THE SOCIAL ORIGINS OF MODERN SCIENCE

BOSTON STUDIES IN THE PHILOSOPHY OF SCIENCE

Editors

ROBERT S. COHEN, *Boston University*
MARX W. WARTOFSKY†, *(Editor 1960–1997)*

Editorial Advisory Board

THOMAS F. GLICK, *Boston University*
ADOLF GRÜNBAUM, *University of Pittsburgh*
SYLVAN S. SCHWEBER, *Brandeis University*
JOHN J. STACHEL, *Boston University*

VOLUME 200

EDGAR ZILSEL

THE SOCIAL ORIGINS OF MODERN SCIENCE

Foreword by
JOSEPH NEEDHAM

Introduction by
DIEDERICK RAVEN & WOLFGANG KROHN

Edited by
DIEDERICK RAVEN
Utrecht University, Utrecht, The Netherlands

WOLFGANG KROHN
Bielefeld University, Bielefeld, Germany

and

ROBERT S. COHEN
Boston University, Boston, U.S.A.

KLUWER ACADEMIC PUBLISHERS
DORDRECHT / BOSTON / LONDON

A C.I.P. Catalogue record for this book is available from the Library of Congress.

ISBN 0-7923-6457-0

Published by Kluwer Academic Publishers,
P.O. Box 17, 3300 AA Dordrecht, The Netherlands.

Sold and distributed in North, Central and South America
by Kluwer Academic Publishers,
101 Philip Drive, Norwell, MA 02061, U.S.A.

In all other countries, sold and distributed
by Kluwer Academic Publishers,
P.O. Box 322, 3300 AH Dordrecht, The Netherlands.

Printed on acid-free paper

All Rights Reserved
© 2000 Kluwer Academic Publishers
No part of the material protected by this copyright notice may be reproduced or
utilized in any form or by any means, electronic or mechanical,
including photocopying, recording or by any information storage and
retrieval system, without written permission from the copyright owner.

Printed in the Netherlands.

Edgar Zilsel (1819-1944)

Picture taken circa 1930, reproduced with the kind permission of Paul Zilsel. The picture in Edgar's right hand is titled 'Das Duel im Pardo'. Most likely it is a reproduction of a painting that used to be attributed to Diego Velázques but probably was executed long after his time. It forms part of Spanish collection of the National Gallery, London, no 1376, title: Landscape with Figures.

TABLE OF CONTENTS

PREFACE	*Robert S. Cohen*	ix
FOREWORD	*Joseph Needham*	xi
ORIGIN OF THE ESSAYS		xv
A NOTE OF THE USE OF ARCHIVE MATERIAL		xvi
ACKNOWLEDGEMENTS		xvii
EDITORIAL POLICY		xviii

INTRODUCTION:
Edgar Zilsel: His Life and Work (1891-1944)
      Diederick Raven and Wolfgang Krohn xix

PHOTO OF EDGAR ZILSEL lxi

PART I: THE SOCIAL ORIGINS OF MODERN SCIENCE

1. The Social Roots of Science	3
2. The Sociological Roots of Science	7
3. The Methods of Humanism	22
4. Remarks on Zilsel's 'The Methods of Humanism' *Paul O. Kristeller*	65
5. The Origins of William Gilbert's Scientific Method	71
6. The Genesis of the Concept of Physical Law	96
7. Copernicus and Mechanics	123
8. The Genesis of the Concept of Scientific Progress and Cooperation	128

PART II: PHYSICAL LAW AND SOCIO-HISTORICAL LAW

9. Problems of Empiricism — 171

10. Physics and the Problem of Historico-sociological Laws — 200

11. Phenomenology and Natural Science — 209

12. Concerning 'Phenomenology and Natural Science' — 214

13. History and Biological Evolution — 216

14. Science and the Humanistic Studies — 221

PART III: APPENDICES

Appendix I: The Sociological Roots of Science — 227

Appendix II: Laws of Nature and Historical Laws — 233

Appendix III: Bibliography of Works Cited by Zilsel — 235

Appendix IV: Bibliography of Edgar Zilsel's Works With a Bibliography of Secondary Literature Concerning Zilsel — 243

Index of names — 253

Index of topics — 259

ROBERT S. COHEN

PREFACE

Edgar Zilsel (1891-1944) lived through the best of times and worst of times, through the renewal of scientific optimism and humane politics, and through the massive social collapse into idolatrous barbarism. With it all, and with his personal and family crises in Vienna and later in America, Zilsel was, I believe, a heroic, indeed a model, scholar of the first half of the 20th century.

He was widely admired as a teacher, at high schools, in workers education, in research tutoring and seminars. He was an original investigator on matters of the methodology of science, and of the history of the sciences. He was a social and political analyst, as a critical Marxist, of the turmoil of Vienna in the 20s. Above all, he achieved so much as a sociological historian who undertook research on two central facts of the early modern world: recognition of the creative individual, and the ideal of genius; and the conditions and realities of the coming of science to European civilization.

His teacher, I think of Heinrich Gomperz the philologist, was distinguished indeed, and so was Leonardo Olschki from whose work on Galileo he learned so much, but Zilsel went far beyond. Thus, why did our science arise, where and when it did, and why not at other times? He was one of a handful of pioneers in this comparative historical sociology of science, seeking to understand the history of the concepts of laws and regularities of nature, of the idea of progress in knowledge, of the specific explanatory terms of mechanism, and so on. In his few years writing in English, his results seemed to be that of a research institute, as this book shows.

His theme was enlightening. Three social strata came together through Renaissance and later: the working artisans, the new post-medieval secular humanists, the higher university scholars. With indefatigable documentation, Zilsel traced how those whose minds were devoted to formal knowledge, known through logic and mathematics, came to collaborate with others who lived by observation and hand-working experience, even by experimentation and common sense causality. Zilsel's story was a study of early capitalism, of society rationalized, mechanized, exploratory, enterprising, and of one central part of its cultural accompaniment, early modern science.

I first came upon Zilsel's works on the sociology of science 50 years ago. This was partly due to my wish for a treatment of natural science within the sociology of knowledge (Mannheim) but beyond the initial studies in that direction which were stimulating but incomplete in scope and in critical depth of their scientific content (Merton, Lukacs, Hessen). To a reader of the essays in this volume, especially to one who follows Zilsel for the first time, his impact may be solid enough, and his results persuasive, but perhaps not so surprising any longer. Indeed, the sociological history of science has flourished these past five decades, both internalist and externalist, but Joseph Needham's happy and welcoming recognition of Zilsel stands today.

Needham has a genial chemical metaphor for the process of science developing, what he calls the 'fusion' of works of the hand and the brain, of Zilsel's three strata, and this is an apt summary of Zilsel's teaching and his challenge. Just what were the social, political, economic, religious, aesthetic and perhaps other factors required for that innovative 'fusion' to occur? Zilsel's method was clearly comparative in plan, across civilizational conditions, but here we can only wonder how far he would have been able to go if the world of his earlier social dream and hope, and the other world of his private stability, had not collapsed.

Lately a friend, Elisabeth Nemeth, brought me Zilsel's essay of 1912 on Mozart and his times, which the author subtitled 'a didactic fantasy'. How lovely to see in this little essay of the young Zilsel the new sensitivity of the first years of that century, not as a scientific inquiry but as a matter of feelings, an essay in which we recognize what reappears throughout Edgar Zilsel's life and works: the deeply emotional and moral dimensions of his search for rationality.

Zilsel's life touched many, beyond his knowing. Paul Kristeller, in his 90^s, wrote to me as 'a great admirer of Zilsel for many decades, and a personal friend'. Also in old age, the leading Chinese empiricist philosopher Tscha Hung (who took his doctorate in Vienna with Schlick) wrote to me from Beijing to aks about the career of his old friend Zilsel with whom he had lost touch; he was especially interested in Zilsel's own 'fusion' of logical empiricism with a historical materialist sociology of science.

When Zilsel continued his personal teaching in Vienna after the murder of Schlick, the little 'Zilsel Circle' bravely met until the early months of 1938. Kurt Gödel gave probably his last lecture to an audience in Vienna at the Zilsel Circle, accepting Zilsel's suggestion to talk about 'the status of consistency questions in logic'; this took place the evening of January 29th. Except for Zilsel, there was likely to be a mathematically unsophisticated audience ... but there was Zilsel!

I did not know Edgar Zilsel. But I know his son Paul, my friend and fellow physics graduate student at Yale, where we received our doctorates in 1948. Paul's understanding of his father's life and work is admirable, and equally his sympathetic resonance for the vicissitudes of Edgar Zilsel's existence. May this preface, and this volume, be an affectionate and respectful tribute.

JOSEPH NEEDHAM

FOREWORD

It is a great privilege and pleasure for me to have been invited to write a foreword for this collection of the papers of Edgar Zilsel. I never met him, and even now I don't know what he looked like, or indeed any details of his personal life. I only recall, and record, with great affection, the intellectual stimulus which I derived from his writings, and for which, together, I am sure, with many others, I shall always be unfeignedly grateful.

My own intellectual odyssey was, to be sure, extremely different from his. Even now I don't really know very much about the "Vienna Circle" of logical positivism, or the philosophies of Mach, Husserl, and Jaspers. I don't appreciate to the full what is meant by Platonic metexis or Kantian schematism and amphiboly; though I can form a rough idea of what excited Zilsel so much as to lead him to devote such a quantity of time and research to the history of the concept of human genius. I never had any real training in philosophy, or in sociology either; all I can say is that I experienced a great fellow-feeling as I read, one after the other, all the things that Zilsel wrote.

My origins were quite different. I came from a medical family, I was a medical student at Caius, and became a working biochemist and embryologist for half my life. Perhaps it was just because of these interests that I developed a fascination for the philosophy of science, and especially the philosophy of biology. The problem of the nature of organic form was always central for me, and it became the most natural thing in the world to trace out the course of human thought about it from Western antiquity onwards, and particularly in the seventeenth century after the scientific revolution. Since my interests included also Christian theology and its history, as well as Marxist philosophy, it was understandable that the problems which interested Zilsel interested me too.

Why then did I get involved with Chinese culture? It was all due to personal friendship. Just over fifty/forty years ago there came to work in the Cambridge laboratories three young Chinese scientists who became my intimate friends. Lu Gwei-Djen, Shen Shih-Chang and Wang Ying-Lai. My family had no connections with China, none of my forbears had been diplomats, traders or missionaries, but for some reason or other these Chinese friends brought about a revolution in me, much greater, I am sure, than any influence which Cambridge could have had on them. The fact that as scientific minds they were so much like my own raised very vividly in my consciousness the historical problem of why modern science had originated in Europe alone, and not in China or India. With their help, I began to learn Chinese, and arrived at the conviction that somehow or other someone was going to have to write a full-dress history of science, technology and medicine in Chinese culture. The scientific minds of these friends were so like my own that they convinced me, without having any intention of doing so, that the factors which had held back modern science in China could not possibly have been intrinsic; in other words, there was nothing in the "Chinese mentality" which could have prevented the rise of modern science there if the conditions had been propitious. The more I read, the more I came to feel that in fact the conditions had not been propitious. I was reminded of something I had read in one of the books of Leonard Tomlinson about certain fungi - when the conditions of humidity and temperature were just right, up they come; if not, they never appear. And so, all through the Second World War (four years of which I spent in China), and after it, I was reading voraciously about the history of China and Chinese civilization. I got to know about the geographical setting of Chinese culture. I came to realise that the type of feudalism in China (if such it was) was totally different from the aristocratic military set-up in Europe, because it was bureaucratic, perhaps also hydraulic. I began to see how important the growth of the city-states had been in Europe, while these did not develop in China. I was conscious of the role of the higher artisanate, which could join with the early experimental scientists to bring about a quite new form of science, involving the mathematisation of hypotheses about nature, and the testing of these by actual experiment. Many other matters also came in, such as attitudes to time, attitudes to authority, attitudes to cooperation, and attitudes to free publication. In a word, it gradually became clear that the scientific revolution had not happened in a vacuum, but that it had been bound up so intimately that no-one could fully detect the connections, with the rise of capitalism and the Reformation. It must then be obvious that anyone with all these ideas in mind would fall upon the writings of Edgar Zilsel with great delight, and extract from them everything which could be useful for this grand *Fragestellung*.

Our *Science and Civilisation in China* series began to be published in 1954, and sure enough, Zilsel's name figures prominently in the early volumes as it certainly will again in the last. In the volume on the history of scientific thought in China, I got much help from Zilsel, especially with regard to the development of the concept of "Laws of Nature", a subject on which he himself had already written admirably.[1] In the volume on the sciences of the heavens and the earth, I ventured to insert, between mathematics and astronomy, a section in which I tried to analyze what was the essential nature of modern science, first appearing at the scientific revolution, and here I was helped by Zilsel again, both in relation to the higher artisanate and the development of commerce and industry. The general conclusion at this stage was that while many of the components of the scientific revolution had already existed in China, at least as markedly and successfully as in Europe, nevertheless that particular fusion of

---

1 *Science and Civilisation in China* (Cambridge: Cambridge University Press, 1954), Volume 2, pp. 534 ff, 539, 542 ff.

mathematics and scientific experiment which occurred in the West could not happen in China. There, I wrote:

> there came no vivifying demand from the side of natural science. Interest in Nature was not enough. Controlled experimentation was not enough. Empirical induction was not enough. Eclipse-prediction and calendar-calculation were not enough - all of these the Chinese had. Apparently a mercantile culture alone was able to do what agrarian bureaucratic civilization could not - bring to fusion point the formerly separated disciplines of mathematics and nature-knowledge.[2]

And, for a third example, in the volume on physics, Zilsel helped us regarding the relation of magnetism and cosmology, particularly with regard to William Gilbert and Petrus Peregrinus.[3]

This is not to say, of course, that I never disagreed in any particular with the conclusion which Zilsel drew. He would never have claimed to have said the last word about the concept of progress, and indeed many other scholars have had much of interest to say in that regard, for example, E.R. Dodds in his analysis of the idea among the writers of Western antiquity. Furthermore, the fact that Zilsel never considered the Asian civilizations, indeed he never claimed to know anything about them, meant that some of his ideas concerning, e.g. the development of scientific co-operation, needed adjustment. It is a fact that such co-operation was customary from early medieval times in China, as witness, for instance, the *Hsin Hsiu Pên Tshao* (Newly Re-organized Pharmacopoeia) of +659. This was worked at by a whole team under imperial commission, which made it indeed by far the oldest of all the official pharmacopoeias, a whole thousand years earlier than the *Pharmacopoeia Londiniensis*. Individual collaborations were also quite common in China, for instance the close working together of the monk I-Hsing and the engineer Liang Ling-Tsan in their invention about +725 of what was perhaps the earliest of mechanical clock escapements. A similar partnership occurred in +1088 when Su Sung and his collaborator, the engineer Han Kung-Lien, set up the great astronomical clock-tower at Khaifêng; and far from wishing to keep secret what they did, they published an elaborate treatise on the whole mechanism, both in its astronomical and horological aspects, entitled *Hsin I Hsiang Fa Yao* (New Design for an Armillary Clock) translated by my collaborators and myself just under nine centuries later.[4] This monumental work embodied the oldest astronomical clock-drive in any civilization, the earliest power-transmitting chain-drive, the earliest bevel-gearing, and a time-annunciator which took account of the unequal night-watches as

---

2 *Science and Civilisation in China* (Cambridge: Cambridge University Press, 1956), Volume 3, pp. 154, 167-168.
3 *Science and Civilisation in China* (Cambridge: Cambridge University Press, 1959), Volume 4, part 2, p. 542.
4 Joseph Needham, Wang Ling and Derek J. de Solla Price *Heavenly Clockwork: The Great Astronomical Clocks of Medieval China* (Cambridge, etc: published in association with the Antiquarian Horological Society at the University Press, 1960).

well as the equal double-hours of the Chinese day and night. Or again, if I am not giving too many examples, there was the *Shêng Chi Tsung Lu* (Imperial Medical Encyclopedia) produced by twelve eminent physicians under the patronage of the emperor in +1118. All this is simply an indication that while it remains perfectly true that the idea of scientific collaboration and co-operation was something new for Europe at the time of the scientific revolution, it was not at all new for the Chinese; yet, once again, it is the case that modern science did not originate in Chinese culture, but only in the European West.

In the last of our volumes (vol. 7), Zilsel will help us again just as he did at the beginning. Much of this volume is already in draft, and one cannot help feeling how delightful it would have been if Edgar Zilsel himself could have read it all through and criticised it for us. First we shall have a thorough investigation of the social and economic background for science, technology and medicine in China. This necessitates a careful characterisation of what exactly traditional Chinese society was; and in the making of this, we must take into account not only what the Western observers and sociologists over the years from François Bernier onwards thought about Chinese society, but also what the modern economic and social historians of China have themselves thought about the past of their own society. Another valuable contribution is in progress on the world-outlook of the Chinese *literati*; how it differed from that traditional in Christendom and how it affected the chances which modern science had of growing up in China. And then there was logic; what part did it play in traditional China, and how did it develop there, helping or hindering? For that matter, was Aristotelian formal logic a help or a hindrance in Europe? The role of religion will also be investigated, both in relation to the *san chiao* of China (Confucianism, Taoism and Buddhism) and the typical religions of the West, deriving from Hebrew monotheism. It can be seen that there is much yet to be done.

This book of collected papers has been planned for many years past, as I believe, and it is a splendid thing that it is now coming to fruition. I am sure that Edgar Zilsel himself would be very happy in the knowledge that his writings have been of such great use to others, especially in directions the importance of which he himself did not, perhaps, always clearly recognise. But the keynote of his work, even when not exactly so stated, was "the comparative", and so it will always be of use to those who want to compare the development of European thought with that in other civilizations. It has been said that if it is very difficult to ascertain why certain developments of thought happened in one place, it is even more difficult to understand why they did not happen in some other place. All we can be sure of, and this is where Zilsel's work is a veritable torch to light the darkness, is that we have to look for the "sociological roots" as well as the purely intellectual ones, of science and technology, whether it be in the West or in the East. *Fiat lux*, we all cry, and Edgar Zilsel's life and work put him among the most notable taperers in the procession of those who seek to understand.

ORIGIN OF THE ESSAYS

The following list indicates the first publication of the different essays included in the present volume.

1. 'The Social Roots of Science', in H. Pauer-Studer, *Norms, Values, and Society (Vienna Circle Institute Yearbook, Vol 2)* (Dordrecht: Kluwer Academic Publishers, 1994), pp. 305-308.
2. 'The Sociological Roots of Science', *The American Journal of Sociology*, 1942, 47, pp. 544-62.
3. 'The Methods of Humanism', *not previously published.*
4. Kristeller, P. O., 'Remarks on Zilsel's 'The Methods of Humanism'', *not previously published.*
5. 'The Origins of William Gilbert's Scientific Method', *Journal of the History of Ideas*, 1941, 2, 1, pp. 1-32.
6. 'The Genesis of the Concept of Physical Law', *The Philosophical Review*, 1942, 51, pp. 245-79.
7. 'Copernicus and Mechanics', *Journal of the History of Ideas*, 1940, 1, pp. 113-8.
8. 'The Genesis of the Concept of Scientific Progress and Scientific Co-operation', *not previously published.* A shortened version was published under the title 'The Genesis of the Concept of Scientific Progress', *Journal of the History of Ideas*, 1945, 6, pp. 325-49.
9. 'Problems of Empiricism', in O. Neurath, R. Carnap & C. Morris (eds.), *Foundations of the Unity of Science: Towards an International Encyclopedia of Unified Science*, Volume II, number 8, pp. 53-94, Chicago & London: The University of Chicago Press, 1941.
10. 'Physics and the Problem of Historico-Sociological Laws', *Philosophy of Science* 1941, 8, pp. 567-79.
11. 'Phenomenology and Natural Science', *Philosophy of Science*, 1941, 8, pp. 26-32.
12. 'Concerning Phenomenology and Natural Science', *Philosophy and Phenomenological Research*, 1941/42, 2, pp. 219-20.
13. 'History and Biological Evolution', *Philosophy of Science*, 1940, 7, pp. 121-8.
14. 'Science and the Humanistic Studies', *not previously published.*

All the previously published essays appear here with the permission of the respective copyright owners, if any. These permissions are most gratefully acknowledged.

A NOTE ON THE USE OF ARCHIVE MATERIAL

We have been able to find (parts of) Zilsel's correspondence with several persons. In alphabetical order these are: H. Evans, H. Feigl, M. Horkheimer, O. Neurath, H. Reichenbach, G. Sarton, M. Schlick. The correspondence of Feigl, Neurath, and Schlick can be found at a number of places. Most of the original letters are in the files on the Vienna Circle in the Rijksarchief of Noord-Holland, Haarlem, the Netherlands.

In referring to material held in these archives we will use the following acronyms:

1. APS/Z, for the Zilsel file held by the American Philosophical Society, Philadelphia, PA.;
2. EC/Z, for the Emergency Committee in Aid of Displaced Scholars, file Zilsel, which is in the New York Public Library, Astor, Lenox, and Tilden Foundations, Manuscripts and Archives Division;
3. EP/Z, for the Evans papers, file Zilsel, which are in the Bancroft Library, University of California, Berkeley;
4. HP/Z, for the Horkheimer papers, file Zilsel, which are in the City Library of Frankfurt am Main;
5. NP/Z, for the Neurath papers, file Zilsel, which are in Haarlem (NL) and Konstanz (D);
6. PF (private file), for material that is not in any manner publicly available;
7. HR, for the Reichenbach papers, file Zilsel, which are in Pittsburgh (USA) and Konstanz (D);
8. Sch/Z, for the Schlick papers, file Zilsel, which are in Haarlem (NL) and Konstanz (D);
9. SP/Z, for the Sarton papers, file Zilsel, which are in the Houghton Library of Harvard University;
10. ZF/UV, for the University of Vienna's file on Zilsel which is held at the University of Vienna.

We kindly acknowledge the permission given by all these libraries and/or institutions to quote from material held in their possession.

ACKNOWLEDGMENTS

A number of people have been instrumental to us in preparing this volume. We especially like to acknowledge the help in a variety of ways of the following individuals: Hans-Joachim Dahms, Stephan Fuchs, Roger Hahn, Toby Huff, Andreas Kamlah, Paul-Oskar Kristeller, Anna-Katherina Mayer, Elizabeth Nemeth, Ben Schreurs, Friedrich Stadler, and Joanna Zilsel. Additionally we want to thank Thomas Kellein and Angela Lampe for their invaluable help in identifying the picture on the photo of Edgar Zilsel reproduced in this book on page lxi. A special word of gratitude to Annie Kuipers and Stephanie Harmon, both at Kluwer Academic Publishers, for their understanding and patience.

A number of librarians have been very helpful in trying to locate archival material. We like to thank especially: Janice Braun, Special Collections Librarian of Mills College, Oakland, CA, Julio Hernández-Delgado, Archivist at Hunter College of the City University of New York, NY, Bill Roberts, Head of the Special Collections of Bancroft Library of the University of California, Berkeley, CA, and Brigite Uhlemann of Konstanz University.

A special word of thanks to Paul Zilsel for his willingness to discuss every detail of his father's life with us as well as for providing us with copies of letters written by his father shortly before he died.

We kindly acknowledge the financial support we received from the *Deutsche Forschungsgemeinschaft* in preparing this volume as well as from the *American Philosophical Society* for a grant that enabled us to do additional archival research at Hunter College, NY, at the University Library of Chicago University, Chicago, Ill. and at the Houghton Library of Harvard University, Cambridge, Mass. in the summer of 1997.

A final word of thanks to our research assistants Kerstin Klein, Jesse Kraai, and Panagiotis Tsiantopoulos for preparing the final MS, the indexes, the list of works cited by Zilsel and translations from German to English.

EDITORIAL POLICY

This edition combines essays previously published with ones that were not made public before. They are written by someone whose command of English has a distinct foreign flavour. Our general editorial policy has been to be deliberately restrictive in modifying Zilsel's English. In case of the essays that were already in the public domain we have only changed obvious typos and misspellings and the references Zilsel made to his own work to references contained in this volume - wherever this was appropriate. The customary practice of Zilsel was to use notes placed at the bottom of the page and not to use endnotes, located at the end of a chapter. In this volume we have followed Zilsel's practice. Only in the essay 'Problems of Empiricism' did Zilsel use endnotes which we have changed into footnotes. In the essays published here for the first time, alterations were only undertaken when we felt that the sense of the sentence was *significantly* hampered by misspellings, typos, or modes of expression. Mistakes in dates were also corrected.

The essays collected in part I were intended to form the basis of a monograph. We have arranged them as if they are a monograph but they still clearly carry with them the fingerprint of having been published in different scholarly journals each of which had their own printing style. Only a minimal attempt has been made to unify the layout of the different essays. We have however compiled a list of reference of works cited by Zilsel. Added are indexes of names and topics.

Whenever necessary we have put an editorial note at the beginning of an essay to provide some background on the essay. In the case of the unpublished essay 'The Genesis of the Concept of Scientific Progress and Cooperation' the note explains how we have incorporated the published shorter version 'The Genesis of the Concept of Scientific Progress'. All our comments are in footnotes and are between [square brackets].

The Editors.

DIEDERICK RAVEN & WOLFGANG KROHN

EDGAR ZILSEL: HIS LIFE AND WORK (1891-1944)

Introduction

Many-sided marginal-man is, in a nutshell, the best characterization of Edgar Zilsel. Everybody acquainted with him was convinced of his brightness and his gifts as a scholar, and all recognized his enormous intellectual powers.[1] Nevertheless, he was marginal to the University of Vienna - his *Habilitationsschrift* (post-doctoral dissertation) was repeatedly rejected - and he was subsequently not capable of obtaining a teaching position there. As he detested the misuse of scientific hypotheses for party-political goals, he became marginal to the Social Democratic Party. He lost his teaching job and was forced into retirement when the Nazi-sympathizers came to power in Austria in the 1930's. Later in life, during exile in the USA, he remained marginal to the emigrated Frankfurt School even though he was associated with the International Institute of Social Research. He died in obscurity in 1944.

This volume presents Zilsel's later works, all composed during his exile in America (1939-1944). Two highly interesting projects with a common basis are to be found in the later work. The papers collected in part I (pp. 1-170) are related to his project 'on the social origins of modern science', and were originally intended to be the foundations of a book. Part II (pp. 171-226) contains the incompleted work of the second project 'on the concept of law'. Both projects failed to reach completion due to Zilsel's tragic suicide. Zilsel wrote several outlines of both these projects which provide critical insight into his research plans. Translations of the most detailed outline of each project can be found in appendices I and II.[2]

Zilsel's published work 'on the social origins of modern science' is well-known among historians of early modern science as a pioneering and distinctive approach to our understanding of the social factors behind the development of science. We are convinced that these essays, along with two other previously unknown essays, which we have also included in this volume, deserve a wide audience. They touch upon the very reasons why we have modern science in the first place. As such, they are relevant not just for historians of science, but also for philosophers and sociologists of science alike.

---

1 Karl Menger recalls how the logician Jan Lukasiewcz was simply overwhelmed by a lecture Zilsel gave at a meeting in Warsaw and repeatedly said 'What an intellect!', cf. .K. Menger *Reminiscences of the Vienna Circle and the Mathematical Colloquium*, L. Golland, et al (eds.). (Dordrecht, etc.: Kluwer Academic Publishers, 1994), p. 67.
2 Appendix I, this volume, pp. 227-232, appendix II, this volume, pp. 233-234.

Our aim in this introduction is twofold: to provide key biographical data concerning Zilsel's life and to argue that the two sets of essays found in parts I and II of this volume are actually part of one common project: to single out the unique features of modern science and to investigate the sociological conditions of its origins.

Our introduction to Zilsel's work is arranged as follows: section I introduces the reader to Zilsel's life by means of a brief biography. Section II proceeds to a description of Zilsel's two projects. We then move to a discussion of Zilsel's philosophical background in section III. We concentrate in particular on his dissertation (III.1), his post-doctoral dissertation (III.2), his relationship to the Vienna Circle (III.3), and the role of Marxism in his philosophical development (III.4). The material presented in sections II and III is then used to argue for the convergence of Zilsel's two projects in section IV. Section V summarizes the relevance of Zilsel's work for contemporary research.

I. Biography of Zilsel

I.1 Edgar Zilsel in Vienna, 1891-1938

Edgar Zilsel was born in Vienna on August 11, 1891 to the lawyer Jacob Zilsel and his wife Ina Kollmer. He had two older sisters, Wallie and Irma. From 1902 to 1910 Zilsel attended the high school Franz-Joseph-Gymnasium - now the Bundesgymnasium Stubenbastei. Zilsel then went directly to the University of Vienna to study mathematics, physics, and philosophy. His student years were interrupted by his military service, from August 1, 1914 to December 15 of the same year. He attained his Ph.D. with the dissertation *A Philosophical Investigation of the Law of Large Numbers and related Laws* (1915).[3] He published an extended version of this dissertation two years later under the title *The Application Problem. A Philosophical Investigation of the Law of Large Numbers and its Induction.*[4] In September 1915 he took up a job as an insurance mathematician at an insurance company, but left after a year to try to obtain the necessary qualifications to become a secondary school teacher. Despite not being fully qualified, he took up the post of teacher on February 16, 1917. His final teacher's exam was taken on November 18, 1918 in mathematics, physics, and natural history (*Naturlehre*). Zilsel published his second book in the same year: *The Religion of Genius. A Critical Study of the Modern Ideal of Personality.*[5] This

---

3 *Ein philosophischer Versuch über das Gesetz der grossen Zahlen und seine Verwandten.* For full bibliographical details see the bibliography, appendix IV, pp. 243-253.
4 *Das Anwendungsproblem: Ein philosophischer Versuch über das Gesetz der grossen Zahlen und die Induktion.* This book will be discussed in section III.1 of this introduction, pp.xxxix-xl.
5 *Die Geniereligion. Ein kritischer Versuch über das moderne Persönlichkeitsideal.* This book is somewhat of an anomaly in Zilsel's œuvre. It is more a journalistic work than a scholarly book. For a penetrating analysis of this work, see E. Nemeth, '"Wir Zuschauer" und das "Ideal der Sache". Bemerkungen zu Edgar Zilsels Geniereligion', *Lecture Series/Vorträge des Institut Wiener Kreis*, 1997, 5, pp. 157-78.

book polemically addressed the role of the cult of genius in modern society in the form of an investigation into its social and historical roots.

Zilsel married the schoolteacher Dr. Ella Breuer on February 19, 1919 after he had established himself as a teacher at a high school in Vienna. Ella Breuer taught English and German at a women's high school. Their only child, Paul, was born on May 6, 1923. Zilsel was allowed paid leave to take up a position at the *Verein Volksheim Wien*, now the Adult Education Center 'Ottakring', to teach philosophy and physics. The *Verein* ('association') was comprised of several institutes dedicated to adult education (*Volkshochschule*), and was sponsored by the City of Vienna and the socialist trade unions. Zilsel was awarded this position by the city education counselor (*Stadtschulrat*) Hartmann, who was the head of the Viennese Department of Education, "in appreciation of the applicant's particular suitability for this form of education".[6] In addition to his work at the *Volksheim*, Zilsel also became engaged in 1924 as a teacher trainer (*Lehrerbildung*) at the Pedagogical Institute of the City of Vienna.[7] His various duties included chairing the physics department, the philosophy department (beginning 1927/28), as well as being in charge of the apparatus of the physics laboratory. He taught a wide range of topics including lectures on Heidegger, Jaspers, Spinoza, as well as 'Space and Time in Philosophy and Physics', 'The Sociological Essays of Max Weber', 'An Introduction to Psycho-analysis', 'World Religion', 'The Spiritual Life of Natural Peoples', and 'Mysticism and Science'.[8]

Zilsel regularly published essays in academic journals, and was an active contributor to the social democratic journal *Der Kampf* and the newspaper *Arbeiterzeitung*. In the summer of 1923 he submitted his *Habilitationsschrift* which was rejected. The rejection of his *Habilitationsschrift* prevented Zilsel from pursuing a career at the University of Vienna.[9] In 1926 he published his most important book from this period, *The Development of the Concept of Genius: A Contribution to the Conceptual History of Antiquity and Early Capitalism*,[10] which was based on his *Habilitationsschrift*. The book was well received.[11] Shortly thereafter, Zilsel was recommended by the physicist Philipp

---

6 J. Dvořak, *Edgar Zilsel und die Einheit der Erkenntnis.* (Wien: Löcker Verlag, 1981), p. 20.
7 Except for the Winter semester 1924/25, Summer 1925, and Winter 1929/30; cf. C. M. Götz, and T. Pankratz, 'Edgar Zilsels Wirken im Rahmen der Wiener Volksbildung und Lehrertforbildung', pp. 467-73 in *Wien-Berlin-Prag: Der Aufstieg der Wissenschaftlichen Philosophie; Zentenarien Rudolf Carnap - Hans Reichenbach - Edgar Zilsel*, R. Haller & F. Stadler (eds.). (Wien: Verlag Hölder-Pichler-Tempsky, 1993), p. 470.
8 A complete list of courses Zilsel taught at various institutes can be found in F. Stadler, *Studien zum Wiener Kreis: Ursprung, Entwicklung und Wirkung des Logischen Empirismus im Kontext.* (Frankfurt: Suhrkamp, 1997), pp.805-16.
9 The *Habilitation* requires then as now a second larger work following the dissertation. The net result of a *habilitation* exam was that it provided the applicant with a licence, a *venia legendi*, to teach at the university. As this episode was so pivotal in Zilsel's development, we will discuss it in section III.3 of this introduction, pp. xlv-xlvii.
10 *Die Entstehung des Geniebegriffes: Ein Beitrag zur Ideengeschichte der Antike und des Frühkapitalismus*.
11 A list of these reviews may be found in our bibliography of Zilsel, see appendix IV, pp. 243-253.

Frank for the new Chair of philosophy at the German university in Prague, Czechoslovakia.[12] Zilsel's Vienna Circle colleagues Hans Reichenbach and Rudolf Carnap also contended for the Chair Carnap was to eventually obtain.

In the wake of the coup d'etat initiated by Dollfuss in 1934, Zilsel was briefly taken into custody and lost his job. He thereafter worked part-time at various schools until he found a full-time teaching job at a high school in Vienna. Following Austria's *Anschluss*, Zilsel was forced into compulsory retirement on account of his Jewish origin.

1.2 Zilsel in the USA 1939-1944

It is unclear when the Zilsel family left Vienna for London. In a letter to Otto Neurath, dated January 17, 1939, Zilsel wrote: "We indeed did manage to escape the Nazi jailhouse and have been living safely in England for some time now".[13] Very little is known about his time in England. He left England via Southampton on March 26 to arrive in New York on April 4. His sixteen year old son Paul stayed in England until the summer to prepare for his final exams.

Within a very short time after his arrival in New York Zilsel was able to establish contact with Max Horkheimer, the director of the International Institute of Social Research (IISR) - the emigrated Frankfurt School. Although they did not have the necessary means to support Zilsel, they did actively assist his efforts to find such.[14] These first few months were spent with the time-consuming work of writing project proposals and obtaining recommendations.[15] Zilsel was able to make a little money by privately tutoring classical Greek to an émigré he met on the boat.

In June 1939, Zilsel received a Rockefeller Fellowship to work on the origins of early modern science. He was tremendously excited about the prospect of working full time on the project. To Neurath he wrote: "Good news: I obtained

---

12 The assisstence of Frank is mentioned by Zilsel in his own undated CV, written propably shortly after he arrived in the USA in April 1939. We have been able to locate two copies of this CV, one is in correspondence between Zilsel and Horkheimer and the other in the Zilsel file of the Emergence Committee in Aid of Displaced Foreign Scholars. We will use the acronym HP/Z followed by a date to refer to the Zilsel Horkheimer correspondence which is located in the City Library of Frankfurt am Main (Horkheimer papers) and we will use the acronym EC/Z followed by a date to refer to the Zilsel file of the Emergence Committee which is located in the New York Public Library, collection Emergence Committee, box 38. All archive sources with material on Zilsel are listed on page p. xvi.

13 Zilsel to Neurath, quote from the Neurath Papers held in Haarlem, The Netherlands, file Zilsel, NP/Z for short. If necessary the acronym NP/Z will be followed by a date.

14 In the Institute's official correspondence with outside funding agencies its main line was: "[t]he present financial situation of this Institute does not permit us as yet to offer Dr. Zilsel a permanent position with us" (HP/Z May 29, 1940). As background the following should be mentioned. The Institute was founded from the funds of Weil Foundation and had just done badly with its investments in the stock market. Horkheimer, in his function as director, ordered a cut back in spending. Understandable or not, it did not prohibit Horkheimer to set aside $ 50.000 for himself. Critical remarks of members of the IISR on Horkheimers style of managing the funds of the Weill Foundation are quoted in R. Wiggershaus, *The Frankfurt School: Its History, Theories and Political Significance*. (Cambridge: Polity Press, 1994), p. 261.

15 In another letter to Neurath (NP/Z, May 8, 1939), Zilsel mentions that he received "very good" letters of recommendation from Carnap, Feigl, Gomperz, and Reichenbach. We have seen only those by Carnap and Reichenbach.

the Rockefeller fellowship, and am thus taken care of for a year. I've never in my whole life had the opportunity to scientifically work without a secondary job. This is indeed a beautiful country".[16]

In September 1939, he presented a paper entitled 'The Social Roots of Science' to the '5th Congress for the Unity of Science' held at Cambridge, Mass.[17]

Zilsel did not have any money of his own to live on and needed a job or other financial support. After Hitler came to power in 1933, a number of relief agencies had been set up in the United States of America to support the émigré scholars. Their aim was "to save learning, not to provide personal help for individual scholars".[18] Grants awarded by them provided the émigré scholar with an opportunity to present himself to American scholars - through publications or otherwise. The initial funding was for a year and could in some cases be renewed for another year. After that, however, the scholar was expected to have found a job. In Zilsel's case it was renewed. Any further funding was to be given only if Zilsel affiliated himself with an institution other than the IISR. In this he was apparently unsuccessful.[19]

It is very difficult to trace Zilsel's activities after the summer of 1941. Two things are clear: his scholarly output dried out, and he was somehow able to secure a part-time job at Hunter College in New York. Although we have no idea when he started his job at Hunter, we know that he was assigned to teach a course entitled 'Mechanics, Molecular Physics, Heat and Sound'.[20] In the spring

---

16 NP/Z, July 27, 1939.
17 This essay is the first English statement of Zilsel's project 'on the social origins of modern science'. It was presented at the 5th International Congress for the Unity of Science held at Harvard University, Cambridge, Mass., September 3-9, 1939. The papers presented at this conference were originally intended to be published in the *Journal of Unified Science* - the successor to *Erkenntnis*. The original MS of the journal issue was sent to Holland to be printed, but was destroyed during the Nazi invasion of Holland; cf. C. Morris, 'On the History of the International Encyclopedia', *Synthese*, 12, 1960, 517-21, p. 520. Some of the papers were published with the reprint of *Erkenntnis*, vol 8, pp. 386-437. The Zilsel MS was discovered by Friedrich Stadler among the Neurath papers, Haarlem, the Netherlands and published in H. Pauer-Studer, *Norms, Values, and Society (Vienna Circle Institute Yearbook, Vol 2)* (Dordrecht: Kluwer Academic Publishers, 1994), pp. 305-308. We gratefully acknowledge Stadler's assistance to republish it in this volume, see pp. 3-6. A complete list of all the speakers present at the Harvard conference is given in Stadler, op. cit., 1997, pp. 429-33.
18 This was the stated aim of the Emergency Committee in Aid of Displaced Foreign Scholars and the Rockefeller Foundation. A useful source of information on the Emergency Committee is S. P. H. Duggon and B. Drury, *The Rescue of Science and Learning: The Story of the Emergency Committee in Aid of Displaced Scholars*. (New York: Macmillan, 1948) ; the quote is from p. 19.
19 Except for teaching an introductory course in philosophy for eight weeks which he gave at City College, New York. (HP/Z, June 27, 1940). Still, Zilsel was able to secure grants from various agencies. The complete picture of all the grants he was awarded is as follows: for the academic year 1939-1940 both the Emergency Committee in Aid of Displaced Foreign Scholars and the Rockefeller Foundation paid $900. For the following year he received from the same agencies $650, supplemented by a grant from the Social Science Research Council of a total of $500. For the academic year 1941/42 he received $600 from both the Emergency Committee and Social Science Research Council. He was granted $600 for the year 1942 from the American Philosophical Society, and $1,000 for the academic year 1942/43 from the Social Science Research Council and $1,000 for the academic year 1943/44 from the American Philosophical Society.
20 The complete course description is: "45 hours lecture or recitation, 45 hours laboratory work. 4 credits. Motion, Newton's laws of motion, work and energy, mechanics of rigid bodies, elasticity, mechanics of fluids, temperature, quantity of heat and calorimetry, kinetic-molecular theory of matter,

of 1943 Zilsel secured a grant of $1,000 from the American Philosophical Society to finish his work 'on the social origins of modern science'. Shortly thereafter, on what appears to be Lynn White, Jr.'s initiative, Zilsel was offered a job at Mills College in Oakland, California.[21] Like Hunter College, Mills College was an all-female institution.

I.3 Zilsel at Mills College (August 1943 - Spring 1944)

Lynn White, Jr., President at Mills from 1943 to 1958 and a historian with an interest in 'technology and invention in the Middle Ages and Renaissance',[22] was familiar with Zilsel's recently published historical work,[23] and worked on similar subjects. Mills was at that time a "small and impoverished institution"[24] which could not afford a specialist solely assigned to teach the history of science. White nevertheless indicated to Zilsel that he would be given the opportunity to teach a course on the history of science.[25] The initial offer of a half-time job was, as White puts it, "to teach physics".[26] We also know that Zilsel initially spent a lot of time "cleaning up my laboratory and reassembling and repairing apparatus".[27] In the February of the following year however, Zilsel would write to his son Paul "having escaped the worst of difficulties, I no longer take the lab so seriously".[28]

---

wave motion and sound". A similar course was given by another teacher, a Prof. Turner. Source: Calendar of Hunter College for the year 1943, Physics Section, p. 38.

21 In the files of the Emergency Committee in Aid of Displaced Foreign Scholars we found a memorandum written by Betty Drury, the executive secretary of the committee, dated May 22, 1943 which makes clear that Lynn White, Jr. had come to the office of the committee to discus the possibilities of securing a stipend for Zilsel if Zilsel would move to Mills College. This memo seems to indicate that the Emergency Committee did not play an active role in Zilsel's move to Mills but that it was White's own initiative. This is confirmed by the closing paragraph of the memo: 'President White called up again this morning to say that Zilsel was unwilling to have the Emergency Committee approached. If he went to Mills Colege at all he preferred to go independently' (EC/Z; May 22, 1943).

22 The quote is from White's letter of February 16, 1939, to Herbert Evans in which he "accepts with enthusiasm" the invitation to join the History of Science Dinner Club. The main source of our information on this club stems from the Evans Papers which are held at the Bancroft Library of the University of California, Berkeley. We use the acronym EP followed by the date of the letter in our references to this collection. For a useful general assessment of the Dinner Club see Roger Hahn, 'Berkeley's History of Science Dinner Club. A Chronicle of Fifty Years of Activity ', *Isis*, 1999, 90, pp. 182-91.

23 When White wrote to Evans (EP, August 28, 1943) to excuse himself for not being able to attend the History of Science Dinner Club meeting of September 4, he mentioned that Zilsel would join his faculty at Mills, and added "You have doubtless seen some of his recent articles, the best of which I think was that in the *Journal of the History of Ideas* a couple of years ago on the background of Gilbert's scientific method".

24 White in private correspondence to one of the editors.

25 Based on information given by Paul Zilsel. It is clear from some of the correspondence we have seen between White and the leading historian of science at Harvard University, George Sarton, that White called upon Sarton to find a young and talented historian of science to teach at Mills after the war. This person was to presumably replace the vacancy created by Zilsel's death.

26 White to Evans (EP, August 28, 1943). Later, in the already mentioned letter to one of us, White was to recollect - mistakenly we would argue in light of his letter to Evans - that he invited Zilsel to Mills College to teach mathematics.

27 Zilsel to Evans, October 21, 1943.

28 Edgar Zilsel to Paul Zilsel, February 3, 1944. Source: private file with the authors. For material that

The grant in Spring 1943 from the American Philosophical Society had given Zilsel the financial freedom to resume work on his book on the emergence of modern science. In a letter to Paul he indicates that he was working a great deal, as the lack of social intercourse forced him to delve into his work.[29] The essay 'The Genesis of the Concept of Scientific Progress and Scientific Cooperation' is the fruit of this period.[30]

Zilsel was apparently not very happy at Mills College. Due to White's time-consuming administrative duties as president of the college, Zilsel hardly ever had a chance to exchange ideas with him. Zilsel's wife gave him serious cause for concern, not only because of her unstable mental condition, but also because she did not want to follow him to California. Nor did he see much of his son Paul, who studied physics at the University of Wisconsin. All in all, Zilsel found Mills College a very lonely place. All of his fellow teachers were married and went home for lunch or dinner, while he had to remain on campus. He found having to eat with the "girls in the hall" embarrassing.[31] He never managed to adjust to Mills very well and regarded it as "quite provincial".[32]

The one thing that really excited him was the monthly meeting of the History of Science Dinner Club.[33] White suggested that Zilsel should become a member in a letter dated August 28, 1943. Of the 22 members of the Club, Zilsel was acquainted with roughly half.[34] Zilsel was introduced by Evans during the first meeting he attended and was "immediately and unanimously invited to become a member of the club". In a letter dated September 23, in which Evans asks for

---

is not in any manner publicly available, we use the acronym PF followed by the date of the letter.
29 Ibid. His own words are "*So verkrieche ich mich also in meiner Arbeit*".
30 E. Zilsel,'The Genesis of the Concept of Scientific Progress and Scientific Cooperation', this volume, pp. 128-170.
31 In his own words "*ich [empfinde] das Essen mit den girls in der hall langsam immer mehr als etwas komisch und ungewöhnlich*". (Edgar Zilsel to Paul Zilsel, PF, February 3, 1944)
32 Ibid.
33 The Club began meeting in 1933 upon the initiative of Herbert Evans, a biology professor at the University of California, Berkeley and director of the Institute of Experimental Biology. Evans was the secretary of the club. He organized all the meetings which were held on the second Tuesday of the month and made detailed minutes of them. In this he went to great lengths, writing to people to ask them if they could put their remarks and comments on paper. Edward Strong, also a member of the club, would later describe Evans as "the moving spirit of the group" (cf. E. W. Strong *Philosopher, Professor and Berkeley Chancellor 1961-65*. Interviews conducted by H. Nathan, Oral History Project, Berkeley, Cal. 1992, p. 81). The club would meet at 6 p.m. at the O' Neil Room of the Faculty Club, where they would have dinner at the price of $ 1.00. Another ritual was that every member was invited to "bring a book". Before the speaker of the night was given the floor, all these new books - preferably on the history of science - were briefly discussed.
Members of the club at the time were R. Aitken, W. Blasdale, E. Brunswik, C. Camp, E. Essig, A. Foster, S. Holmes, F. Johnson, F. König, C. Kofoid, V. Lenzen, R. H. Lowie, W. Meyer, F. Mood, A. Pabst, C. Porter, H. Reed, E. Strong, H. Torrey, L. White, Jr.. Only Strong and White had a real professional interest in the history of science. Strong's doctoral dissertation *Procedures and Metaphysics* (Berkeley, Cal.: University of California Press, 1936) was very close to Zilsel's own philosophical and historical position on central arguments in the history of science. The amateur nature of the club would continue for a very long time. Kuhn, when he joined the club in the fifties, objected strongly against it and wanted to purge the club of the 'amateurs'.
34 White in a letter to one of the editors. Egon Brunswik was a personal friend of Zilsel from Vienna years

a more complete title of the paper Zilsel agreed to deliver on May 9 of the following year, he writes: "The members of the History of Science Dinner Club greet your appearance in our midst for the current year with great satisfaction".[35]

The Club greatly appreciated Zilsel's contribution. In a letter to Zilsel dated October 13, Evans writes: "All our members have spoken warmly of the profit we are enjoying and will continue to enjoy from your participation in our little club".[36] Zilsel replied in a letter dated October 21 that the remark "is too kind" and continues: "It is certainly I who profits from and enjoys this participation most". He says further:

> It means, therefore, much to me that I found a place of scientific stimulation and an opportunity to meet eminent scholars in the club. Compared to that, the admission of one more person can hardly seem very important to the members of your circle.

Still, Zilsel must later have become deeply depressed, for on the night of Friday, March 11, 1944 he failed to return to his place of residence. Instead, he stayed in his office and wrote three letters, one addressed to Lynn White, Jr., one to his son Paul, and a third note he put on his desk. Having done that, he "fashioned a pillow from excelsior and his jacket, took poison and then reclined on the floor awaiting death - his hands in his pockets".[37] The note on his desk nearby read:

> No fuss, please!
> Just inform Dr. French, don't tell anybody else, please. Keep silent, please!
> Nobody must know of the suicide, everybody must be told that I died through a traffic accident.
> No students must see the body.
> Please, please, don't try to wake me up again.
> I am sorry to have inconvenienced you.
> Thank you.

There was a postscript to this note

> If the janitor finds me, he may keep the $10 bill as compensation for the shock.

The money was on his desk.
Zilsel was found by French, the Dean of the faculty, who had been alerted that Zilsel had not turned up at home.[38]

---

35 EP, Sep. 23, 1943.
36 EP, October 13, 1943.
37 Oakland Tribune, Sunday March 12, 1944, p. A7, col. 1.
38 Zilsel's self-choosen death is six years to the day after the Austrian *Anschluss* with Hitler Germany.

II. Zilsel's Two Projects

II.1 Two Copious Manuscripts

In this section we describe the two projects Zilsel was working on during his exile in America and show how they can be seen as a continuation and elaboration of themes which Zilsel had developed in his earlier work.

In June 1939, Zilsel informed George Sarton in a letter that he had arrived in America with "two copious manuscripts, on Natural and Historical Laws the one, on the Social Roots of Science the other, [which] in spite of preparations lasting years are not quite finished yet and therefore are not published".[39] These two manuscripts were never published. We neither know where they are, nor to what extent the manuscripts were finished. Most likely these two 'manuscripts' were closer to organized research notes than draft versions of books. We are confident that the papers Zilsel published and the unpublished materials we were able to find can be related in a fairly straight way to these two 'manuscripts'.

Zilsel enclosed a "sketch of the planned work The Sociological Roots of Science" in one of his very first letters to Horkheimer at the IISR.[40] Mention is made in the same letter of a second project on 'Natural Laws and Historical Laws' of which, Zilsel says, a sketch had yet to be made. Zilsel sent an abstract (*Inhaltsskizze*) of each of the projects two days later. He also points out that because "I have been working on both subjects with only short interruptions for eleven years, and it would please me greatly, if I could finish at least one of the MSs". In an attempt to adapt his project to the goals of the IISR, Zilsel indicated his preference for the project on the social roots of science "as the other subject leads rather deeply into problems of physics. I am uncertain as to whether these might be too physicalistic for your institute".[41]

Several versions of both projects can be found in the papers of the Zilsel file in the Horkheimer Archive. All of them are undated, and their chronological order is unclear. There is one two-page outline of the project 'on the concept of law' in German and there are five outlines of his project 'on the social origins modern of science'. Two of them are German, one five pages long and the other eight. The remaining three are in English, all of the same length (three pages). Two of these are identical, and the third contains corrections to the English. Although these project outlines constitute the only written material we have on Zilsel's project 'on the social origins of modern science', we do know that he gave a lecture to the *Wiener Kulturwissenschaftlichen Gesellschaft* entitled 'The Genesis of Science - A Sociological Problem'[42] in 1930[43], and a further lecture

---

39 Zilsel to Sarton, June 3, 1939. We quote from the Sarton papers held at the Houghton Library of Harvard University and use the acronym SP followed by the date of the letter.
40 HP/Z; April 19, 1939.
41 HP/Z; April 21, 1939.
42 'Die Entstehung der Wissenschaft - Ein soziologisches Problem'.
43 Zilsel to Reichenbach, HR 013-38-31, May 2, 1930. We have not been able to find the MS used for

on the 'Origins of Modern Science' to the philosophy department of the Volksheim Leopoldstadt in 1936.[44]

II.2 Project I: The Social Roots of Modern Science

We now intend to take a closer look at the extended version of Zilsel's project 'on the social roots of modern science' (a full translation of this text is given in appendix I, see pp. 227-232). Zilsel starts by noting a problem for undertaking a comparative study of the emergence of science: it only developed in the western world. We are thus not in the position to compare the development of science in one culture with that of another non-Western culture. His answer to this dilemma is to assume that all cultures are subject to one general scheme of development. Although Zilsel's terminology varies slightly all schemes comprise the stages: 1. magic, 2. theology, 3. humanism, and 4. science.[45] With this scheme in hand Zilsel believed that he possessed the necessary tool for conducting a historical and sociological comparison of non-scientific and scientific cultures. Above all, the goal was to single out those characteristic social structures which could be correlated with their respective knowledge-systems. These characteristics could then be compared with the preceding and subsequent stages of the historical scheme, thereby highlighting the social roots of the production of knowledge in a way which could be proved or refuted through a comparison with the historical development of other cultures. This comparison of social structures and their corresponding knowledge-systems is best seen in Zilsel's comparison of science and humanism.[46] In the project outline, Zilsel stipulates five key characteristics of the scientific spirit, namely, 1. worldly interests, 2. causal interests, 3. reliance upon independent thinking instead of authority and tradition, 4. aiming at quantification, and 5. empiricism. These are compared and contrasted not only with humanism, but with the other stages of historical development as well. For Zilsel, the scientific validity of his approach was guaranteed by the possibility of comparing this singular development in European history with a general development to be found in all cultures.

The main factor which in Zilsel's eyes brought forth the age of science from humanism, and which was lacking in the historical developments of all the other cultures of the world, was capitalism. Humanism, like science, is a mundane and rationally oriented activity, but it has a contemptuous opinion of manual labor. Capitalism is the main force behind the changing attitude towards manual work.

---

this lecture. In quoting from the Reichenbach papers, which are located at the Archives of Scientific Philosophy in the Twentieth Century, Pittsburgh, Penn. we will use the acronym HR followed by the inventory control number of that collection, followed by the date of the letter.

44 As is seen from the list of Zilsel's activities at the *Wiener Volkshochschule* given by Stadler, op. cit., 1997, p. 805-17.

45 Zilsel uses the concept of humanism not only to denote the Italian based cultural period in the early Renaissance but it covers every cultural movement carried by *literati* with a bend to rational and mundane learning.

46 Cf., in particular, 'The Methods of Humanism', this volume, pp. 22-64.

Yet "Why didn't science unfold fully in Antiquity?" and "Why didn't it develop in China?", where limited forms of capitalism existed? These questions, posed at the end of his exposition, are typical of the cross-cultural comparisons which Zilsel would have presumably further pursued. His conclusion offers us a brief description of the results and current state of his project:

> Attempt to solve the main problem: Science originates in urban cultures, money economy, market economy, and competition so that first the magical and theological mode of thinking is destroyed and the mode of thinking of certain groups of superior artisans rises to the level of the literary educated upper class. If rough manual labor is mainly carried out by slaves, this rise is hampered.

II.3 The Table of Contents of Zilsel's Planned Book

In addition to the various project outlines, we have been able to locate three 'Tables of Contents' in which Zilsel sketched the nature of his planned book. These allow us to obtain an impression of the historical materials, plan, and state of the planned book. Let us begin then by taking a closer look at Zilsel's last (February 23, 1943) 'Table of Contents'[47]:

I. The Precursors of Modern Science:
(1) Scholasticism.
(2) Humanism.

II. The immediate predecessors of science:
(1) Technology in nascent capitalism.
(2) Liberal and mechanical arts. The disdain of manual labor and its implications for experimentation and dissection.
(3) Superior manual laborers (artists, artists-engineers, makers of nautical and makers of musical instruments, navigators, map-makers, surveyors, gunners, surgeons).
(4) The writings of superior artisans 1500-1600.

III. The emergence of science 1570-1620:
(1) Learned literature on the mechanical arts before 1600;
(2) Representatives of the mechanical arts adopt methods of the scholars (the map-maker Ortelius, the book-keeper and military engineer Simon Stevin, the surgeon Ambrois Paré).
(3) Academically trained scholars adopt methods of the mechanical arts (William Gilbert, Galileo, Bacon, Harvey).

IV. The Rise of the Quantitative Spirit:
(1) The technological background (clocks, book-keeping, monetary reform, surveying, gunnery, mechanics).
(2) Mathematics and its relation to commerce, military engineering, technology, and painting 1300-1600.

V. The Genesis of the Concept of Physical Law.

VI. The Special Position of Astronomy:
(1) Astronomy as a liberal art. Astrology, calendar reform.

---

47 Quoted from the Zilsel file held by the American Philosophical Society, Philadelphia, PA. We will use the acronym APS/Z followed by a date in our references to this collection.

(2) Practical astronomy and navigation.

VII. The Genesis of the Idea of Progress and Scientific Co-operation:
(1) Virtual absence of co-operation and the ideal of progress in classical antiquity.
(2) Scholasticism and Humanism.
(3) The superior artisans of the 15th and 16th centuries.
(4) Sociology of the first learned societies.[48]

Our interpretation of this table is informed by Zilsel's desire to explicate the concept of the 'scientific spirit'. This scientific spirit, or ethos as the current sociological phrase would be, is comprised of the five characteristics he mentioned in his project outline (see above section II.2). For each of these features his problem was to identify their respective social origins. Zilsel's radical suggestion is that the origin of experimental work is to be found outside academia with the craftsman, artisans, surgeons, instrument makers, surveyors, navigators and all those who earned their living by getting their hands dirty. The *pièce de résistance* of the book in our view is chapter III. Here Zilsel has to argue, as he puts it in the abstract of his most famous essay 'The Sociological Roots of Science', that "Science was born when, with the progress of technology, the experimental method eventually overcame the social prejudice against manual labor and was adopted by rationally trained scholars".[49] He also had to demonstrate the counterpart of this process: superior artisans adopting the formal methods of rationally trained scholars. This leaves Zilsel with a number of problems. Most importantly, he had to explain why there was this crossing of social barriers in the late 15th and early 16th century. This is the topic of chapter II. Zilsel singles out the emergence of capitalism as the prime social motor behind the rise of science. In his view, capitalism is responsible for the changing attitudes towards manual labor, and this is taken to explain the erosion of the social boundaries between those who work with their hands and those who work with their brain. Chapter I was to present the two main immediate predecessors of the scientific spirit, humanism and scholasticism. Its title mirrors the development sequence of Comte's classical positivism. The content table of the first two chapters leaves the reader with the impression that science was the inevitable next step waiting to happen, if only the inhibiting social factors could be unlocked.

---

48 The Table of Contents given by Dahms in his 'Edgar Zilsels Projekt "The Social Roots of Science" und seine Beziehungen zur Frankfurter Schule', pp. 474-500 in *Wien-Berlin-Prag: Der Aufstieg der Wissenschaftlichen Philosophie; Zentenarien Rudolf Carnap - Hans Reichenbach - Edgar Zilsel*, R. Haller & F. Stadler (ed.). (Wien: Verlag Hölder-Pichler-Tempsky, 1993), p. 485 is dated by Zilsel June 22, 1941. When that Table of Contents is compared to the one Zilsel composed about 18 months later, three differences come to the fore. The first is that chapter I in the earlier version comprised three subsections, Magic being the third. The second is that in the earlier version no mention was made of a chapter dealing with progress and scientific co-operation - although the topic was mentioned in the project outline - see appendix I. The third difference is that the earlier version had as chapter VII 'The society of European early capitalism, of classical antiquity, and of China'. In the later version this chapter is dropped and a chapter on progress and scientific co-operation is added.
49 This volume, p. 7.

This explains why science arose at the time and place it did. But the argument so far only touches upon one feature of the scientific spirit: experimentation. We interpret chapters IV-VII as being planned to explain the social origins of the other features of the scientific spirit or ethos. Chapter IV picks up the eminent example of mathematics which is not an experimental activity but a formal science. Zilsel's treatment of the quantitative nature of modern science needs to explain where the origins of this feature are to be located and how it was incorporated into the new science. Another important example of a science not based on experiments is that of astronomy. Chapter VI was planned to describe the "somewhat different way, sociologically, [in which] modern astronomy developed".[50] In his Copernicus essay, Zilsel comments briefly on the sociological difference between the astronomer and the mechanic as exemplified in the figures of Copernicus and Galileo:

> The difference between Copernicus and Galileo is not a difference of individual psychology only, and even less can it be explained by the mere difference of time. Kepler, who was a contemporary of Galileo was, as is generally known, at least as Pythagorean and thought at least as teleologically as Copernicus. There rather seems to be a difference between astronomy and mechanics as to their historical evolution and sociological origins. The very first astronomers were Babylonian priests and this connection with priesthood was never quite interrupted... A more extensive inquiry, however, than could be given in this short note on Copernicus, would be necessary to verify this sociological explanation.[51]

Zilsel hints at the religious, astrological, and calender-making functions of astronomy which associated astronomers with the class of priests. He also points out that 'practical astronomy' was linked to navigation. One of the primary reasons for the sociological difference between mechanics and astronomy is due to the difference between observation (passive looking) and experimentation (active manual work).[52] According to Zilsel, the success of astronomy in spite of its lack of experimentation is an extraordinary case which arises from the fact

> that in our solar system superimposed effects belong to very different orders of magnitude and therefore can be separated comparatively easily. The solar system is exceptionally well isolated and the sun surpasses by far all planets in mass. Were the solar system continually bombarded by heavy meteorites or, what is the same, were it passing through a dense star-cluster, and were

---

50 *Ibid.*
51 'Copernicus and Mechanics', this volume p. 126.
52 Zilsel has this to say on why experiments are so essential to empirical science. "Mere observation is a passive affair. It means but 'wait and see' and often depends on chance. Experiment, on the other hand, is an active method of investigation. The experimenter does not wait until events begin, as it were, to speak for themselves; he systematically asks questions. Moreover, he uses artificial means of producing conditions such that clear answers are likely to be obtained. Such preparations are indispensable in most cases. Natural events are usually compounds of numerous effects produced by different causes, and these can hardly be separately investigated until most of them are eliminated by artificial means. There is, therefore, in all empirical sciences a distinct trend toward experimentation" ('Problems of Empiricism', this volume, p. 171-199).

Jupiter's mass of the same order of magnitude as that of the sun, Copernicus, Kepler, and Newton would not have achieved much.[53]

In a similar vein, we see the other chapters dealing with features of the scientific spirit which are not simply covered by the general acceptance of the experimental method, i.e. lawfulness of nature, progress, and co-operation.

II.4 Completion to be Expected within a Year
How far advanced was Zilsel's project? Let us begin by comparing the Table of Contents given in the project sketch of 1943 with a chronologically-ordered list of the essays published in this volume:

'The Social Roots of Science', 1939, unpublished;
'Copernicus and Mechanics', 1940;
'History and Biological Evolution', 1940;
'The Origins of William Gilbert's Scientific Method', 1941;
'Phenomenology and Natural Science', 1941;
'Physics and the Problem of Historico-Sociological Laws', 1941;
'Concerning "Phenomenology and Natural Science"', 1941;
'Problems of Empiricism', 1941;
'Science and the Humanistic Studies', 1941, unpublished;
'The Genesis of the Concept of Physical Law', 1942;
'The Sociological Roots of Science', 1942;
'The Methods of Humanism', 1942, unpublished;
'The Genesis of the Concept of Scientific Progress and Scientific Co-operation', 1944, unpublished;
'The Genesis of the Concept of Scientific Progress', 1945;

Of these essays the following can be directly connected to the Table of Contents:

---

[53] This volume, p. 176. Zilsel's hint that astronomy developed differently from the other sciences is not without its significance for the historiography of the emergence of modern science and our understanding of Zilsel's argument. Alexandre Koyré was one of Zilsel's harshest critics - see for example the first two chapters of Koyré's *Metaphysics and Measurement* (Yverdon, etc: Gordon & Breach, 1992 [1968]). Central to Koyré's argument is that the new science of Galileo developed without any role of experiments or experience or sense-perception. It was achieved by "pure unadulterated thought" (*Metaphysics and Measurement*, p. 13), i.e. Koyré claims that theoretical imagination had primacy over experience. This stress on the sole role of reason in the creation of the new science should be seen against the background of Koyré's deep interest in the way astronomy was transformed during the 'scientific revolution' - see for example his *The Astronomical Revolution: Copernicus -Kepler-Borelli* (New York: Dover, 1992). But it is now well established - see the various contributions to the Koyré special issue of the journal *History and Technology* of 1987 - that Koyré's picture is extreme and one-sided. It is an artifact of his limiting himself to astronomy. As Kuhn pointed out in his review of *Metaphysics and Measurement* ('Alexandre Koyré & the History of Science', *Encounter*, 1970, 34, pp. 67-9), Koyré's neglect of the entire Baconian movement with its emphasis on experiment, instrumentation, utility, and the study of crafts would have been a "disaster" if applied to the development of chemistry, electricity, or magnetism (p. 69).

'Copernicus and Mechanics';
'The Origins of William Gilbert's Scientific Method';
'The Genesis of the Concept of Physical Law';
'The Genesis of the Concept of Scientific Progress and Scientific Co-operation'
'The Methods of Humanism'.

It must first be noted that Zilsel's best known essay 'The Sociological Roots of Science' was not projected to be included in the book plan. This should hardly be surprising as it was meant as a summary of the main argument of the book and there was no need for it to appear as a chapter heading.

The inescapable conclusion, however, must be that the material which Zilsel did manage to publish does not even come close to meeting the aspirations found in the 'Table of Contents' described above. What we do have is part of Chapter III, section 3 (academically trained scholars who experiment), in the form of his essay on Gilbert.[54] His 1942 essay on 'Physical Law' looks like the whole of Chapter V. The unpublished essay 'The Methods of Humanism' can easily be recognized as being part of section I, 'The Precursors of Modern Science'.[55] Dahms has interpreted the Copernicus essay to be chapter VI (astronomy).[56] But for reasons already set out we believe it should at best be seen as

---

54 The Gilbert essay is generally considered to be one of Zilsel's best. He himself was far from pleased with the published version. The publication history of this essay is as follows. In early 1940 an essay of 55 type-written pages titled "The Origin of William Gilbert's Scientific Method" was accepted for publication by *Osiris* (editor: George Sarton, Harvard). But as Zilsel informed Horkheimer (HP/Z, May 24, July) "Since *Osiris* is printed in Belgium and its further existence is questionable, negotiations about its publication in the *Journal of the History of Ideas* (editor Prof. Lovejoy) are pending. *Osiris* is devoted to the history of the natural sciences, JHI however to the general history of the mind. Therefore, a rewriting of the essay had become necessary and is already done". The rewritten version is the version Zilsel published. From his correspondence with Sarton a far more dramatic picture emerges. Since Sarton ceased to publish *Isis* after 1940, *Osiris* had become the only place where the MS could be published. Zilsel underestimated the length of the MS and worked on it longer than he first indicated to Sarton - in effect up till mid February 1940. Sarton quickly accepted the MS but indicated that it would not be published before the end of 1941. This was very unwelcome news for Zilsel who was in a hurry to get US scholars to take notice of his work as soon as possible. Hitler Germany invaded the Netherlands, Belgium, and France in May 1940 and WWII began in earnest on the European western front. Zilsel queried Sarton on the sixth of July whether there still was a chance that the MS might be published "after the European catastrophe" since he has an opportunity to publish it elsewhere, "but I ought to shorten it so considerably that it would be damaged. I should prefer, therefore, to publish it in your periodical if possible". Sarton answers (July 9) Zilsel that "Your paper is the only *Osiris* paper which had not been mailed to Belgium at the time of the catastrophe" and he returned it, "for I cannot guarantee early publication, and I may be unable to publish it at all". Hence the situation was as follows: the published Gilbert essay is not the essay originally intended by Zilsel but a drastically shortened version of a longer MS - which we have been unable to trace - and Zilsel was not pleased by the shortening. We know (SP/Z, March 7, 1940) that of the original 55 pages sections 9-11, "contain the main results on Gilbert himself" and that the "hypotheses on the origin of modern science" are explained in section 12. This section is clearly no longer in the Gilbert essay, whether it was to constitute the bulk of Zilsel's 'Social Roots' essay of 1942 is impossible to say but is our best guess.
55 Directly beneath the title of this essay stands (in the typescript version) "I/2: The Methods of Humanism". This suggests that Zilsel intended to use this essay as the second section of chapter one of his book.
56 H.-J. Dahms, op. cit. p. 485.

only part of that chapter. The essay on 'Scientific Progress', published after his death, and of which we publish an extended version in this volume, is easily recognized as Chapter VII.

In the summer of 1941, Zilsel judged the state of his project as follows:

> Collection of the material is completed in the following fields: Galileo, Stevin, Francis Bacon, Harvey, the Italian mechanics and mathematicians before 1600, the distinction between liberal and mechanical arts and the social prejudice against liberal and mechanical arts and the social prejudice against manual labor, the artist-engineers, the English mathematicians in the 16th century, the makers of nautical and geodetical instruments, double-entry-bookkeeping and the quantitative spirit.
> Collection of the material is nearly completed in the following fields: Kepler, the "scientific" literature composed before 1600 by authors without academic training i.e. by superior craftsmen, the early advocates of the idea of scientific co-operation, application of the scholastic method to mundane matters.
> Considerable gaps still exist in the following fields: late scholasticism, beginnings of the quantitative method in the late middle ages, geographers and map-makers in the 16th century, the literature on navigation, monetary reform and the quantitative spirit, German, French, and Spanish authors in the 15th and 16th centuries, the Italian academies in the 16th century.[57]

As a final remark he writes: "Completion of the whole study can be expected by July 1942".[58] In light of the discrepancy between the plan of the book and the state of the (un)published essays this is most striking. This discrepancy makes us uncertain when it comes to judging the feasibility of his plan and his perhaps overly optimistic approach to it. We have no idea how much of the plan was an illusion and how much could actually be completed in one year. We do, however, feel that some scepticism is justified.

One difficulty in interpreting Zilsel's plan is that the purpose of the report must be taken into account. It was most likely written to persuade outsiders to provide funds for his research. Hence there is bound to be an element of impression management at work. This is at least how we read Zilsel's remarks on how his work was being carried out. He writes:

> after the leading ideas had been formulated and a considerable amount of source material had been collected the whole study was divided into chapters and subsections. This stage was reached before 1939. As far as possible the single sections are to be completed successively. However it was not feasible to strictly adhere to this plan. Very numerous authors must be read and often one author contributes material to several sections at the same time.

The tension between feasibility and illusion reappears when we are able to get another glimpse of Zilsel's project 'on the social origins of modern science'. We find this in an exchange of letters between Princeton University Press and Zilsel in which Zilsel once again gave the impression that his book was close to being finished. In a letter dated August 16, 1943, Datus C. Smith, Jr., director of

---

57 HP/Z, June 22, 1941.
58 We know from the copies of later grant applications that he always asked for money for one year. This most likely reflects the fact that shortly after the great influx of émigrés in 1938 the relief agencies were no longer willing to commit themselves for longer than one year.

Princeton University Press, inquires whether Zilsel was "already committed to a publisher" and whether "the manuscript is in shape to show".[59] The reason for his interest is that a "mutual friend, Albert T. Lauterbach", had informed Smith "a little about your work on the social origins of modern science". Smith urged Zilsel to submit the MS to Princeton U. P. "as this kind of history of science is of special interest to us, and from what Mr. Lauterbach tells me of your work I feel sure that you are making a major contribution". Zilsel submitted a prospectus in a letter dated Sept. 26, which we have not seen. Smith acknowledged receiving the prospectus in a response to Zilsel dated October 7, 1943. He let Zilsel know that he was delighted to read "the prospectus of your really exciting book project", and that he will present it to the Editorial Board in about two weeks' time. Smith reported the results of that meeting to Zilsel in a letter dated October 21, 1943. He writes: "Although it was no surprise to me, it is nevertheless a pleasure to report that this project excited the undisguised enthusiasm of everyone on the Board". Smith adds that this should not be interpreted as an official commitment on the part of the P.U.P. to publish the MS - for that they would need the MS "or a generous portion of it". He has "no hesitation in saying, however, that if the manuscript fulfils the promise of the prospectus - and your Princeton admirers are confident that it will - publication would be immediately approved". He indicates that, as soon as the MS is ready to show, "I believe quick action can follow". We also hear about the title of the projected book in this letter: *The Genesis of the Modern Scientific Method*. As we have not been able to track down the prospectus, it is difficult to establish what it contains and in what way it is similar to the earlier project reports we have. What is clear however is that Zilsel gave the impression to others that his book was nearly finished. Zilsel also gave this impression to the American Philosophical Society. In his application for a research grant of October 28, 1941, he indicated that he would be able to finish the book within a year, and in an application of February 28, 1943, he is even more adamant. After acknowledging that he had "underestimated the duration of the investigation" in his application of 1941, he now claimed that "completion by the fall 1944 can be expected practically with certainty".[60] Zilsel strongly believed that he could finish his book, but, from what we are able to publish in this volume, it is clear that a number of parts are missing, for example, the section on Scholasticism, chapter II, and large parts of chapter III.

II.5 Physical Law and Historical Law
We now move to a discussion of Zilsel's second project 'on the concept of law' and Zilsel's outline of it. A full translation of this outline is given as in appendix II, pp. 233f. The projected book, entitled *Origins and Transformation of the*

---

[59] This and some other letters related to this correspondence between Zilsel and Princeton University Press were kindly made available to us by Paul Zilsel (PF).
[60] APS/Z, February, 28. 1943.

Concept of Natural Law, would consist of six chapters and an introduction.[61] Zilsel presents the aim of the book as follows:

> To clarify the claim that there are "laws" in history and sociology through an analysis of the concept of a law of nature with examples taken from modern science. It shall at the same time be shown that there isn't any fundamental difference between history and sociology on the one hand, and science on the other, which would make the search for laws in history and sociology completely hopeless.

Obviously, to simply explicate the concept of historical and sociological laws does not imply their existence. Since the state of the humanities did not allow Zilsel to cut the epistemological controversy short by pointing at examples of valid laws, he was forced to take those scholars seriously who denied the possibility of lawlike statements in humanistic research.[62] This project 'on the concept of law' aimed at doing just that in a fundamental and thorough way.

Zilsel begins in chapter two with a discussion of various ways of discovering and explaining laws, and goes on to discuss the properties of scientific laws. Need they be precise and predictive? Both features depend on the availability of 'isolated systems', which rarely exist in nature or history. Only when the parameters of a system can be controlled can precision and prediction be achieved. Many natural sciences share a similar state with the humanities in their inability to analyze their subject of research as a closed system, and establish deductive laws. As such, the condition of the humanities, or the geo-sciences, may be viewed as immature "data-gathering".

Zilsel proceeds, in chapter three, to examine the distinction between 'micro- and macro-laws'. Historical laws need to be macro-laws. This has several implications: the macro-variables may be rather different from the micro-variables; the statistical significance of the laws depends on the availability of comparable cases; the impact of micro states may effectively blur the macro states of a system. These implications directly aim at the heart of the essential question of whether historical and sociological laws are possible at all. It could, for example, be argued that the interferences between micro and macro levels of analysis are so heavy that no sound statistical relations can be found. As we shall show in section III, Zilsel was convinced that only a direct search for such relations would be able to solve the problem. The next chapter addresses the difference between time-dependent and time-independent laws. Clearly, the first type relates to the historical dimension of cultural development (Dollo's law is mentioned as an example from biology).[63] The second type comprises sociological

---

61 In HP/Z, undated.
62 In his essays he mentions the philosophers Wilhelm Dilthey, Heinrich Rickert, Wilhelm Windelband, and the sociologists Georg Simmel, Werner Sombart, Rudolf Stammler, Alfred Weber, Max Weber. For a thorough analysis of the background of this debate see K. C. Köhnke, *The Rise of Neo-Kantianism: German Academic Philosophy between Idealism and Positivism*. (Cambridge: Cambridge University Press, 1991)
63 Dollo's law, named after the Belgian paleontologist Louis Dollo (1857 - 1931), states that evolution is irreversible, i.e. that evolution never returns to a previous condition. Richard Dawkins explains

correlations between coexisting phenomena. Chapter five addresses the causality crisis of quantum physics and tackles the problem of free will. The physicist Pascual Jordan had tried to build an analogy between the uncertainty principle of quantum physics and man's free will. Zilsel was sceptical about this analogy.[64] He seems to argue here that, even if we take free will for granted, this would - as such - not impair statistical effects. Without doubt, this is an intriguing problem. Unfortunately, we do not know how Zilsel intended to answer it.

If we look at the essays published in this volume that are not related to Zilsel's project 'on the social origins of modern science' we still have:

x: 'History and Biological Evolution', 1940;
xi: 'Phenomenology and Natural Science', 1941;
xii: 'Physics and the Problem of Historico-sociological Laws', 1941;
xiii: 'Concerning Phenomenology and Natural Science', 1941;
xiv: 'Science and the Humanistic Studies', 1941;
xv: 'Problems of Empiricism', 1941.

The third essay can easily be seen as covering many of the topics mentioned in Chapter 3. The same applies to the first essay. 'History and Biological Evolution'[65] was based on a paper entitled 'Biology and History' which Zilsel presented at the University of London when he was in London waiting for his visa to enter the USA. In a letter to Horkheimer, Zilsel points out that this

> essay is a polemic against Wells' popular world history,[66] although the name is not mentioned. It tries to prove that even from a purely natural scientific point of view, history *cannot* be conceived of as a period of human phylogenesis, rather, history reveals its own characteristic laws. Indeed, it wants to show that this separation can even be carried out with purely scientific means.[67]

Dollo's law as follows: "Dollo's law is really just a statement about the statistical improbability of following exactly the same evolutionary trajectory twice (or indeed any particular trajectory), in either direction. A single mutational step can easily be reversed. But for larger numbers of mutational steps, even in the case of the biomorphs with their nine little genes, the mathematical space of all possible trajectories is so vast that the chance of two trajectories ever arriving at the same point becomes vanishingly small" (*The Blind Watchmaker* [London: Penguin books, reprinted with an appendix, 1991], p.94).

64 Cf. E. Zilsel, 'P. Jordans Versuch, den Vitalismus quantenmechanisch zu retten', *Erkenntnis*, 1935, 5, pp. 56-65.

65 This essay is a shortened version of the considerably longer 'Geschichte und Biologie, Überlieferung und Vererbung' which was originally published in *Archiv für Sozialwissenschaft und Sozialpolitik*, 1931, 65, pp. 475-524.

66 The reference is to H.G. Wells' (1866-1946) popular history textbook, *The Outline of History* (1920, German edition 1928). *The Outline* grew out of Wells's belief that human history could not be understood without reference to biology, and the book was intended as a history of the species man. As Wells pointed out to a critic, the principles for inclusion in the Outline were 'the contribution an individual had made to the growing centralization of power, or to the increased concern for the future of the species'. See David C. Smith, *H. G. Wells: Desperately Mortal. A Biography*. (New Haven, London: Yale University Press, 1986), pp. 245-267, quote is from p. 257. We thank Anna Mayer for her help in clarifying some of the background of this remark by Zilsel.

67 HP/Z, December 20, 1939.

The second paper takes issue with what Zilsel considers to be the remarkable thing about phenomenology, namely, that it supplants causal investigation with *Wesensschau*. As such, phenomenology only tries to revive, in Zilsel's eyes, premodern and prescientific modes of investigation. The fourth essay is a rebuttal of the commentary which W. Cerf published in reply to Zilsel's essay on phenomenology. 'Science and the Humanistic Studies' is an unpublished paper[68] which shows that Zilsel was still actively engaged with this MS in 1942. The essay 'Problems of Empiricism' will be discussed below.

III. Zilsel's Philosophical Background

We now turn to the relationships between Zilsel's project 'on the social origins of modern science' and his project 'on the concept of law'. We maintain that the central thrust of Zilsel's philosophical program may be found in the relations between these two projects. However, in order to appreciate the richness of this research program, we must first take a closer look at various facets of his earlier work.

a) His first book, *The Application Problem* (1916), attempts to solve the fundamental philosophical problem of applying statistics to an 'irrational nature'. Methodologically, Zilsel operates with the Kantian conviction that the lawfulness of nature ultimately rests in our attributing certain fundamental features to nature which result from our own conditions of cognition. This early attempt to make the philosophy of nature (*Naturphilosophie*) contingent upon a theory of knowledge (*Erkenntnistheorie*) is discussed in section III.1.

b) Zilsel's second major book *On the Development of the Concept of Genius* (1926), an elaboration of his 'post-doctoral dissertation' (*Habilitationschrift*), presents a sociological analysis of the genius personality cult which dominated Renaissance humanism and, in Zilsel's eyes, continued to pervade the ideals of his time. More important for our analysis is that this book aimed to find and explain the empirical laws which governed the development of the concept of genius. Zilsel's commitment to this project and its methodology were at stake when the University of Vienna denied him the right to teach at the professorial level by refusing this work. Section III.2 describes this conflict.

c) Zilsel's interest in the sociological factors underlying cultural history and the history of science led him to assert the role of the humanities within the Vienna Circle's unity of science program. Zilsel's position with respect to this program and his relation to the Vienna circle are analyzed in section III.3.

---

68 Zilsel used this as a basis for his presentation at the Sixth International Congress for the Unity of Science held at the University of Chicago, 2-6 September, 1941.

d) Critical to Zilsel's entire philosophical program is his commitment to Marxism. Zilsel used Marxism as a philosophical framework which would, in principle, be able to explain historical developments with as much of a claim to truth as scientific knowledge. He refused, however, to use Marxism as a political dogma. He rather understood it as a set of hypothetical assumptions which was only empirically provable. Section III.4 describes Zilsel's ideas on Marxism.

(e) Finally, we interpret the essay 'Problems of Empiricism'[69] as a concise example of the confluence of Zilsel's historical ('on the social origins of modern science') and philosophical work ('on the concept of law'). The latter essay illustrates the extent to which Zilsel's project 'on the social origins of modern science' was meant as an empirical study supporting the case for a historical science oriented to a search for laws as argued for in his project 'on the concept of law'. (section III.5).

III.1 The Fundamental Dualism or How can Rational Law describe Irrational Nature?

Zilsel fibst grappled with the role of the concept of law in his book *The Application Problem*. The main topic of the book is a riddle posed by the so-called law of large numbers. The law states what at first glance seems to be a rather truistic statement of probability theory, namely that "with a large number of repeated throws of a chance game ... the relative frequency almost equals the mathematical probability."[70] Nature, however, could be rather different. She could produce frequencies quite different from the expected result. It is therefore not at all trivial to ask why the law of large numbers is applicable at all. Zilsel construed this problem as being part of a wider one: how can rational mathematical constructions apply to a vague and irrational nature? This is what Zilsel termed 'the application problem'.

Since it is neither natural law nor mathematical principle, Zilsel found the epistemological and ontological status of the law of large numbers to be in need of explanation. Zilsel ventured a Kantian solution based upon contemporary theories of statistics and induction. Kant's transcendental solution to the Humean problem of relating deductive reasoning and inductive sense experience was to declare the principle of causality a precondition of knowledge. Either we understand the mutual relations of things in terms of causes and effects, or we don't understand anything.

The critical aspect of Zilsel's thinking that developed out of this analysis of the relationship between the rational laws of probability and empirical causal laws of nature led him to accept that there are general philosophical problems

---

69 Otto Neurath asked Zilsel to write this essay in a letter dated 10. 1. 1938 (NP/Z). He was working on the MS for roughly two years interrupted of course by his and his family's moves to London and New York.
70 Zilsel, *The Application Problem*, p. 3. Apparently Zilsel's analysis made a deep impression in mathematical circles, cf. Stadler, 1997, op. cit., p. 802. Hans Hahn (*Monatshefte für Mathematik und Physik*, 1917, 27/8, p. 37-8) however wrote a critical review, claiming that Zilsel's mathematical argument was fundamentally flawed. For Zilsel's reaction see note 71.

which are related to all sciences but not solvable within any one specific science. In opposition to the proponents of logical positivism within the Vienna Circle - Otto Neurath, Moritz Schlick, Rudolf Carnap, Richard von Mises - Zilsel believed that the discussion of such problems was fruitful and should not be denounced as metaphysical '*Scheinprobleme*'.[71] At the same time, however, he never believed in the capacity of philosophy to solve fundamental problems independent of empirical research and distrusted philosophy as an independent discipline. Zilsel was very outspoken in his desire to unite these fundamental philosophical problems with the contemporary problems presented in empirical research. He particularly despised all attempts by "schoolmasters ... who would separate ... philosophy from the empirical disciplines".[72] This philosophical position is perhaps clearest in the following quote:

> the remaining unsolved and fundamental philosophical problems can only be discussed in a fruitful manner if the results and methods already made in the empirical sciences are taken into account ... [and as] Ernst Mach and Henri Poincaré [have shown] it is at present only possible to fruitfully discuss philosophical problems in that intimate connection to living science which characterized the classical philosophy of the 17th century.[73]

Zilsel's subsequent research on the concept of genius arose within the context of this commitment to the "living sciences". He began to direct the 'application problem' toward the statistical analysis of history and culture. How is the 'nature' of society to be perceived if it is subordinate to the law of large numbers? Is this 'nature' fundamentally different from that of the natural sciences? Philosophers of culture would strongly emphasize the difference which Zilsel ventured to deny.

III.2 Zilsel and his failed Habilitationsschrift

On July 10, 1923, Zilsel submitted his two-volume *Contributions to the History of the Concept of Genius*[74] to the Philosophy Department of the University of

---

[71] In an undated letter to Moritz Schlick, in which Zilsel responded to Schlick's paper 'Die Kausalität in der gegenwärtigen Physik' ('Causality in Contemporary Physics', *Die Naturwissenschaften*, 1931, 19, pp. 145-62), he insisted upon these views. In a long and careful argumentation he makes his point that the presupposition of a "specific constellation of nature" (15) is needed in order to apply statistical laws to nature. (SchP/Z). In a letter to Zilsel Reichenbach responded: "I completely agree that positivism suppresses certain things which are simply there; I consider it wrong to declare these *Scheinprobleme*, but prefer to raise the question as to how to conceive the concept of the scientifically-statable (*wissenschaftlich-Sagbaren*) so that these otherwise suppressed things might be comprehended". (HR 013-38-15, July 4, 1932. Quoted by permission of the University of Pittsburgh. All rights reserved). Reichenbach and Zilsel had corresponded on probability and induction since 1925. The 'spirit' of their exchange may be grasped by the following remark of Zilsel: "By the way, my *Application Problem* contains a lot of mistakes. To make mistakes seems to adhere to the essence of philosophy; one only wishes to have a philosophical method in which right and wrong are discernable at all, and in which the mistakes are discovered as quickly as possible". (HR 016-24-06, May 22, 1925. Quoted by permission of the University of Pittsburgh. All rights reserved.)
[72] Zilsel, 'Philosophische Bemerkungen', in *Der Kampf*, 1929, 22, pp.178-86; republished in E. Zilsel *Wissenschaft und Weltanschauung: Aufsätze 1929-1933*, K. Acham and G. Mozetič (eds.), 1992, pp.31-44, p. 40.
[73] Ibid, p. 39.
[74] *Beiträge zur Geschichte des Geniebegriffes*.

Vienna, and applied for a *venia legendi*[75] for the subject of philosophy. The first part of this work was entitled 'The Classical Roots of the Concept of Genius'[76] and the second part simply 'The Renaissance'. The examination committee consisted of eight people - Robert Reininger, Karl Bühler, Moritz Schlick, Richard Meister, Heinrich Gomperz, Julius Schlosser, Rudolf Wegscheider, and Felix Ehrenhaft. Later in life, Zilsel would describe himself as "a pupil of H. Gomperz under whose direction I studied especially ancient philosophy and civilization and learned to use philological and historical methods". From what we know, he was always very close to Schlick, with whom he had "worked intensely on the fundamentals of modern physics"[77], and whose obituary he wrote for the journal *Die Naturwissenschaften* after a mentally disturbed student shot Schlick in 1936. Even with his close supporters on the committee, however, Zilsel found himself in a difficult situation. Both Gomperz and Schlick asked Zilsel on separate occasions to withdraw his application, and after some objections he did so on June 3, 1924.

If we look at the various positions of the committee members, two substantial differences concerning philosophy and politics stand out. Philosophically, there was a clash between the different strands of traditional philosophy (including German idealism, Catholic philosophy, and neo-romanticism) and a new philosophy based on formal logic, language analysis, and recent developments in physics. Among German-speaking countries the battle lines were perhaps most clearly drawn in Vienna. The battle began in 1895 when a new chair for the philosophy of the exact sciences was established and occupied by the positivist physicist-philosopher Ernst Mach. His successors were Ludwig Boltzmann, Adolf Stöhr, and from 1922 on, Moritz Schlick. Opposed to this development were the representatives of the traditional conservative and Catholic philosophy. Although the established positions of the conservative faction differed in many respects, they had united views on several issues where knowledge politics played a significant role, among them the conviction that philosophy should be regarded as the 'queen of the sciences'. They defended the speculative and synthesizing power of philosophy and were opposed to any attempt to turn philosophy into a rigorous scientific enterprise. They emphasized a sharp division between philosophy and the empirical disciplines, and did not consider the new logic of Frege, Russell, and Whitehead nor the new metamathematics of Hilbert to be a part of philosophy. Politically, there was a clash between the democratic alliance of the new scientific philosophers like Carnap, Reichenbach, Schlick, Neurath and, directly opposed, the conservative, neo-romantic, and anti-socialist faction.

---

75 Upon acceptance of a *Habilitaion* a *venia legendi* may be obtained with which one may teach in the given area on a free-lance basis. Generally, it also serves as a license to apply for a university professorship.
76 *Die antiken Wurzeln des Geniebegriffes*.
77 Both quotes are from the curriculum vitae in HP/Z, undated.

Robert Reininger (who had a chair in the History of Philosophy) and Richard Meister (who had a chair in Pedagogy) were the two members of the committee who denied Zilsel's work "any value as a professional piece of philosophy". Reininger[78] considered "the writing on the concept of genius as unsatisfactory"; the second part "has nothing to do with philosophy" and is a "mere collection of materials". Meister claimed to have found a number of fundamental mistakes in the methodology of the work, amongst them "that everything is focused on economics". He also made a point of stating that he had nothing against the candidate on personal grounds, but that he had problems with his "one-dimensional and rationalistic" approach in which "everything is construed in economic terms". Without taking sides in this dispute,[79] it is possible to say that Reininger and Meister did have a point. The MS they were presented with was to a large extent a philological and historical analysis of the genealogy of the concept of genius supplemented by a set statements specifying under which socio-economic and institutional conditions the concept of genius could develop. Even today, the MS would not easily qualify as a thesis in philosophy. Zilsel's fight for the acceptance of his *Habilitationsschrift* was fought and lost in a dispute over what constitutes the subject matter of philosophy proper.

Zilsel had very outspoken ideas about exactly this point, and he was by no means willing to compromise. When Schlick at one point asked Zilsel to withdraw the MS and submit a new, more philosophically oriented *Habilitationsschrift*, Zilsel firmly refused. In his reply to Schlick, Zilsel demonstrates his strong commitment to stick to his research program:

> In continuing my work in philosophy and physics on phenomena of chance and large numbers in inanimate nature, my interests, in recent years, have turned toward the application of natural scientific methods to the humanities as well as toward the disclosure of fairly exact laws concerning the events in these fields. This research area shall occupy my mind for a longer period of time. I have already collected rather extensive material, especially regarding the history of the concept of genius. The results of the finished parts are formulated with respect to this material. How long the final completion of the other parts will occupy me, I, of course, cannot say today. I could not, however, justify to myself that the direction of my scientific work be influenced by any considerations other than by the problems themselves, my interest in them, and my previous work.

---

78 There may have been a tension between Zilsel and Reininger dating back to the publication of Zilsel's book *The Religion of Genius* (1918). In the author's summary (*Selbstanzeige*) of it (*Kantstudien*, 1919/20, 24, pp. 165/6) Zilsel mentions Reininger as one of the authors upon whom he had orientated his work. Zilsel's criticism of the 'cult of genius' may have been taken badly by Reininger for whom it was a very important idea.

79 The conflict needs not only be seen in the context of a larger inner-university fight between positivist and traditional philosophers but also be placed within a special controversy concerning school reform. Meister was a vigorous opponent of the school reform movement in which Zilsel was very much involved. Zilsel's scientifically based rationalism clashed with Meister's *a priori* intuitionism in legitimating authority - both in science and in politics. More details can be found in F. Stadler, 'Aspects of the Social Background and Position of the Vienna Circle at the University of Vienna', pp. 51-77 in *Rediscovering the Forgotten Vienna Circle: Austrian Studies on Otto Neurath and the Vienna Circle*, T. E. Uebel (ed.). (Dordrecht, etc.: Kluwer Academic Publishers, 1991), esp. pp. 59-61.

Furthermore, I would not have called my analysis of the history of the concept of genius, the outcome of eight years of work, a *Habilitation* thesis, if, in my judgement, it could not withstand scientific criticism. A withdrawal of the present application appears to me to amount to a revocation of my intent. I would not be ready to withdraw it unless scientific reasons convince me of the untenability of my judgement.[80]

However, Zilsel withdrew his application for the *venia legendi* in a letter to the Dean on the third of November 1924 and refused to present another manuscript "which belonged to the most narrow definition of philosophy", as had been requested by the evaluating committee. At whatever the cost to his career, Zilsel would stubbornly champion his understanding of philosophy. In his letter to the Dean, he writes:

I have, as seen in my previous work, approched philosophy, not coincidentally, not from the studies of historical literature. I have rather tried to develop my philosophy of nature and history with the help of physical and historical factual material in the hope of serving philosophy better than I would were I to cut her off from the fruitful ground of the individual sciences. It appears very improbable that I should fundamentally change my method in completing my planned philosophical work; it is thus improbable that the mentioned hopes for my future work can be fulfilled in the near future. I nevertheless find myself forced to follow the advice given to me with an apparently benevolent intention. I thus withdraw my application[81].

A combination of philosophical and political factors prevented the acceptance of Zilsel's work.[82] However, he did rework the MS - to what extent is unclear -

---

80 The complete text of this letter is given by Dvořak, op. cit., 1981, p. 130-1, n. 11.
81 This letter, d.d . November 3, 1924, is in the Zilsel file at the University of Vienna; we will use the acronym ZF/UV for references to this file.
82 As to the scientific merits of Zilsel's MS, the following is relevant. In order to find a way out of the situation the commission negotiated a compromise. It decided to seek an additional opinion from three external experts. The committee agreed upon three distinguished philosophers of high reputation though differing orientation - Ernst Cassirer (Berlin), Adolf Dyroff (Bonn), and Heinrich Scholz (Kiel). They were asked "whether according to your opinion the work can be appreciated as a piece of philosophy, and whether you believe that its ways of thought (which are obviously located in a boundary area of the history of the humanities) provide a sufficient basis for qualifying its author as a teacher in philosophy". (Letter from the Dean, May 10, 1924) Cassirer was clearly impressed and indicated that "in general I received from it the most favourable impression. (...) I learned many new things with respect to the content of the work as well as its method guiding its subject". On the crucial question of whether the study had philosophical merit his answer was an unequivocal yes. He wrote: "For even in the richness of historical details, which are indispensable in such a work, the selection of the material and the manner of weighing and questioning is consistently determined by general topics related to the history of the humanities and the philosophy of culture. I believe therefore, that the work in question - in line with Zilsel's former works, which I highly esteem as valuable contributions to logic and philosophy of science - is to be considered as a fully valid basis for a philosophical habilitation". Heinrich Scholz gave a lukewarm response. Scholz originally was a philosopher of religion who had turned to the new logic and to a rigid style of applying it to traditional philosophical problems. Though he admitted possessing a concept of philosophy radically different from that of Zilsel's, he did not doubt the philosophical value of the MS. Adolf Dyroff, a philosopher of culture who had the chair for 'catholic philosophy' at the University of Bonn supported Zilsel's critics: "The statistics are not without value ... but they need to be more strongly tied to the heart of the subject matter". Dyroff tried to build a golden bridge. "So it would be deplorable, if the great intellectual power and sophistication of the author would be lost for the university", and suggested that Zilsel should be allowed to revise the MS. With respect to the controversial point as to whether the MS could be rated as a proper piece of philosophical work, these experts represented exactly the variety of opinions already present on the committee. (All quotes are from ZF/VU). The complete text of the Cassirer letter can also be found in Dvořak, op. cit., p. 131, n.

and published it in 1926 as *The Development of the Concept of Genius*. It became his most influential book and was reviewed by, among others, Benedetto Croce and Georg Lukacs.[83] Zilsel developed in this work the new empirical methods necessary for what he saw as the goal of his empirical historical research: the discovery of historical laws. Historical material about beliefs, attitudes, professional activities, and economic structure are collected and grouped together in each chapter. At the end of each chapter 'results' are formulated, followed by 'provisional explanations'. At the end of the book a 'final result' is formulated: 'Laws on the concept of genius'.[84] The structure of these laws is a thesis-like summary in which all information on spatial and temporal locations is omitted so as to attain the character of a general 'if-then' proposition. The crux of the scientific validity of these laws rested upon their testability through cross-cultural comparisons. They were meant as provisional scientific hypotheses, to be confirmed, rejected, or modified through research. He considered the analysis of 'renaissances' in different cultures and times as a suitable field for testing the laws he had tentatively formulated. It is exactly this point that offended his enemies. Zilsel's method carried a dramatic shift of perspective in its wake: People, professional work, and cultural periods would lose their uniqueness and come to be seen as mere variables in a temporal development. Even if the resulting laws only barely met the minimum theoretical requirements of lawlike statements, they were of paradigmatic significance to Zilsel for establishing a socio-historical science that is fueled by cooperation and applies comparative cross-cultural methods. In his essay 'Physics and the Problem of Historico-Sociological Laws'[85] he pursues this topic and provides examples of the kinds of laws he was looking for in his book on *The Development of the Concept of Genius*.

III.3 Zilsel and the Vienna Circle

Zilsel's attempt to combine philosophical analysis with detailed historical research directed toward the tentative articulation of general laws caused his *Habilitation* to falter, and, to a certain extent, it also distanced him from the central figures of the Vienna Circle. In a letter to Reichenbach, Zilsel objected to "the content free methodology and logic of science as practiced today".[86] His target was explicitly, though not exclusively, the practice of members of the Vienna Circle to write abstract philosophy. To some extent this criticism is surprising. Zilsel was, after all, part of the Moritz Schlick discussion-circle, and as a member of the *Verein Ernst Mach*,[87] he was one of the founding members of the Vi-

---

14.
83 For a complete list of the reviews we have been able to locate of this book see appendix III, pp. 235-242.
84 *The Development of the Concept of Genius*, pp. 101ff, 209f, 323ff.
85 This volume, pp. 200-208.
86 HR 013-38-31, May 2, 1930. Quoted by permission of the University of Pittsburgh. All rights reserved.
87 See *Erkenntnis*, Vol. 1, 1930, p. 70.

enna Circle. His observation however that its members were predominantly interested in the methodology of the sciences and not in the study of new fields is not completely unjustified. For Zilsel, the Vienna Circle could be ironically characterized as an empirical school without empirical research. His critical evaluation of the methodology of the Vienna Circle is particularly clear in two book reviews: Max Adler's *Textbook of the Materialist Conception of History* (1930),[88] is criticized by Zilsel for having presented only "three concrete examples from real history" in the entire book.[89] In a review of Otto Neurath's book *Empirical Sociology*,[90] Zilsel noted that "the book has no intrinsic interest in the living content of sociology - in any case, a lot less than in the promotion of the basic logical ideas of the Vienna school of philosophy. Thus, in this 'empirical sociology' fertile empiricism withdraws behind logic".[91] Zilsel considered himself an exception, not only due to his question as to whether history is a potential field for studying laws, but also in terms of epistemological discourse.

An exchange between Zilsel and Reichenbach concerning an essay submitted by Zilsel for the first volume of *Erkenntnis*[92] also sheds some light on this issue. Zilsel's essay was on the relationship between history and biology.[93] Reichenbach was short of space in the journal and by letter of October 16, 1930, proposed a number of abridgements. Zilsel opposed such an action as "such a cut would affect the core of such a densely styled work". He was particularly opposed to the elimination of the historical examples.

> There is no other way than to describe the historical facts and examples. Were I to omit the examples, nothing but a formal program would be left. This may appeal to some readers with purely natural scientific interests, but it would be scientifically sterile and not convince any expert. I regard the customary content-free methodology and logic of science, as practiced today, to be so detrimental that I certainly could not delete the material content of my own work and publish merely the "principal considerations".[94]

---

88 *Lehrbuch der materialistischen Geschichtsauffassung*.
89 E. Zilsel, 'Partei, Marxismus, Materialismus, Neukantianismus', *Der Kampf*, 1931, 24, pp. 213-220, p. 21; republished in *Wissenschaft und Weltanschauung*, op. cit., 1992, pp. 88 - 98.
90 *Empirische Soziologie*, the bulk of this book is translated in O. Neurath *Empirical Sociology*, M. Neurath and R. S. Cohen, (eds.) (Riedel: Dordrecht & Boston 1973.). The German edition was published in 1931 by Springer Verlag.
91 E. Zilsel 'Review of O. Neurath *Empirische Soziologie*', *Der Kampf*, 1932, 25, pp. 91-94, p. 93, republished in *Wissenschaft und Weltanschauung*, op. cit., pp 145-9. Apart from his methodological queries, Zilsel did not have the highest opinion of Neurath's book. Reichenbach asked him to review it in *Erkenntnis* (HP 013-38-18, November 18, 1931) and Zilsel replied: "By the way, Neurath's book does not seem to me to be worked out very well and I don't believe *Erkenntnis* would lose a lot if a review were omitted". (HR 013-38-17, December 5, 1931. Quoted by permission of the University of Pittsburgh. All rights reserved.)
92 Reichenbach was the main editor of the new journal.
93 The MS at issue was published in 1931 as 'Geschichte und Biologie, Überlieferung und Vererbung', *Archiv für Sozialwissenschaft und Sozialpolitik*, 65, pp. 475-524; republished in *Wissenschaft und Weltanschauung*, op. cit., pp 101-44. The later fate of this essay was already mentioned in section II.5.
94 HR 013-38-22, October 18, 1930. Quoted by permission of the University of Pittsburgh. All rights reserved.

His dislike of abstract philosophical argument and his belief in accurate case studies does not imply any serious doubts about the validity of the basic ideas of logical empiricism. Zilsel most certainly did agree with them. He however did harbor doubts as to whether the plan for a unified science could be advanced on the basis of a theory-of-science-program alone. The core of his criticism was directed against Carnap's construction of a spatio-temporal universal language.[95] He made the point that such a language would still allow the possibility of more than one unified science. Therefore, one had to raise the "far more difficult question"

> whether all areas of contemporary specialized sciences can be linked together through unified laws. This is an empirical problem ... Laws should first be found in the field of the socio-cultural sciences (Representatives of the present-day *"Geistes"-Wissenschaften* object, above all, to such laws). The discovery of cultural laws would bring about a far more interesting unification of the sciences than the spatio-temporal universal language.[96]

Zilsel realized in his own historical research how difficult the task of linking the social with the natural sciences was. This made him sceptical about the usefulness of formulating a *program* of unity as compared to generating unity by means of empirical research. The program of the Vienna Circle as it was formulated in 1929 by Neurath, Carnap, and Hahn was vague in regard to the unity of the social and natural sciences. 'Unity' was conceived as something which was to be found through the "search for a formalized neutral system, ... for a universal system of concepts".[97] The all-important question as to what these basic scientific terms represent for the study of cultural and social phenomena and how they fit into such a system is touched upon rather carelessly. The program simply states: "The object of history and economics are people, things and their arrangement".[98] As to Zilsel's focal question concerning the possibility of laws in all realms of knowledge, the manifesto expresses exactly Zilsel's conviction: "Only step by step can the advancing research of empirical science teach us in what degree the world is lawful".[99] But this very sentence stands, interestingly enough, under the subtitle 'Foundations of physics'.

---

[95] Due to lack of space we will not deal with this exchange of ideas at any great length. We are referring to Zilsel's reply, 'Bemerkungen zur Wissenschaftslogik', (*Erkenntnis*, 1932, 3, 2/3, pp. 143-61) to R. Carnap's essay 'Die Physikalische Sprache als Universalsprache der Wissenschaft', (*Erkenntnis* 1931, 2, (5/6), pp. 432-65). Carnap in turn replied in his 'Erwiderung auf die vorstehenden Aufsätze von E. Zilsel und K. Duncker', (*Erkenntnis*, 1932, 3, 2/3, pp. 177-88). For an extensive analysis see H. Rutte, 'Zu Zilsels erkenntnistheoretischen Ansichten in der Phase des Wiener Kreises', pp. 447-66 in *Wien-Berlin-Prag: Der Aufstieg der Wissenschaftlichen Philosophie; Zentenarien Rudolf Carnap - Hans Reichenbach - Edgar Zilsel*, R. Haller & F. Stadler (ed.). (Wien: Verlag Hölder-Pichler-Tempsky, 1993).
[96] 'Bemerkungen zur Wissenschaftslogik', op. cit., p. 154.
[97] Quoted from: Otto Neurath et. al., *Wissenschaftliche Weltauffassung - der Wiener Kries*, in: O. Neurath, *Wissenschaftliche Weltauffassung, Sozialismus und Logischer Empirismus*, R. Hegselmann (ed.), (Frankfurt: Suhrkamp, 1979), pp. 81-102, p. 87. An English translation of the Vienna Circle manifesto can be found in Neurath *Empirical Sociology*, op. cit.
[98] Ibid., p. 98.
[99] *"Nur Schritt für Schritt weiter dringende Forschung der Erfahrungswissenschaft kann uns darüber belehren, in welchem Grade die Welt gesetzmäßig ist."* Ibid., p. 95.

Our point is not to insinuate any serious unbridgeable opposition between the Vienna Circle program and Zilsel.[100] On the contrary, Zilsel tried to contri-bute by incorporating the social and historical sciences into the scientific enterprise. If there was a significant difference between Zilsel and the Vienna Circle, it was that he did not believe that the unity of the natural and the social sciences could ever be accomplished by logical analysis. The only significant difference between the Vienna Circle and Zilsel was that Zilsel did not believe that a program based on logical analysis and language construction could help in uniting the social and natural sciences. In the following section we want to contextualize this difference by showing that Zilsel did not choose the 'Scientific World View' as it was articulated in the manifesto of the Vienna Circle as his frame of reference, but in the Marxist philosophy of history and society.

IV. Marxism as a Comprehensive Philosophy

The bond between logical empiricism and the socialist movement gradually established itself after World War I. Common to both were anti-metaphysical attitudes, a non-elitist education system, planning of economic progress, and a plebiscitarian democratic order. Several members of the Vienna Circle had socialist ideas or were even active in the socialist movement: Carnap, Hahn, and Neurath, the latter in particular, emphasized the "internal bonds" to the scientific worldview and tried "to strive for a reform of the economic and societal relations".[101] Reichenbach, Joergensen, and Frank held similar opinions.

Zilsel became a member of the Austrian Social Democratic Party in 1918[102], and quickly found a role within the Workers Education Movement in Vienna. He dedicated his book *The Development of the Concept of Genius* to the People's University of Vienna. Zilsel, as other members of the Vienna Circle, considered the socialist movement to be the enactment of the rationalist moderniza-

---

100 It seems likely that there must have been a conflict of some sort between Zilsel and Neurath over the exclusion of specific interests from the Vienna Circle, to which Zilsel was opposed. This is reflected in a letter of Neurath from 1935 and Zilsel's reply to it. Neurath writes: "Even if you don't want to be counted as member of the Vienna Circle you belong to the scientism movement *per definitionem*, as you also believe that *only* scientific statements are meaningful..." (NP/Z, March 20, 1935) to which Zilsel answered: "I find this impression gravely unpleasant, and I ask you to communicate this to our common friends. Verbally, written, and printed I have emphasized again and again, that factual objections one time to Carnap, the other to Frank or Hahn or Schlick or yourself concern *irrelevant details* as compared to our anti-metaphysical common basic attitude. I have never marked off myself *scientifically*, but rather only found, some 6 years ago, certain external structures of the organization to be not without danger". (NP/Z, March 24, 1935) The conciliatory tone of this statement may have been caused by the increased strength the common enemy, the fascist ideology, had gained. Similar problems with Neurath's strategies in forming the Vienna School are reported from the mathematician Karl Menger. Cf. R. Leonard, 'Ethics and the Excluded Middle: Karl Menger and Social Science in Interwar Vienna', *Isis*, 1998, 89, pp. 1-26, see esp. p. 9-11.
101 Neurath, *Wissenschaftliche Weltauffassung, Sozialismus und logischer Empirismus*, op. cit., p. 85.
102 Cf. *Austrian Labour Information*, Nr. 24, March-April 1944, p. 9.

tion of science and society as it was proposed by logical empiricism. For Zilsel, however, Marxism held deeper attractions. As he did not believe in the methodologically founded program of the unity of science, he was compelled to look for a more material approach to unity. In a 1929 paper, he said of Marxism that it

> represents a melting together of natural, scientific, and historical-sociological ideas ... which is indispensable for the entire body of philosophical theory ... From Marxist socialism ... one can learn that, from all regular natural processes, that of history is the most complicated.[103]

This is to say that Zilsel's philosophy was first and foremost Marxist. The notion of law features strongly in both Marxism and in the philosophy of science of the Vienna Circle. For Zilsel, Marxism served as a model for a law-governed conception of history and presented hypotheses on the derivation of social structures from a few formative principles.

> The search for laws is rather that condition which has intellectually separated European culture, for four hundred years, from all other cultures that have ever arisen at the most fundamental level ... Thus, for four hundred years, laws have been gaining ground. This and only this is Europe, is modern, is science.[104]

In the same essay, he wrote: "If there exists a problem today which is unresolved and which joins the sciences together, that is thus truly philosophical, then it is indeed the problem of historical laws".[105] And he did not shy away from assigning an affective connotation to the concept of law: "In the need to find something holy, a socialist could feel newly enriched in front of law-governed history".[106] At the same time however, Zilsel opposed a party dogmatism that attempted to bind party membership and political activity to an unconfirmed theory of historical materialism.

> Precisely because I consider Marx's theory of history, in its most radical form, to be correct, I resist its misuse ... Turning it into a party dogma does equal damage to party and theory. Every scientist is aware of the enormous difficulties that necessarily attach themselves to any far-stretched theory. As a scientist, I in no way want to be a member of a party that really expects 400,000 members to agree with an incredibly demanding theory - one which can only be confirmed through the most careful studies. Such a thoughtless party must sooner or later founder politically.[107]

Marxism is "incredibly demanding" if it is not only taken as a general comprehensive view of societal development, but also as a theory capable of producing

---

103 E. Zilsel, 'Philosophische Bemerkungen', *Der Kampf*, 1929, 22, pp. 178-186, p. 186; republished in *Wissenschaft und Weltanschauung*, op. cit., pp. 31-44.
104 E. Zilsel, 'Soziologische Bemerkungen zur Philosophie der Gegenwart', *Der Kampf*, 1930, 23, pp. 410-24, p. 421 republished in *Wissenschaft und Weltanschauung*, op. cit., pp. 88-98.
105 'Soziologische Bemerkungen zur Philosophie der Gegenwart', p. 411.
106 E. Zilsel, 'Partei, Marxismus, Materialismus, Neukantianismus', *Der Kampf*, 1931, 24, pp. 213-20, p. 214f.
107 Ibid., p. 214.

that very conceptual precision, founded on hard material evidence, which the principles of logical positivism were calling for. Zilsel's research program, as seen against this Marxist background, demanded the development of a historical theory which proceeded by means of a recursive integration of empirical laws discovered by historical and comparative research, as well as the construction of hypotheses capable of statistical falsification.

V. The Confluence of Zilsel's two Projects

Zilsel claimed a special role for philosophical analysis within the scientific enterprise, because, as he puts it, there exists

> a great number of very general statements which are neither tautological nor *Scheinsätze* and which can nevertheless not be comprehended by the customary special branches of science because they equally belong to all disciplines... In accordance with the historically given meaning of the word one is accustomed to call such statements philosophical. It is thus by no means necessary to deny philosophy of any sort of content, or to relegate her to the mere 'activity' of clarifying concepts. There are by all means substantive, general, and binding fundamental and boundary questions in the framework of science which, although they can't be comprehended by any one science, can be studied in an exact manner.[108]

The fundamental and boundary questions Zilsel kept asking since he wrote his dissertation were focused upon the relevance of the concept of law in science. His post-doctoral dissertation on the 'Concept of Genius' narrowed the problem down to analysing the conditions of the existence of law-like explanations in the socio-historical sciences. The discussions in the Vienna Circle made him think explicitly about the unity of the natural and the socio-historical sciences. Marxism provided him with a theoretical framework which allowed him to generate hypotheses for empirical research.

Seen against this background, the relation between the two projects is that the project 'on the social origins of modern science' is the case study for the project 'on the concept of law'. This interpretation squares with the only early statement we have been able to find by Zilsel on the nature of his project 'on the social roots of modern science'. When Reichenbach asked Zilsel to offer an essay for the new journal *Erkenntnis* Zilsel replied in a positive way:

> I would most like to work out for print a talk I recently gave at the *Wiener kulturwissenschaftliche Gesellschaft* (*Society for Cultural Studies, Vienna*). The work is directed toward exact natural science in a twofold sense: firstly, because it considers historical processes themselves as natural processes and attempts to connect them by statistical laws; and secondly, because it treats the origins of the exact sciences, while it understands the so-called humanities, as they are practiced today, as remnants of a prescientific age[109].

---

108 'Bemerkungen zur Wissenschaftslogik', op. cit., p. 154/5.
109 HR 013-38-31, May 2, 1930. Quoted by permission of the University of Pittsburgh. All rights reserved.

In this early statement from 1930 both projects are intimately bound together. The quote furthermore indicates how he planned to relate statistical analysis with an epistemological claim and a broader sociological explanation. To make this central thrust in Zilsel's work as clear as possible, we shall quote two longer passages from essays published in this volume. The first concerns the use of the comparative method in historical sociology. It is especially noteworthy how Zilsel links this comparative method to 'finding and verifying causal explanations'.

> The rise of science is usually studied by historians who are primarily interested in the temporal succession of the scientific discoveries. Yet the genesis of science can be studied as a sociological phenomenon, too. The occupations of the scientific authors and of their predecessors can be ascertained. The sociological function of these occupations and their professional ideals can be analyzed. The temporal succession can be interrupted and relevant sociological groups can be compared to analogous groups in other periods and other civilizations - the medieval scholastics with Indian priest-scholars, the Renaissance humanists with Chinese mandarins, the Renaissance artisans and artists with their colleagues in classical antiquity. *Since, in the sociology of culture, experiments are not feasible, comparison of analogous phenomena is virtually the only way of finding and verifying causal explanations. It is strange how rarely investigations of this kind are made.* As the complex intellectual constructs are usually studied historically only, so sociological research for the most part restricts itself to comparatively elementary phenomena. *Yet there is no reason why the most important and interesting intellectual phenomena should not be investigated sociologically and causally.*[110]

The second comment concerns the manner in which a causally oriented sociohistorical science needs to be carried out:

> Among sociologists there are today various schools and many controversies; some schools even disregard the investigations of most of the others. It might be generally agreed that sociology is an empirical science. It is based on observation and comparison, if not yet on experiment. As it does not deal with individuals but investigates groups and mass phenomena, the general sociological statements which appear in still rather uncritical forms in Hegel and Comte must be based on careful and complete collection of material if reliable results are to be achieved. It is here that statistical and sometimes even quantitative methods were successfully introduced in sociology. They were, however, largely applied to quite elementary phenomena and their use frequently resulted in mere collections of material. Causal and comprehensive sociological theories, based on statistics, are still lacking.[111]

Taken together, these three comments allow us to characterize the relation between Zilsel's two projects as follows. Zilsel's more general goal was to demonstrate the feasibility of the notion of law in the humanities. Given his own philosophical position, he could not just argue the case in general. He had to show, through detailed socio-historical analysis, that causal historical and comparative research was possible. His project 'on the social origins of modern science' was an attempt to search for the laws that would causally explain how the concept of law came to dominate our understanding of nature and the reasons

---

110 'The Sociological Origins of Modern Science', this volume, p. 7-21, emphasis added.
111 'Problems of Empiricism', this volume, p. 194; unless otherwise indicated all subsequent quotes are from this essay, and page references will be given in the main text.

for its lack of use in the humanities. His project 'on the concept of law' intended to demonstrate that there are no meaningful restrictions preventing one from doing so. The former project, in turn, demonstrates the fruitfulness of doing so.

The most explicit essay which Zilsel was able to produce on what we term the confluence of his two projects is his 'Problems of Empiricism'. This comprehensive analysis of the development and meaning of the concept of law is paralleled by an investigation into the social origins and development of modern science.[112] With respect to the project 'on the concept of law', the essay inspects the sources of the deterministic version of causal law in the 17th century and pursues the major changes that lead to the concept of law based on statistical probability in the late 19th and early 20th century. Zilsel explains the development of the concept of law within the context of changing technological conditions, economic demands, and the respective epistemological implications. Furthermore, he touches upon the application of the concept of law in various disciplines. Thus, the most important components of his project of law are present in this essay, even if he concedes the omission of certain aspects related to the "deductive side of theoretical knowledge" (p. 199).

This essay also takes its start from Zilsel's project 'on the origins of modern science'. Summarizing the results of his earlier essays, Zilsel describes how the experimental method and the concept of law compounded to form the conviction of a necessary and complete inner-connectedness of all natural phenomena in the metaphysics of the mechanistic worldview. Three characteristics are central for the mechanistic worldview: (a) differentiation of reality into the real world of mechanical laws and the world of appearance; (b) reduction of all empirical laws to mechanical laws; (c) an introspective psychology of knowledge.

The concluding section ('The Decline of the Mechanical Conception of Nature') describes how these characteristics fell apart and gave way to a new concept of law, a non-reductionist view of reality, and an epistemology no longer based on psychology. The essay thus provides hints as to how Zilsel would have studied the socio-cultural conditions of the *origins* of modern science, of the spread of its spirit into various disciplines, as well as the conditions of the *decline* of some of its essential tenets. Technological change is one of the most important conditions: "With the inventions of the fifteenth and sixteenth century the technology of the Middle Ages was completely revolutionized" (p. 172). Later he says "The breakdown of mechanistic physics took place during a period of complete revolution in technology" (p. 198). He was convinced that "man is inclined to conceive natural processes after the pattern of how he himself influences nature" (p. 199). The dominance of mechanical technology led

---

112 A similar survey is given in Zilsel's essay 'Das mechanistische Weltbild und seine Überwindung' ('The Mechanistic World View and its Overcoming'), *Der Atheist*, 1932, 6, 9, pp. 129-31. This essay, however, was not addressed to an academic audience, but rather aimed at explaining the revolutionary changes in the history of physics to a lay public.

to the ideal of the causal deterministic law. The rise of electrical and chemical industries helped establish a non-reductionist conception of causal law.

The essay also deals with the issue as to why the human sciences followed a separate path of development. Zilsel discusses the various origins of psychology, political science, history, economics, and sociology. He connects the practical needs to which they responded to their research ideals. None of them, at least in their early stages, were seriously interested in the experimental method and in causal laws, but were orientated upon rational norms, progress of humanity, and class interests. Zilsel also touches upon the

> sociological schools which deny the possibility of any sociological laws. They maintain that in social research causes and laws have to be replaced by 'types', by 'understanding' - that is empathy, by 'wholeness', entelechies and values (p. 194).

He calls this position a "rebellion against causality" and summarizes all of the objections he had developed in his earlier essays against it. The differences between the origins of science and those of the humanities provide a causal explanation of their different methods and ideals.

His project 'on the social origins of modern science' explains the origins of science *and* the non-participation of the humanities in its development. Still, the anti-naturalists, defenders of hermeneutics, and introspectionists could be right in denying a common epistemology to all sciences grounded 'on the concept of law'. That is the reason why the very idea of the unity of science is at stake if the concept of law is not applicable to the humanities. As long as the metaphysical architecture of the mechanistic world view dominated the concept of law, the perspectives for theory development in the humanities on the basis of empirical laws would have been rather ridiculous - one could know in advance that the gulf between the fundamental concepts of mechanics and the theoretical terms of culture, history, and society could not be bridged by any sort of reductionism. Zilsel's opponents thus did have a point in defending themselves against reductionism. Yet Zilsel suggests that since the metaphysics of the mechanistic world view are now obsolete, the traditional arguments against developing the humanities into a law orientated science no longer pertain. Zilsel's recurring argument against the epistemologists of the human sciences[113] is that they proceed on the basis of a false understanding of the natural sciences. They wrongly believe that the natural sciences are still governed by the concept of the mechanical causal law (and insofar rightly resist accepting such an approach in the humanities). Unfortunately, they developed ideologies which rejected the concept of law completely for the social sciences, and declared the heuristic means of empathy and understanding as 'ultimate goals' (p. 194). But a more modern, non-reductionist, and statistical conception of law could well serve as a framework for the social sciences. The "empirical methods of causal research have, in all sciences, proved to be so fruitful that we shall not rashly give up

---

113 See note 62 for the people Zilsel takes as his opponents in this debate.

hope of finding them successful in the field of sociology too" (p. 195).[114] Zilsel's project 'on the social origins of modern science' was intended to provide an example of this kind of social research. The project 'on the concept of law' tried to clarify the epistemological changes of the concept of law and the methodological conditions of its application in the social sphere. The focal point that unites both projects under one common perspective is the programmatic question Zilsel kept asking for 30 years: the empirical as well as theoretical/conceptual clarification of what it means to say "that there are also 'laws' in history and sociology".[115] The need for a more general description required in an encyclopedic essay gave him the liberty to combine the perspectives and leitmotifs of both of his projects.

VI. The Fragmentary State of Zilsel's Research Program and its Prospects

This leaves us with the final question we want to raise. Zilsel could not complete either of his projects. Was the fragmentary state of his work the result from the poor circumstances of his life in exile, or did he, the longer he worked out his program, run into serious inherent problems? Unfortunately, the archive material we have collected so far provides almost no evidence to base a plausible answer on.[116] A point of departure could be to speculate that the permanent pressure to design projects and to offer results in the business of grant application widened the gap between the announced completion of the book and the actual state of finished pieces of work. We will present a few suggestions regarding this situation. A more far-reaching speculation we would like to offer, is that the more historical material he accumulated for formulating law-like rela-

---

114 It should be noted that Zilsel's empirical hypotheses about the origins of the humanistic theory of knowledge include the normative and legitimatory functions of early political sciences, historical studies, and economics. Cf. 'The Methods of Humanism', this volume, p. 22-64.
115 The quote is from the *Problemstellung* of Zilsel's project, 'Laws of nature and historical laws', this volume, p. 233.
116 As far as we have been able to establish, the suicide of Zilsel was to all who were close to him not only a great emotional shock but also an unexpected event: no one had seen it coming. Prof. David French, Dean of the Science Faculty at Mills College informed the *Emergency Committee* as follows: "It grieves me very much to inform you that Dr. Zilsel took his own life when he was suffering a period of depression that was unbearable to him. I have been unable to discover any very evident immediate cause of this tragedy. Dr. Zilsel had made many friends here, had entered into the activities of the campus, was getting ahead with his writing, and was assured of a position at the college for the coming year. He maintained a cheerful manner with students and colleagues, but it was known to a few intimates and to his physician that he was deeply disturbed in spirit. I can think only that the prolonged bitter experience and suffering finally overwhelmed his so sensitive nature" (EC/Z, April 19, 1944; the file does not contain French's letter but the quoted fragment was included in what looks like item 2 of the agenda of the subcommittee on application which was held May 2, 1944). Lynn White, in personal correspondence with one of us, had expressed similar shock and amazement when he heard about Zilsel's suicide - he was in Cleveland at the time. For his son Paul, as well, the death of his father came totally unexpected. This makes venturing into the reasons of Zilsels suicide a hazardous enterprise. Was it that he could not cope with the prospect of having to start a new life in such a bourgeois place as Mills College, or was it the realization that his intellectual program was falling apart, or was it something else? We will probably never be able to tell.

tions, the more difficult it became to give structure to it. We assume that his strong belief in the necessity to search for empirical correlations made him neglect the equally important theoretical work.

In June 1941, Zilsel wrote that a "chapter on humanism and its conformities with and differences from science is nearly ready for the press".[117] In his grant application for the American Philosophical Society of February 28, 1943, Zilsel wrote: "I underestimated the duration of this investigation in my application of 1941". He explained his delay thus: "In the meantime so much further material was collected that completion by the fall can be expected practically with certainty".[118] In both instances the essay did not reach the press. A similar claim from 1941: "Completion of the whole study can be expected by July 1942"[119] did not turn out to be correct. Surely this remark cannot be taken at face value. One could easily conform to the view that Zilsel would probably have needed a few more months or even another year to finish his book. But we do not observe him working on rendering the final manuscript. Instead he is still fighting his case against the misguided self-understanding of the humanities. We recently came across an outline of a presentation Zilsel gave at the 6th Congress for the Unity of Science held in 1942 at Chicago. It is published here for the first time.[120] The paper is entitled 'Science and the Humanistic Studies' and its central argument is that scientists and historians "connect" their data differently. For historians, "temporal succession of the states of the same single object" is the focus of attention, for scientists, "natural laws".[121] While this acknowledgment shows some sensitivity to the point his opponents were making, he maintains his position to restrict the field of historical phenomena that can be scientifically investigated to those parts where comparative and statistical methods can be used. The paper ends:

> Analogous processes in as many different cultures as possible must be compared. And the single facts in each culture must be collected and *worked up by statistical methods*. Success or failure of prediction is the decisive test of the correctness of a historical and sociological law.[122]

Here again Zilsel puts weight on the empirical side of his program. Socio-historical facts should be 'worked up by statistical methods' to establish the existence of 'historical and sociological laws'. But he does not offer any clue as to

---

117 HP/Z, June 22, 1941.
118 APS/Z, February 28, 1943.
119 HP/Z, June 22, 1941.
120 We found this outline among the papers C. Morris donated to the University of Chicago Library. Cf. Department of Special Collections, file: Unity of Science Movement, Box 3, folder 4. A complete list of all the speakers present at the Chicago conference is given in Stadler, op. cit., 1997, pp. 433-6.
121 Zilsel takes the concept of causality as a subspecies of the law concept; universal implications are the logical expression of natural laws. See also W. Krohn, 'Edgar Zilsel zur Methodologie einer exakten Geisteswissenschaft', pp. 257-75 in *Philosophie, Wissenschaft, Auklärung. Beiträge zur Geschichte und Wirkung des Wiener Kreises*, H-J. Dahms (ed.). (Berlin & New York: W. de Gruyter, 1985) and W. Krohn, 'In Search of Laws: Edgar Zilsel, the Vienna Circle, and Marxian Tradition', *Unpublished MS/Paper presented to the 1993 annual History of Science Society Meeting*.
122 'Science and the Humanistic Studies', this volume, p. 223, emphasis added.

how to arrange them in a theoretical framework. With respect to physics he obviously accepted the necessity of an 'internal construction of science'.[123] But in his socio-historical research, as far as we can see, he hesitated in his work on a theoretical architecture. In his essays from the 30s he took over the Marxist theory and accepted it as a suitable framework for deriving testable hypotheses. One of his most fruitful hypotheses was the assumption that superior artisans and other practitioners had been operative in developing the epistemic principles of causal explanation and methodical experimentation. This is now known as the Zilsel thesis.[124] But in the final years of his life, we do not see him relating his findings to this or any other substantial theoretical framework, except in rather cursory and scattered remarks.[125] Without doubt, Zilsel was sceptical about a premature search for deductive theories in history.[126] But can any attempt to understand and explain a complex process such as the emergence of modern science be successful if it dispenses with a recursive interaction between theory and data? Is Zilsel's one-sided decision to collect and correlate data in provisional laws responsible for his difficulties in completing his book?

The reconstruction of the relations between Zilsel's two projects has perhaps led us away from the interests which historians and sociologists of science usually identify with Zilsel and the Zilsel-thesis.[127] We have tried to show that both projects were part of a more comprehensive research interest which Zilsel had pursued since the beginning of the 1920s: the relevance of the concept of law for modern science and modern society. It was an extremely ambitious program which comprehended sociological, epistemological, and methodological theory as well as empirical historical research.

The pioneering work of Zilsel's research on the relationships between mathematical practitioners, superior artisans, engineers, surgeons on the one hand and scholarly trained members of various traditional disciplines on the other can today more easily be appreciated than in his own times. Historians, sociologists, and philosophers of science have vastly demonstrated the impact of experimentation, intervention, instrumentation, or in one word, of practice on scientific knowledge formation.[128] Zilsel's attempt to understand the institutional and cog-

---

123 'Problems of Empiricism', this volume, p. 193.
124 Cf. S. Shapin in *Dictionary of the History of Science*, W.F. Bynum et al. (eds.). (London & Basingstoke: Macmillan, 1981), sv Zilsel-thesis, p. 450.
125 The scattered remarks can usually be found at the end of the essays. The most detailed one is at the end of the programmatic essay 'The Sociological Roots of Science', this volume, p. 18/19.
126 'The Problem of Historico-Sociological Law', this volume, p. 196/197.
127 Based on a conference on the Zilsel thesis held in Berlin in the spring of 1998 we are currently editing a separate volume dealing with this aspect of Zilsel's work.
128 Cf. D. Gooding, et al. (eds.) *The Uses of Experiment: Studies in the Natural Sciences.* (Cambridge: Cambridge University Press, 1989); D. Gooding, *Experiment and the Making of Meaning: Human Agency in Scientific Observation and Experiment.* (Dordrecht: Kluwer Academic Press, 1990); I. Hacking, *Representing and Intervening: Introductory Topics in the Philosophy of Natural Science.* (Cambridge: Cambridge University Press, 1983); A. Pickering, *The Mangle of Practice: Time, Agency, and Science.* (Chicago & London: University of Chicago Press, 1995); S. Shapin & S. Schaffer, *Leviathan and the Air-pump: Hobbes, Boyle, and the Experimental Life.* (Princeton, N. J.: Princeton University Press, 1985.)

nitive dynamics which brought these complementary sources together in terms of lawlike relations can now be established upon a much firmer empirical basis. Furthermore, the computer has opened up new possibilities for modelling the structural dynamics of historical changes.[129] Quantitative statistical research has since become fully established in the social sciences and has found its way into historical research (less though in cultural studies). But this does not necessarily imply that the idea of the unification of the social and the natural sciences, which Zilsel tried to establish by giving the concept of law a central position, has gained general acceptance today. On the contrary, talk is now openly of the disunity of science.[130] To subclassify the adherents of qualitative methods as remnants of a pre-scientific state of research, as Zilsel liked to do, would be encountered with even more scepticism than by that of 50 years ago.

The contemporary controversy concerning the legitimacy of a causal sociological explanation of the development of science has reached new heights in the so-called 'science wars'.[131] Zilsel intended to contribute to the unity of the sciences "by generating it empirically", and by fighting against qualitative heuristics as ultimate goals. The so-called strong program in the sociology of knowledge, which has adopted Zilsel's causal approach,[132] has triggered a long-lasting debate between a constructivist and a realist interpretation of scientific knowledge. Philosophy and politics are now just as intertwined as they were when Zilsel defended his program early in his career at the University of Vienna. He did not live to put the pieces of his work together, but - as we hope this collection will serve to demonstrate - the vital questions of his intellectual endeavor continue to persevere.

---

129 Cf. P. Ahrweiler and N. Gilbert. *Computer Simulations in Science and Technology Studies*. (Berlin & Heidelberg: Springer, 1998).
130 For a useful survey see P. Galison and D. J. Stump (eds.). *The Disunity of Science: Boundaries, Context, and Power*. (Stanford, Cal.: Stanford University Press, 1996).
131 Cf. P. R. Gross and N. Levitt. *Higher Superstition: The Academic Left and Its Quarrels with Science*. (Baltimore, Maryland & London: The Johns Hopkins University Press, 1998, 2nd. ed.) and the subsequent controversy. For more on this controversy see: www.members.tripod.com/ScienceWars/.
132 The first tenet of the strong program is that explanations should be causal, cf. D. Bloor *Knowledge and Social Imagery*. (Chicago & London: University of Chicago Press, 1991, 2nd. ed.), p. 7.

References

Ahrweiler, P. and N. Gilbert. 1998. *Computer Simulations in Science and Technology Studies*. Berlin & Heidelberg: Springer.

Bloor, D. 1991. *Knowledge and Social Imagery*, 2nd. ed., Chicago & London: University of Chicago Press.

Carnap, R. 1931. 'Die physikalische Sprache als Universalsprache der Wissenschaft', *Erkenntnis* 2, 5/6, pp. 432-65.

———. 1932/33. 'Erwiderung auf die vorstehenden Aufsätze von E. Zilsel und K. Duncker', *Erkenntnis* 3, 2/3, pp. 177-88.

Dahms, H.-J. 1993. 'Edgar Zilsels Projekt "The Social Roots of Science" und seine Beziehungen zur Frankfurther Schule', pp. 474-500 in *Wien-Berlin-Prag: Der Aufstieg der Wissenschaftlichen Philosophie; Zentenarien Rudolf Carnap - Hans Reichenbach - Edgar Zilsel*, R. Haller & F. Stadler (ed.). Wien: Verlag Hölder-Pichler-Tempsky.

Dawkins, R. 1991. *The Blind Watchmaker*. London, etc: Penguin Books.

Duggon, S. P. H. and B. Drury. 1948. *The Rescue of Science and Learning: The Story of the Emergency Committee in Aid of Displaced Scholars*. New York: Macmillan.

Dvořak, J. 1981. *Edgar Zilsel und die Einheit der Erkenntnis*. Wien: Löcker Verlag.

Galison, P. and D. J. Stump (eds.). 1996. *The Disunity of Science: Boundaries, Context, and Power*. Stanford, Cal.: Stanford University Press.

Gooding, D. 1990. *Experiment and the Making of Meaning: Human Agency in Scientific Observation and Experiment*. Dordrecht, etc.: Kluwer Academic Publishers.

Gooding, D. et al (eds.). 1989. *The Uses of Experiment: Studies in the Natural Sciences*. Cambridge: Cambridge University Press.

Götz, C. M. and T. Pankratz. 1993. 'Edgar Zilsels Wirken im Rahmen der Wiener Volksbildung und Lehrerfortbildung', pp. 467-73 in *Wien-Berlin-Prag: Der Aufstieg der Wissenschaftlichen Philosophie; Zentenarien Rudolf Carnap - Hans Reichenbach - Edgar Zilsel*, R. Haller & F. Stadler (eds.). Wien: Verlag Hölder-Pichler-Tempsky.

Gross, P. R. and N Levitt. 1998. *Higher Superstition: The Academic Left and Its Quarrels with Science*, 2nd. ed. Baltimore, Maryland & London: The Johns Hopkins University Press.

Hacking, I. 1983. *Representing and Intervening: Introductory Topics in the Philosophy of Natural Science*. Cambridge: Cambridge University Press.

Hahn, H. 1917. 'Review of E. Zilsel *Das Anwendungsproblem* (Leipzig, Barth, 1916)', *Monatshefte für Mathematik und Physik*, 27/8, pp. 37-8.

Hahn, R. 1999. 'Berkeley's History of Science Dinner Club. A Chronicle of Fifty Years of Activity ', *Isis* 90, pp. 182-91.

Köhnke, K. C. 1991. *The Rise of Neo-Kantianism: German Academic Philosophy between Idealism and Positivism*. Cambridge: Cambridge University Press.

Koyré, A. 1992. *The Astronomical Revolution: Copernicus - Kepler - Borelli*. New York: Dover.

———. 1992. *Metaphysics and Measurement*. Yverdon, etc.: Gordon & Breach.

Krohn, W. 1985. 'Edgar Zilsel zur Methodologie einer exakten Geisteswissenschaft', pp. 257-75 in *Philosophie, Wissenschaft, Auklärung. Beiträge zur Geschichte und Wirkung des Wiener Kreises*, H- J. Dahms (ed.). Berlin & New York: W. de Gruyter.

———. 'In Search of Laws: Edgar Zilsel, the Vienna Circle, and Marxian Tradition', *Unpublished MS, Paper Presented to the 1993 Annual History of Science Society Meeting*.

Kuhn, T. S. 1970. 'Alexandre Koyré & the History of Science (Review of A. Koyré's *Metaphysics and Measurment: Essays in the Scientific Revolution* [London: Chapman & Hall, 1968])', *Encounter* 34, pp. 67-9.
Leonard, R. J. 1998. 'Ethics and the Excluded Middle: Karl Menger and Social Science in Interwar Vienna', *Isis* 89, 1, pp. 1-26.
Morris, C. 1960. 'On the History of the International Encyclopedia', *Synthese*, 12, 517-21.
Nemeth, E. 1997. '"Wir Zuschauer" und das "Ideal der Sache". Bemerkungen zu Edgar Zilsels Geniereligion', *Lecture Series/Vorträge des Institut Wiener Kreis* 5, pp.157-78.
Neurath, O. 1931. *Empirische Soziologie. Der wissenschaftliche Gehalt der Geschichte und Nationalökonomie*. Wien: Springer Verlag.
Neurath, O. et al. 1979. 'Wissenschaftliche Weltauffassung - Der Wiener Kreis', pp. 81-102 in O. Neurath, *Wissenschaftliche Weltauffassung, Sozialismus und logischer Empirismus*, R. Hegelsmann (ed.). Frankfurt: Suhrkamp.
Pickering, A. 1995. *The Mangle of Practice: Time, Agency & Science*. Chicago & London: University of Chicago Press.
Rutte, H. 1993. 'Zu Zilsels erkenntnistheoretischen Ansichten in der Phase des Wiener Kreises', pp. 447-66 in *Wien-Berlin-Prag: Der Aufstieg der Wissenschaftlichen Philosophie; Zentenarien Rudolf Carnap - Hans Reichenbach - Edgar Zilsel*, R. Haller & F. Stadler (ed.). Wien: Verlag Hölder-Pichler-Tempsky.
Schlick, M. 1931. 'Die Kausalität in der gegenwärtigen Physik', *Die Naturwissenschaften* 19, pp. 145-62.
Shapin, S. 1981. 'Zilsel-Thesis', p. 450 in *Dictionary of the History of Science*, London & Basingstoke: Macmillan.
Shapin, S. and S. Schaffer. 1985. *Leviathan and the Air-Pump: Hobbes, Boyle and the Experimental Life*. Princeton, N. J.: Princeton University Press.
Smith, D. C. 1986. *H. G. Wells: Desperately Mortal. A Biography*. New Haven & London: Yale University Press.
Stadler, F. 1991. 'Aspects of the Social Background and Position of the Vienna Circle at the University of Vienna', pp. 51-77 in *Rediscovering the Forgotten Vienna Circle: Austrian Studies on Otto Neurath and the Vienna Circle*, T. E. Uebel (ed.). Dordrecht, etc.: Kluwer Academic Publishers.
———. 1997. *Studien zum Wiener Kreis: Ursprung, Entwicklung und Wirkung des Logischen Empirismus im Kontext*. Frankfurt am Main: Suhrkamp.
Strong, E. W. 1936. *Procedures and Metaphysics: A Study in the Philosophy of Mathematical-Physical Science in the Sixteenth and Seventeenth Centuries*. Berkeley, Cal.: University of California Press.
———. 1992. *Philosopher, Professor and Berkeley Chancellor 1961-65*. Interview conducted by H. Nathan, Oral History Project Berkeley, Cal..
Wells, H. G. 1920. *The Outline of History: Being a Plain History of Life and Mankind*. London: Newnes.
Wiggershaus, R. *The Frankfurt School: Its History, Theories and Political Significance*. Cambridge, Mass.: MIT Press, 1994.
Zilsel, E.1915. *Ein philosophischer Versuch über das Gesetz der grossen Zahlen und seine Verwandten*. Dissertation Universität Wien.
———. 1916. *Das Anwendungsproblem. Ein philosophischer Versuch über das Gesetz der grossen Zahlen und die Induktion*. Leipzig: Barth.
———. 1918. *Die Geniereligion. Ein kritischer Versuch über das moderne Persönlichkeitsideal*. Wien-Leipzig: Braumüller

———. 1919. 'Selbstanzeige of *Die Geniereligion*', *Kantstudien* 24, pp.156-7.
———. 1926. *Die Entstehung des Geniebegriffes: Ein Beitrag zur Ideengeschichte der Antike und des Frühkapitalismus*. Tübingen: Mohr (Paul Siebeck).
———. 1929. 'Philosophische Bemerkungen', *Der Kampf* 22, pp.178-86.
———. 1930. 'Soziologische Bemerkungen zur Philosophie der Gegenwart', *Der Kampf* 23, pp. 410-24.
———. 1931. 'Geschichte und Biologie, Überlieferung und Vererbung', *Archiv für Sozialwissenschaft und Sozialpolitik* 65, pp. 475-524.
———. 1931. 'Partei, Marxismus, Materialismus, Neukantianismus', *Der Kampf* 24, pp. 213-20.
———. 1932. 'Review of O. Neurath *Empirische Soziologie* (Wien: Springer, 1931)', *Der Kampf* 25, pp. 91-4.
———. 1932. 'Das mechanistische Weltbild und seine Überwindung.', *Der Atheist* 6, 9, pp. 129-31.
———. 1932/33. 'Bemerkungen zur Wissenschaftslogik', *Erkenntnis* 3, 2/3, pp. 143-61.
———. 1935. 'P. Jordans Versuch, den Vitalismus quantenmechanisch zu retten', *Erkenntnis* 5, pp. 56-65.
———. 'Copernicus and Mechanics', this volume, pp. 123-127.
———. 'The Genesis of the Concept of Scientific Progress and Scientific Co-Operation', this volume, pp. 128-170.
———. 'History and Biological Evolution', this volume, pp. 216-220.
———. 'The Methods of Humanism', this volume, pp.22-64.
———. 'Physics and the Problem of Historico-Sociological Laws', this volume, pp. 200-208.
———. 'Problems of Empiricism', this volume, pp. 171-199.
———. 'Science and the Humanistic Studies', this volume, pp. 221-226.
———. 'The Sociological Roots of Science', this volume, pp. 7-21.

Edgar Zilsel (1891-1944)
Oil painting by a unknown artist in Austrian Expressionist style painted during World War I when Zilsel was in his early twenties. Picture is reproduced with the kind permission of Paul Zilsel.

PART I

THE SOCIAL ROOTS OF MODERN SCIENCE

1

THE SOCIAL ROOTS OF SCIENCE*

Fully developed, science is to be found only in modern European-American civilization. As its development began in early capitalism we shall have to study the period from the end of the Middle Ages until 1600. Results obtained by ancient mathematicians, astronomers, and physicists and by medieval Arabic physicians have greatly influenced the beginning of science in modern Europe. We shall not discuss this influence, but the social and economic conditions which made it possible.

Some general characteristics of early capitalistic society which are the necessary conditions for the rise of science are well known. Early capitalistic society is a society of trading and manufacturing townsmen. Therefore theology recedes, worldly and empirical thinking advances. Technology progresses rapidly in this period (period of inventions, machines). This sets tasks to mechanics and chemistry and furthers thinking in general. Economic competition dissolves the collective feudal society and especially the medieval guilds. This destroys the collective-mindedness and traditional thinking of the Middle Ages, furthers individual thinking and is the presupposition of scientific criticism. Early capitalistic economy proceeds rationally, calculates, and measures (bookkeeping, machines). This furthers the rise of rational scientific methods. It can be shown that the mathematical writings from 1300 to 1600 are intimately connected with the needs of tradesmen and bankers on the one hand, of architects, craftsmen, and military engineers, on the other.

In order to understand the rise of science in greater detail we have to distinguish three strata of intellectual activity in the period from 1300 to 1600:
(1) At the *universities* of this period theology and scholasticism still rule. The university scholars were trained to think rationally, they liked rational distinctions, divisions, and disputations, but were scarcely interested in experience. They relied on authorities and, therefore favored quotations and comments. If they were at all concerned with mundane and natural events, they did not search into causes, but endeavoured to explain the aims, purposes, and

---

* [This essay is the first English statement of Zilsel's project 'on the social origins of modern science'. It was presented at the 5th International Congress for the Unity of Science held at Harvard University, Cambridge, Mass., September 3-9, 1939. This MS was discovered by Friedrich Stadler among the Neurath papers, Haarlem, the Netherlands and published in H. Pauer-Studer, *Norms, Values, and Society (Vienna Circle Institute Yearbook, Vol 2)* (Dordrecht: Kluwer Academic Publishers, 1994), pp. 305-308. We gratefully acknowledge Stadler's assistance to republish it in this volume. Eds.]

meanings of the phenomena. The universities were scarcely influenced by humanism in this period.

(2) The first representatives of mundane learning were not scientists but secretaries and officials of municipalities, princes, and the Pope (14th century). They became the fathers of *Humanism*. Their aims were mastery of writing and speech and perfection of style. In the following centuries the humanists lose in large part their official connections and became free *literati* dependent on princes, noblemen, and bankers as patrons. Their aims remain unchanged, their pride of memory and learning, their passion for fame even increase. They acknowledge certain ancient writers as patterns of style and are bound to these mundane authorities almost as strictly as the theologians are to their religions ones. Also humanism proceeds rationally. It develops the methods of scientific philology, but it neglects causal research and is more interested in form than in content, more in words than in things.

Both university-scholars and humanists despise the uneducated lower classes. Both, therefore, wrote and spoke only Latin. Both especially despise manual labor and distinguish between liberal and mechanical arts: only professions which do not require manual work are considered to be worthy of well-bred men. The medical doctors, therefore, content themselves with commenting on the medical writings of antiquity; the surgeons who operate and dissect belong with the barbers and midwives. *Literati* are much more highly esteemed than artists. In the 14th century, the latter are not separated from whitewashers and stonedressers, but very slowly gain social esteem by stressing their relations to learning (perspective needs geometry) and literature. The inventors and discoverers, being craftsmen and mariners, are scarcely mentioned by the humanistic *literati*. Those men to whom, from to-day's point of view, the culture of the Renaissance owes the most important achievements, the artists, the inventors, and the discoveries, entirely recede into the background in contemporary literature.

(3) Beneath both the university-scholars and the humanistic literati there were some groups of superior *craftsmen* who needed more knowledge for their work than their colleagues did. The most important of them may be called artist-engineers, for not only did they paint their pictures, cast their statues, and build their cathedrals, but also constructed lifting-gears, earthworks, canals and sluices, guns and fortresses, found new pigments, detected the geometrical laws of perspective, and invented new measuring tools for engineering and gunnery. Many of them wrote diaries and papers in Italian on their achievements; the best known among them is Leonardo da Vinci (1452-1519). Related to them are the surgeons (painting needs knowledge of anatomy) and the constructors of musical instruments (Zarlino). These superior craftsmen invent, experiment, dissect. They already develop considerable theoretical knowledge in the fields of mechanics, chemistry, metallurgy, geometry, anatomy, and acoustics. However, since they had not learned how to proceed systematically their achievements form a collection of isolated discoveries. They are the immediate predecessors of science. The two components of scientific method were still

separated: methodical training of intellect was preserved for upper-class learned people, for university-scholars and humanistic *literati*; experiment and observation were left, more or less, to plebeian workers. Real science is born when, with the progress of technology, the experimental method of the craftsmen overcomes the prejudice against manual work and is adopted by rationally trained university-scholars. This is accomplished with Galileo (1564-1642).

Galileo's relations to technology, to military engineering, and the artist-engineers are often underrated. When he studied medicine at the University of Pisa, mathematics was not taught there at all. He learned mathematics privately from Ostilie Ricci who was a teacher of the *Accademia del Disegno*, a school for artists and artist-engineers. As a young professor of mathematics and astronomy at the University of Padua, he lectured privately on mechanics and engineering and established working-rooms in his private house where craftsmen were his assistants - the very first university-laboratory. He started his researches with studies on pumps, on the regulation of rivers, and on the construction of fortresses. His first printed publication describes a new measuring tool for military purposes. His detection of the law of falling bodies is intimately connected with the needs of gunnery. The shape of the curve of projection had often been discussed by the gunners of his time. Galileo was the first one who was able to solve this problem. From 1610 onwards he wrote only in Italian, no longer in Latin. This also shows his relations to the lower ranks of society, his aversion to university-scholars and humanists.

The same opposition against both humanism and scholasticism is to be found with Bacon of Verulam (1561-1626). Bacon feels enthusiastic about the achievements of the great navigators, the inventors, and the craftsmen of his period. He proclaims their work as a model for the scholars. Bacon did not make any important discovery in the field of natural science and his writings abound with scientific mistakes. But he is the very first writer who realizes the importance of methodical scientific research for the advancement of human civilization.

Relations with the ranks of craftsmen and surgeons can be shown also in the case of Gilbert (1540-1603) and Harvey (1578-1657).

The humanists call themselves 'dispensers of glory' (*dispensatores gloriae*): by their writings they make their patrons and at the same time themselves famous. Their professional ideal is individual fame. Bacon substitutes two new aims of intellectual activity: 'domination of nature' by means of science and 'advancement of learning'. In his *Nova Atlantis* he depicts an ideal state in which technical and scientific progress is reached by planned collaboration of scientists, each using and contouring the researches of his predecessors and fellow workers. These scientists are the rulers of *Nova Atlantis*. They form a staff of officials, organized in accordance with the principle of division of labor. Bacon's *Utopia,* is suggested partly by the progress of division of labor in the field of contemporary economics, partly by the progress of rational organization of contemporary government. Early capitalistic society tends not only to individualism (economic competition) but also to rational organization (military organization, public administration).

The idea that scientists must collaborate in order to bring about the progress of civilization is essential to modern science. Neither disputing scholastics nor *literati* greedy of glory are scientists.

Bacon's idea is entirely new and can not be found either in antiquity or in the Renaissance. Somewhat similar ideas appear in the same period with Descartes and Campanella. It can be shown that Bacon's *Nova Atlantis* and greatly influenced the foundation of learned societies. In 1654 the Royal Society was founded in London, and in 1663 the Academie Française in Paris. In 1664 the *Proceedings of the Royal Society* appeared for the first time. Since this period the collaboration of scientists in scientific periodicals, societies, institutes and organizations has steadily advanced.

Result: In the period from the end of the Middle Ages until 1600 the university scholars and the humanistic *literati* are rationally trained but they do not experiment as they despise manual labor. Many more or less plebeian craftsmen experiment and invent but lack methodical rational training,. About 1600, with the progress of technology, the experimental method is adopted by rationally trained scholars of the educated upper class. Thus, the two components of scientific search are united at last: modern science is born. The whole process is embedded in the advance of early capitalistic economy which weakens collective-mindedness, magical thinking, traditions, and the belief in authority, which furthers mundane, rational, and causal thinking, individualism and rational organization.

2

THE SOCIOLOGICAL ROOTS OF SCIENCE

ABSTRACT. In the period from 1300 to 1600 three strata of intellectual activity must be distinguished: university-scholars, humanists, and artisans. Both university-scholars and humanists were rationally trained. Their methods, however, were determined by their professional conditions and differed substantially from the methods of science. Both professors and humanistic literati distinguished liberal from mechanical arts and despised manual labor, experimentation, and dissection. Craftsmen were the pioneers of causal thinking in this period. Certain groups of superior manual laborers (artist-engineers, surgeons, the makers of nautical and musical instruments, surveyors, navigators, gunners) experimented, dissected, and used quantitative methods. The measuring instruments of the navigators, surveyors, and gunners were the forerunners of the later physical instruments. The craftsmen, however, lacked methodical intellectual training. Thus the two components of the scientific method were separated by the social barrier: logical training was reserved for upper-class scholars; experimentation, causal interest, and quantitative method were left to more or less plebeian artisans. Science was born when, with the progress of technology, the experimental method eventually overcame the social prejudice against manual labor and was adopted by rationally trained scholars. This was accomplished about 1600 (Gilbert, Galileo, Bacon). At the same time the scholastic method of disputation and the humanistic ideal of individual glory were superseded by the ideals of control of nature and advancement of learning through scientific co-operation. In a somewhat different way, sociologically, modern astronomy developed. The whole process was imbedded in the advance of early capitalistic society, which weakened collective-mindedness, magical thinking, and belief in authority and which furthered worldly, causal, rational, and quantitative thinking.

Were there many separate cultures in which science has developed and others in which it is lacking, the question about the origin of science would generally be recognized as a sociological one and could be answered by singling out the common traits of the scientific in contrast to the nonscientific cultures. Historical reality, un-fortunately, is different, for fully developed science appears once only, namely, in modern Western civilization. It is this fact that obscures our problem. We are only too inclined to consider ourselves and our own civilization as the natural peak of human evolution. From this presumption the belief originates that man simply became more and more intelligent until one day a few great investigators and pioneers appeared and produced science as the last stage of a one-line intellectual ascent. Thus it is not realized that human thinking has developed in many and divergent ways - among which one is the scientific. One forgets how amazing it is that science arose at all and especially in a certain period and under special sociological conditions.

It is not impossible, however, to study the emergence of modern science as a sociological process. Since this emergence took place in the period of early European capitalism, we shall have to review that period from the end of the Middle Ages until 1600. Certain stages of the scientific spirit, however, developed in other cultures too, e.g., in classical antiquity and, to a lesser degree, in some oriental civilizations and in the Arabic culture of the Middle Ages. Moreover, the scientific and half-scientific cultures are not independent of each other. In modern Europe the beginnings of science, particularly, have been greatly influenced by the achievements of ancient mathematicians and astronomers and medieval Arabic physicians. We shall, however, discuss not this influence but the sociological conditions which made it possible. We can, necessarily, give but a sketchy and greatly simplified analysis of this topic here. All details and much of the evidence must be left to a more extensive exposition at another place.

I

Human society has not often changed so fundamentally as it did with the transition from feudalism to early capitalism. These changes are generally known. Even in a very brief exposition of the problem, however, we must mention some of them, since they form necessary conditions for the rise of science.

1. The emergence of early capitalism is connected with a change in both the setting and the bearers of culture. In the feudal society of the Middle Ages the castles of knights and rural monasteries were the centers of culture. In early capitalism culture was centered in towns. The spirit of science is worldly and not military. Obviously, therefore, it could not develop among clergymen and knights but only among townspeople.

2. The end of the Middle Ages was a period of rapidly progressing technology and technological inventions. Machines began to be used both in production of goods and in warfare. On the one hand, this set tasks for mechanics and chemistry, and, on the other, it furthered causal thinking, and, in general, weakened magical thinking.

3. In medieval society the individual was bound to the traditions of the group to which he unalterably belonged. In early capitalism economic success depended on the spirit of enterprise of the individual. In early feudalism economic competition was unknown. When it started among the craftsmen and tradesmen of the late medieval towns, their guilds tried to check it. But competition proved stronger than the guilds. It dissolved the organizations and destroyed the collective-mindedness of the Middle Ages. The merchant or craftsman of early capitalism who worked in the same way as his fathers had was outstripped by less conservative competitors. The individualism of the new society is a presupposition of scientific thinking. The scientist, too, relies, in the last resort, only on his own eyes and his own brain and is supposed to make

himself independent of belief in authorities. Without criticism there is no science. The critical scientific spirit (which is entirely unknown to all societies without economic competition) is the most powerful explosive human society ever has produced. If the critical spirit expanded to the whole field of thinking and acting it would lead to anarchism and social disintegration. In ordinary life this is prevented by social instincts and social necessities. In science itself the individualistic tendencies are counterbalanced by scientific co-operation. This, however, will be discussed later.

4. Feudal society was ruled by tradition and custom, whereas early capitalism proceeded rationally. It calculated and measured, introduced bookkeeping, and used machines. The rise of economic rationality furthered development of rational scientific methods. The emergence of the quantitative method, which is virtually non-existent in medieval theories, cannot be separated from the counting and calculating spirit of capitalistic economy. The first literary exposition of the technique of double-entry bookkeeping is contained in the best textbook on mathematics of the fifteenth century, Luca Pacioli's *Summa de arithmetica* (Venice, 1494); the first application of double-entry bookkeeping to the problems of public finances and administration was made in the collected mathematical works of Simon Stevin, the pioneer of scientific mechanics (*Hypomnemata Mathematica* [Lyon, 1608]), and a paper of Copernicus on monetary reform (*Monetae cudendae ratio* [composed in 1552] is among the earliest investigations of coinage. This cannot be mere coincidence.

The development of the most rational of sciences, mathematics, is particularly closely linked with the advance of rationality in technology and economy. The modern sign of mathematical equality was first used in an arithmetical textbook of Recorde that is dedicated to the "governors and the rest of the Companio of Venturers into Moscovia" with the wish for "continualle increase of commoditie by their travel" (*The Wetstone of Witte* [London, 1557]). Decimal fractions were first introduced in a mathematical pamphlet of Stevin that begins with the words: "To all astronomers, surveyors, measurers of tapestry, barrels and other things, to all mintmasters and merchants good luck!" (*De thiende* [Leyden, 1585]). Apart from infusions of Pythagorean and Platonic metaphysics, the mathematical writings of the fifteenth and sixteenth centuries first deal in detail with problems of commercial arithmetic and, second, with the technological needs of military engineers, surveyors, architects, and artisans. The geometrical and arithmetical treatises of Piero della Francesca, Luca Pacioli, and Tartaglia in Italy, Recorde and Leonard Digges in England, Dürer and Stifel in Germany, are cases in point. Classical mathematical tradition (Euclid, Archimedes, Apollonius, Diophantus) could be revived in the sixteenth century because the new society had grown to demand calculation and measurement.

Even rationalization of public administration and law had its counterpart in scientific ideas. The loose state of feudalism with its vague traditional law was gradually superseded by absolute monarchies with central sovereignty and rational statute law. This political and juridical change promoted the emergence of the idea that all physical processes are governed by rational natural laws established by God. This, however, did not occur before the seventeenth century (Descartes, Huyghens, Boyle).[2]

II

We have mentioned a few general characteristics of early capitalistic society which form necessary conditions for the rise of the scientific spirit. In order to understand this development sociologically, we have to distinguish three strata of intellectual activity in the period from 1300 to 1600: the universities, humanism, and labor.

At the universities theology and scholasticism still predominated. The university- scholars were trained to think rationally but exercised the methods of scholastic rationalism which differ basically from the rational methods of a developed economy. Tradesmen are interested in reckoning; craftsmen and engineers in rational rules of operation, in rational investigation of causes, in rational physical laws. Schoolteachers, on the other hand, take an interest in rational distinction and classifications. The old sentence, "bene docet qui bene distinguit", is as correct as it is sociologically significant. Schoolteaching, by its sociological conditions, produces a specific kind of rationality, which appears in similar forms wherever old priests, intrusted with the task of instructing priest candidates, rationalize vague and contradictory mythological traditions of the past. Brahmans in India, Buddhist theologians in Japan, Arabic and Catholic medieval scholastics conform in their methods to an astonishing degree. Jewish Talmudists proceeded in the same way, though, not being priests by profession, they dealt with ritual and canon law rather than with proper theological questions. This school rationality has developed to a monstrous degree in Brahmanic Sankhya-philosophy (sankhya means "enumeration").

As a rule the specific scholastic methods are preserved when theologians, in the course of social development, apply themselves to secular subject matters. Thus in Indian literature Brahmans who had entered the service of princes discussed politics and erotics by meticulously distinguishing and enumerating the various possibilities of political and sexual life (Kautilya, Vatsyayana).[3] In a somewhat analogous way the medieval scholastics and the European university-scholars before 1600 indulged in subtle distinctions, enumerations, and disputations. Bound to authorities, they favored quotation and uttered their opinions for the most part in the form of commentaries and compilations. After the thirteenth century mundane subject matters were treated by scholars, too, and, as an exception, even experience was referred to by some of them. But when the Schoolmen were at all concerned with secular events they did not, as a

---

2 Cf. Edgar Zilsel, 'The Genesis of the Concept of Physical Law', this volume pp. 96-122.
3 Cf. M. Winternitz, *Geschichte der indischen Literatur* (Leipzig, 1920), III, 509 ff., 536 ff.

rule, investigate causes and, never, physical laws. They endeavored rather to explain the ends and meanings of the phenomena. Obviously, the occult qualities and Aristotelian substantial forms of scholasticism are but rationalizations of prescientific, magic, and animistic teleology. Thus, till the middle of the sixteenth century the universities were scarcely influenced by the development of contemporary technology and by humanism. Their spirit was still substantially medieval. It seems to be a general sociological phenomenon that rigidly organized schools are able to offer considerable resistance to social changes of the external world.[4]

The first representatives of secular learning appeared in the fourteenth century in Italian cities. They were not scientists but secretaries and officials of municipalities, princes, and the pope looking up with envy to the political and cultural achievements of the classical past. These learned officials who chiefly had to conduct the foreign affairs of their employers became the fathers of humanism. Their aims derive from the conditions of their profession. The more erudite and polished their writings, the more eloquent their speeches, the more prestige redounded to their employers and the more fame to themselves. They therefore strove after perfection of style and accumulation of classical knowledge. In the following centuries the Italian humanists lost in large part their official connections. Many became free literati, dependent on princes, noblemen, and bankers as patrons. Other were engaged as instructors to the sons of princes, and several got academic chairs and taught Latin and Greek at universities. Their aims remained unchanged, and their pride of memory and learning, their passion for fame, even increased. They acknowledged certain ancient writers as patterns of style and were bound to these secular authorities almost as strictly as the theologians were to their religious ones. Though humanism also proceeded rationally, its methods were as different form scholastic as from modern scientific rationality. Humanism developed the methods of scientific philology, but neglected causal research and was ignorant of physical laws and quantitative investigation. Altogether it was considerably more interested in words than in things, more in literary forms than in contents. Humanism spread over all parts of western and central Europe. Though the professional conditions and intellectual aims of the humanists outside Italy were somewhat more complex, on the whole their methods were the same.[5]

---

4 Pierre Duhem has brought into prominence the fourteenth-century Ockhamists of the University of Paris (Buridan, Oresme, and others) and has attempted to vindicate for them scientific priority to Copernicus and Galileo. Though knowledge of late scholasticism has been greatly furthered by Duhem's investigation of the Paris Schoolmen, he has considerably overrated their "anticipations" of modern physical and astronomical ideas. He singles out the scarce and rather extrinsic conformities with modern natural science and omits the abundance of differences. Duhem's opinion has been uncritically adopted by many followers.

5 It seems to be a rather general sociological phenomenon that, where there are professional public officials, secular learning first appears in the form of humanism. In China also after the dissolution of feudalism in the period of Confucius, a group of literati officials developed who were chiefly interested in perfection of style and who acknowledged certain ancient writings as literary models. In the following period admission to civil service was made dependent on examinations regarding literary style and

The university-scholars and the humanistic literati of the Renaissance were exceedingly proud of their social rank. Both disdained uneducated people. They avoided the vernacular and wrote and spoke Latin only. Further, they were attached to the upper classes, sharing the social prejudices of the nobility and the rich merchants and bankers and despising manual labor. Both, therefore, adopted the ancient distinction between liberal and mechanical arts: only professions which do not require manual work were considered by them, their patrons, and their public to be worthy of well-bred men.

The social antithesis of mechanical and liberal arts, of hands and tongue, influenced all intellectual and professional activity in the Renaissance. The university-trained medical doctors contented themselves more or less with commenting on the medical writings of antiquity; the surgeons who did manual work such as operating and dissecting belonged with the barbers and had a social position similar to that of midwives. Literati were much more highly esteemed than were artists. In the fourteenth century the latter were not separated from whitewashers and stone-dressers and, like all craftsmen, were organized in guilds. They gradually became detached from handicraft, until a separation was effected in Italy about the end of the sixteenth century. In the period of Leonardo da Vinci (about 1500) this had not yet been accomplished. This fact appears rather distinctly in the writings of contemporary artists who over and over again discussed the question as to whether painting and sculpture belong with liberal or mechanical arts. In these discussions the painters usually stressed their relations to learning (painting needs perspective and geometry) in order to gain social esteem. Technological inventors and geographical discoverers, being craftsmen and seamen, were hardly mentioned by the humanistic literati. The great majority of the humanists did not report on them at all. If they mentioned them, they did so in an exceedingly careless and inaccurate way. From the present point of view the culture of the Renaissance owes its most important achievements to the artists, the inventors, and the discoverers. Yet these men entirely recede into the background in the literature of the period.[6]

Beneath both the university-scholars and the humanistic literati the artisans, the mariners, shipbuilders, carpenters, foundrymen, and miners worked in silence on the advance of technology and modern society. They had invented the mariner's compass and guns; they constructed paper mills, wire mills, and stamping mills; they created blast furnaces and in the sixteenth century

---

knowledge of antiquity. In China even calligraphy belonged to the formal requirements of higher education, Chinese writing characters being more complicated than European ones. Secular scribes, proud of their profession and learning and bound to ancient models, can be found also in ancient Egypt and the neo-Parthian empire. In classical antiquity, there was an abundance of rhetors, grammarians, philologists, and philosophers rather resembling the humanistic literati of the Renaissance. Yet lack of professional civil servants in the republican period prevented development of a perfect correspondence.

6 On the prestige of the literati, artists, inventors, and discoverers cf. Edgar Zilsel, *Die Entstehung des Geniebegriffes: Ein Beitrag zur Ideengeschichte der Antike und des Frühkapitalismus* (Tübingen, 1926), pp. 130-175, and 176 f. (statistical evidence).

introduced machines into mining. Having outgrown the constraints of guild tradition and being stimulated to inventions by economic competition, they were, no doubt, the real pioneers of empirical observations, experimentation, and causal research. They were uneducated, probably often illiterate,[7] and, perhaps for that reason, today we do not even know their names. Among them were a few groups which needed more knowledge for their work than their colleagues did and, therefore, got a better education. Among these superior craftsmen the artists are most important. There were no sharp divisions between painters, sculptors, goldsmiths, and architects; but very often the same artist worked in several fields, since, on the whole, division of labor had developed only slightly in the Renaissance. Following from this, a remarkable professional group arose during the fifteenth century. The men we have in mind may be called artist-engineers, for not only did they paint pictures, cast statues, and build cathedrals, but they also constructed lifting engines, canals and sluices, guns and fortresses. They invented new pigments, detected the geometrical laws of perspective, and constructed new measuring tools for engineering and gunnery. The first of them is Brunelleschi (1377-1446), the constructor of the cupola of the cathedral of Florence. Among his followers were Ghiberti (1377-1466), Leone Battista Alberti (1407-1472), Leonardo da Vinci (1492-1519), and Vanoccio Biringucci (d. 1538) whose booklet on metallurgy is one of the first chemical treatises free of alchemistic superstition. One of the last of them is Benvenuto Cellini (1500-1571), who was a goldsmith and sculptor and also worked as military engineer of Florence. The German painter and engraver Albrecht Dürer, who wrote treatises on descriptive geometry and fortifications (1525 and 1527), belongs to this group. Many of the artist-engineers wrote - in the vernacular and for their colleagues - diaries and papers on their achievements. For the most part these papers circulated as manuscripts only. The artist-engineers got their education as apprentices in the workshops of their masters. Only Alberti had a humanistic education.

The surgeons belonged to a second group of superior artisans. Some Italian surgeons had contact with artists, resulting from the fact that painting needs anatomical knowledge. The artificers of musical instruments were related to the artist-engineers. Cellini's father, for example, was an instrument-maker, and he himself was appointed as a pope's court musician for a time. In the fifteenth and sixteenth centuries the forerunners of the modern piano were constructed by the representatives of this third group. The makers of nautical and astronomical instruments and of distance meters for surveying and gunnery formed a fourth group. They made compasses and astrolabes, cross-staffs, and quadrants and invented the declinometer and inclinometer in the sixteenth century. Their measuring-instruments are the forerunners of the modern physical apparatus.

---

7 Cf. the statistical data on population and number of schoolchildren in the chronicle of Giovanni Villani (X, 162 [fourteenth century, Florence]) and J.W. Adamson, "The Extent of Literacy in England in the 15th and 16th Centuries", *Library*, X (4th ser., 1930), 167.

Some of these men were retired navigators or gunners.[8] The surveyors and the navigators, finally, were also considered as representatives of the mechanical arts. They and the map-makers are more important for the development of measurement and observation than of experimentation.

These superior craftsmen made contacts with learned astronomers, medical doctors, and humanists. They were told by their learned friends of Archimedes, Euclid, and Vitruvius; their inventive spirit, however, originated in their own professional work. The surgeons and some artists dissected, the surveyors and navigators measured, the artist-engineers and instrument-makers were perfectly used to experimentation and measurement, and their quantitative rules of thumb are the forerunners of the physical laws of modern science. The occult qualities and substantial forms of the scholastics, the verbosity of the humanists were of no use to them. All these superior artisans had already developed considerable theoretical knowledge in the fields of mechanics, acoustics, chemistry, metallurgy, descriptive geometry, and anatomy. But, since they had not learned how to proceed systematically, their achievements form a collection of isolated discoveries. Leonardo, for example, deals sometimes quite wrongly with mechanical problems which, as his diaries reveal, he himself had solved correctly years before. The superior craftsmen, therefore, cannot be called scientists themselves, but they were the immediate predecessors of science. Of course, they were not regarded as respectable scholars by contemporary public opinion. The two components of scientific method were still separated before 1600 - methodological training of intellect was preserved for upper-class learned people, for university-scholars, and for humanists; experimentation and observation were left to more or less plebeian workers.

The separation of liberal and mechanical arts manifested itself clearly in the literature of the period. Before 1550 respectable scholars did not care for the achievements of the nascent new world around them and wrote in Latin. On the other hand, after the end of the fifteenth century, a literature published by "mechanics" in Spanish, Portuguese, Italian, English, French, Dutch, and German had developed. It included numerous short treatises on navigation, vernacular mathematical textbooks, and dialogues dealing with commercial, technological, and gunnery problems (e.g. Étienne de la Roche, Tartaglia, Dürer, Ympyn), and various vernacular booklets on metallurgy, fortification, bookkeeping, descriptive geometry, compass-making, etc. In addition there were the unprinted but widely circulated papers of the Italian artist-engineers. These books were diligently read by the colleagues of their authors and by merchants. Many of these books, especially those on navigation, were frequently reprinted, but as a rule they were disregarded by respectable scholars. As long as this separation persisted, as long as scholars did not think of using the disdained methods of manual workers, science in the modern meaning of the

---

8 Cf., e.g., the *Oxford Dictionary of National Biography* on the English instrument-makers, Humfrey Cole (d. 1580), William Bourne (d. 1583), and Robert Norman.

word was impossible. About 1550, however, with the advance of technology, a few learned authors began to be interested in the mechanical arts, which had become economically so important, and composed Latin and vernacular works on the geographical discoveries, navigation and cartography, mining and metallurgy, surveying, mechanics, and gunnery.[9] Eventually the social barrier between the two components of he scientific method broke down, and the methods of the superior craftsmen were adopted by academically trained scholars: real science was born. This was achieved about 1600 with William Gilbert (1544-1603), Galileo (1564-1642), and Francis Bacon (1561-1626).

William Gilbert, physician to Queen Elizabeth, published the first printed book composed by an academically trained scholar which was based entirely on laboratory experiment and his own observation (*De Magnete* [1600]). Gilbert used and invented physical instruments but neither employed mathematics nor investigated physical laws. Like a modern experimentalist he is critically-minded. Aristotelism, belief in authority, and humanistic verbosity were vehemently attacked by him. His scientific method derives from foundrymen, miners, and navigators with whom he had personal contacts. His experimental devices and many other details were taken over from a vernacular booklet of the compass-maker Robert Norman, a retired mariner (1581).[10]

Galileo's relations to technology, military engineering, and the artist-engineers are often underrated. When he studied medicine at the University of Pisa in the eighties of the sixteenth century, mathematics was not taught there. He studied mathematics privately with Ostilio Ricci, who had been a teacher at the Accademia del disegno in Florence, a school founded about twenty years earlier for young artists and artist-engineers. Its founder was the painter Vasari. Both the foundation of this school (1562) and the origin of Galileo's mathematical education show how engineering and its methods gradually rose from the workshops of craftsmen and eventually penetrated the field of academic instruction. As a young professor at Padua (1592-1610), Galileo lectured at the university on mathematics and astronomy and privately on mechanics and engineering. At this time he established workrooms in his house, where craftsmen were his assistants. This was the first "university" laboratory

---

9 Peter Martyr (1511, 1530), Peter Apian (1529) Gemma Phrysius (1530), Orontius Finaeus (1532), Nunes (1537, 1546, 1566), George(1530, 1556), Pedro de Medina (1545), Ramusio (1550), Leonard Digges (1556, 1571, 1579), Mercator (1569, 1578, 1594), Benedetti (1575), Guido Ubaldo (1577), Hakluyt (1589), Thomas Hood (1590, 1592, 1596, 1598), Robert Hues (1594), Edward Wright (1599), and others. The high percentage of English authors is striking. They seem to have been interested in the mechanical arts earlier than Continental writers (cf. Francis R. Johnson, *Astronomical Thought in Renaissance England* [Baltimore, 1937]). On the other hand, in the same period a few "mechanics" rose to a scientific level in their activities and their writings: the Dutch engraver and map-maker Abraham Ortelius (1527-98), who became geographer to Philip II of Spain and a scientific cartographer; the French barber-surgeon Ambroise Paré (1510-90), who became surgeon to Henri II of France and the founder of modern scientific surgery; the cashier and bookkeeper of the municipalities of Antwerp and Bruges, Simon Stevin (1548-1620), who became technological and mathematical instructor and adviser to Maurice of Nassau, quartermaster-general of Holland, and one of the founders of modern scientific mechanics.
10 Cf. Edgar Zilsel, 'The Origins of William Gilbert's Scientific Method', this volume pp. 71-95.

in history. He started his research with studies on pumps, on the regulation of rivers, and the construction of fortresses. His first printed publication (1606) described a measuring tool for military purposes which he had invented. All his life he liked to visit dockyards and to talk with the workmen. In his chief work of 1638, the *Discorsi*, the setting of the dialogue is the Arsenal of Venice. His greatest achievement - the detection of the law of falling bodies, published in the *Discorsi* - developed from a problem of contemporary gunnery, as he himself declared.[11] The shape of the curve of projection had often been discussed by the gunners of the period. Tartaglia had not been able to answer the question correctly. Galileo, after having dealt with the problem for forty years, found the solution by combining craftsman-like experimentation and measurement with learned mathematical analysis. The different social origins of the two components of his method - which became the method of modern science - is obvious in the *Discorsi*, since he gives the mathematical deductions in Latin and discusses the experiments in Italian. After 1610 Galileo gave up writing Latin treatises and addressed himself to nonscholars. His greatest works, consequently, are written completely or partially in Italian. A few vernacular poets were among his literary favorites. Even his literary taste reveals his predilection for the plain people. His aversion to the spirit and methods of the contemporary professors and humanists is frequently expressed in his treatises and letters.

The same opposition to both humanism and scholasticism can be found in the works of Francis Bacon. No scholar before him had attacked belief in authority and imitation of antiquity so passionately. Bacon was enthusiastic about the great navigators, the inventors, and the craftsmen of his period; their achievements, and only theirs, are set by him as models for scholars. The common belief that it is "a kind of dishonor to descend to inquiry upon matters mechanical"[12] seems "childish" to him. Induction, which is proclaimed by him as the new method of science, obviously is the method of just those manual laborers. He died from a cold which he caught when stuffing a chicken with snow. This incident also reveals how much he defied all customs of contemporary scholarship. An experiment of this kind was in his period considered worthy rather of a cook or knacker than of a former lord chancellor of England. Bacon, however, did not make any important discovery in the field of natural science, and his writings abound with humanistic rhetoric, scholastic survivals, and scientific mistakes. He is the first writer in the history of mankind, however, to realize fully the basic importance of me- thodical scientific research for the advancement of human civilization.

Bacon's real contribution to the development of science appears when he is confronted with the humanists. The humanists did not live on the returns from their writings but were dependent economically on bankers, noblemen, and

---

11 Letter to Marsili (November 11, 1632), *Opere* (ed. nazionale), XIV, 386.
12 *Novum Organum* I, aph. 120.

princes. There was a kind of symbiosis between them and their patrons. The humanist received his living from his patron and, in return, made his patron famous by his writings. Of course, the more impressive the writings of the humanist, the more famous he became. Individual fame, therefore, was the professional ideal of the humanistic literati. They often called themselves "dispensers of glory" and quite openly declared fame to be the motive of their own and every intellectual activity. Bacon, on the contrary, was opposed to the ideal of individual glory. He substituted two new aims: "control of nature" by means of science and "advancement of learning". Progress instead of fame means the substitution of a personal ideal by an objective one. In his *Nova Atlantis* Bacon depicted an ideal state in which technological and scientific progress is reached by planned co-operation of scientists, each of whom uses and continues the investigations of his predecessors and fellow-workers. These scientists are the rulers of the New Atlantis. They form a staff of public officials organized in nine groups according to the principle of division of labor. Bacon's ideal of scientific co-operation obviously originated in the ranks of manufacturers and artisans. On the one hand, early capitalistic manual workers were quite accustomed to use the experience of their colleagues and predecessors, as is stressed by Bacon himself and occasionally mentioned by Galileo. On the other hand, division of labor had advanced in contemporary society and in the economy as a whole.

Essential to modern science is the idea that scientists must cooperate in order to bring about the progress of civilization. Neither disputing scholastics nor literati, greedy of glory are scientists. Bacon's idea is substantially new and occurs neither in antiquity nor in the Renaissance. Somewhat similar ideas were pointed out in the same period by Campanella and, occasionally, by Stevin and Descartes. As is generally known, Bacon's *Nova Atlantis* greatly influenced the foundation of learned societies. In 1654 the Royal Society was founded in London, in 1663 the Académie française in Paris; in 1664 the *Proceedings* of the Royal Society appeared for the first time. Since this period co-operation of scientists in scientific periodicals, societies, institutes, and organizations has steadily advanced.

On the whole, the rise of the methods of the manual workers to the ranks of academically trained scholars at the end of the sixteenth century is the decisive event in the genesis of science. The upper stratum could contribute logical training, learning, and theoretical interest; the lower stratum added causal spirit, experimentation, measurement, quantitative rules of operation, disregard of school authority, and objective co-operation.[13]

---

[13] The development of modern astronomy took place in a somewhat different way. After the days of the Babylonian priests, the links connecting astronomy with priesthood, calendar-arranging, and religious feasts had never been quite interrupted. Astronomy, therefore, was linked with the idea of celestial sublimity and always belonged to the free arts. As a consequence Pythagorean and nonmechanical animistic ideas are conspicuous in Copernicus and Kepler. Practical astronomy, on the other hand, was linked with navigation, which was interested in exact star positions and measuring instruments. In the period of Newton the metaphysical and astrological spirit was definitely overcome in scientific astronomy.

III

The indicated explanation of the development of science obviously is incomplete. Money economy and co-existent strata of skilled artisans and secular scholars are frequent phenomena in history. Why, nevertheless, did science not develop more frequently? A comparison with classical antiquity can fill at least one gap in our explanation.

Classical culture produced achievements in literature, art, and philosophy which are in no way inferior to modern ones. It produced outstanding and numerous historiographers, philologists, and grammarians. Ancient rhetoric is superior to its modern counterpart both in refinement and in the number of representatives. Ancient achievements are considerable in the fields of theoretical astronomy and mathematics, limited in the biological field, and poor in the physical sciences. Only three physical laws were correctly known to the ancient scholars: the principles of the lever and of Archimedes and the optical law of reflection. In the field of technology one difference is most striking: machines were used in antiquity in warfare, for juggleries, and for toys but were not employed in the production of goods. On the whole, ancient culture was borne by a rather small upper class living on their rents. Earning money by professional labor was always rather looked down upon in the circles determining ancient public opinion. Manual work was even less appreciated. In the same manner as in the Renaissance, painters and sculptors gradually detached from handicraft and slowly rose to social esteem. Yet their prestige never equaled that of writers and rhetors, and even in the period of Plutarch and Lucianus the greatest sculptors of antiquity would be attacked as manual workers and wage-earners. Compared with poets and philosophers, artists were rarely mentioned in literature, and engineers and technological inventors virtually never. The latter presumably (very little is known of them) were superior artisans or emancipated slaves working as foremen. In antiquity rough manual work was done by slaves.

As far as our problem is concerned, this is the decisive difference between classical and early capitalistic society. Machinery and science cannot develop in a civilization based on slave labor. Slaves generally are unskilled and cannot be intrusted with handling complex devices. Moreover, slave labor seems to be cheap enough to make introduction of machines superfluous. On the other hand, slavery makes the social contempt for manual work so strong that it cannot be overcome by the educated. For this reason ancient intellectual development could not overcome the barrier between tongue and hand. In antiquity only the least prejudiced among the scholars ventured to experiment and to dissect. Very few scholars, such as Hippocrates and his followers, Democritus, and Archimedes, investigated in the manner of modern experimental and causal science, and even Archimedes considered it necessary to apologize for constructing battering- machines. All these facts and correlations have already been pointed out several times.

It may be said that science could fully develop in modern Western civilization because early European capitalism was based on free labor. In early capitalistic society there were very few slaves, and they were not used in production but were luxury gifts in the possession of princes. Evidently lack of slave labor is a necessary but not a sufficient condition for the emergence of science. No doubt further necessary conditions would be found if early capitalistic society were compared with Chinese civilization. In China, slave labor was not predominant, and money economy had existed since about 500 B.C. Also there were in China, on the one hand, highly skilled artisans and, on the other, scholar-officials, approximately corresponding to the European humanists. Yet causal, experimental, and quantitative science not bound to authorities did not arise. Why this did not happen is as little explained as why capitalism did not develop in China.

The rise of science is usually studied by historians who are primarily interested in the temporal succession of the scientific discoveries. Yet the genesis of science can be studied also as a sociological phenomenon. The occupations of the scientific authors and of their predecessors can be ascertained. The sociological function of these occupations and their professional ideals can be analyzed. The temporal succession can be interrupted and relevant sociological groups can be compared to analogous groups in other periods and other civilizations - the medieval scholastics with Indian priest-scholars, the Renaissance humanists with Chinese mandarins, the Renaissance artisans and artists with their colleagues in classical antiquity. Since, in the sociology of culture, experiments are not feasible, comparison of analogous phenomena is virtually the only way of finding and verifying causal explanations. It is strange how rarely investigations of this kind are made. As the complex intellectual constructs are usually studied historically only, so sociological research for the most part restricts itself to comparatively elementary phenomena. Yet there is no reason why the most important and interesting intellectual phenomena should not be investigated sociologically and causally.

BIBLIOGRAPHICAL NOTE

The sociological analysis of nascent science must be based primarily on the writings of the scientific authors from 1400 to 1650. The material is very extensive but must be used in its entirety. For the relations of science to technology, commerce, military engineering, and instrument-making the following authors are especially important: Luca Pacioli, Tartaglia, the English mathematicians Recorde and Leonard and Thomas Digges, Stevin, William Gilbert, Galileo, and Francis Bacon. Often (e.g. in Guido Ubaldo) valuable sociological material is contained in the prefaces and dedications. The vernacular writings of the craftsmen, instrument-makers, and navigators are important. The following authors may be mentioned: Ghiberti (*Commentarii* [ca. 1450]), Piero della Francesca (*De prospectiva pingendi* [1484]), Leonardo, Alberti, Biringuccio (*Pirotechnia* [ca. 1540]), Dürer (*Underweysung der Messung* [1525], *Befestigung der stett, schloss und flecken* [1527]), William Bourne (*Inventions or Devices* [1578], *On the properties and qualities of glasses*, in I.O. Halliwell (ed.), *Rara mathematica*), Robert Norman (*The new attractive* [1581], William Borough (*Discourse of the variation of the compass* [1581]), Palissy (*Récepte véritable* [1560], *Discours admirables* [1580]. Strictly speaking, the works also of Tartaglia, Stevin, and Ambroise Paré belong to this group. Also many textbooks on mathematics and treatises on navigation were composed by nonscholars.

The modern literature is of secondary importance. A few works may be mentioned: extensive material on the economy and technology of the period is contained in Werner Sombart, *Modern Capitalism*. On scholasticism: M. Grabmann, *Geschichte der scholastischen Methode* (1909); George Sarton, *Introduction to the History of Science*, Vol. II. Much valuable material on late medieval physics is contained in Pierre Duhem, *Etudes sur Léonard da Vinci* (Paris, 1906) and *Les Origines de la statique* (Paris, 1905).[14] Duhem, however, disregards the differences of the scholastic and the scientific methods and greatly overestimates the results of the Paris Ockhamites. In a special case this has been shown in B. Gunzburg, "Duhem and Jordanus Nemorarius", *Isis*, XXV (1936), 341 ff. On humanism: J. Burckhardt, *The Civilization of the Renaissance*; J.A. Symonds, *The Revival of Learning*; J.E. Sandys, *A History of Classical Scholarship*. On artist-engineers and the influence of the mechanical arts on beginning science: Julius Schlosser, "Materialien zur Quellenkunde der Kunstgeschichte", *Vienna Academy of Science (Phil. hist. Klasse)*, Vols. CLXXVII, CLXXIX, CLXXX, CXCII; Leonard Olschki, *Geschichte der neusprachlichen wissenschaftlichen Literatur*, Vols. I and II; W.E. Houghton, "The History of Trades", *Journal of the History of Ideas*, II (1941), 33 ff.; R.K. Merton, "Science and Technology in the Seventeenth Century", *Osiris*, IV (1938), 360-632. On instrument-makers: important material is contained (but not analyzed) in Robert T. Gunther, *The Astrolabes of the World* (Oxford, 1932). Monographs: E.R.G. Taylor, *Geogr. Journal* (1924), on Jean Rotz, and *ibid.* (1928); on William Bourne. On Galileo and contemporary technology: L. Olschki, *Galilei und seine Zeit* (Halle, 1927); on Stevin: George Sarton, *Isis*, XXI (1934), 241 ff. On scientific co-operation: Martha Ornstein, *The Role of Scientific Societies in the Seventeenth Century* (Chicago, 1938). The various histories of science are generally known. F.R. Johnson, *Astronomical Thought in Renaissance England* (Baltimore, 1937) is the best special study on sixteenth-century science. M. Cantor, *Vorlesungen ueber Geschichte der*

---

14 [Recently translated into English as *The Origins of Statics: The Sources of Physical Theory*. (Dordrecht; etc. Kluwer Academic Publishers, 1991). *Boston Studies in the Philosophy of Science*, Vol 123; Eds.].

Mathematik, still is the best history of mathematics from the sociologists' point of view. Much sociological material is contained in the papers of the author of this article referred to in notes 2, 6, and 10. An excellent bibliography of the modern literature on Renaissance science (including navigation, map-making, nautical and astronomical instruments) is found in *Modern Language Quarterly*, II (1941), 363-401. The bibliography is composed by F.R. Johnson and S.V. Larkey.

3

THE METHODS OF HUMANISM*

The first representatives of worldly learning in the modern era were the Italian humanists. Humanism is older than modern science. Though they conform in some respects - both humanism and science deal with worldly subject matters and proceed rationally - the two intellectual attitudes differ hardly less from one another than science and scholasticism. Just because of this contrast an analysis of humanism can shed light on the characteristics of the scientific spirit. As the methods of the scholastics are understood best through the study of their professional tasks, so the sociological analysis of humanism must start with the occupations and professional aims of its representatives.

1. THE PROFESSIONS OF THE ITALIAN HUMANISTS 1300-1600

The ancestors of the humanists are found among the public officials and secretaries of the late medieval Italian cities. In thirteenth century Italy merchants and artisans, conscious of their worldly interests and their wealth, had arisen. Numerous noblemen had moved from their castles to the cities and in some cities, as in Florence, had even turned to trading like burghers. Feudalism which had settled all public affairs within the framework of the traditional relations between the feudal lords, their vassals, and bondsmen, was disintegrating. The advance of money economy had considerably increased the tasks of public administration and required public officials with rational training and juridical knowledge. Also the intellectual world of the feudal period, being substantially rural, could no longer satisfy the rising townsmen. On the other hand, some ancient traditions still survived in the doctrines of the church and

---

* [This essay has not been previously made public. We know that a first version of this essay was written before the summer of 1941 for in his 'Report on the present state of the study of Dr. Edgar Zilsel on the *Sociological Roots of Science*' of June 22, 1941, Zilsel mentions a MS on humanism and writes "The Chapter on humanism and its conformities with and differences from science is nearly ready for the press" (HP/Z). In his first application to the *American Philosophical Society* (APS) of October 28, 1941, Zilsel again mentions what we take to be the same MS. This time he writes: "The section of the relation ship of the scientific to the humanistic methods (about forty typewritten pages) is nearly ready for the press". For reasons that are unclear to us Zilsel did not publish this MS. In his second application to the APS of February 28, 1943, he writes: "The section on the methods of humanism (71 typewritten pages) . . . [is] ready for the press". We take this to be a reference to the MS published here. Like he did in his essay 'Problems of Empiricism' Zilsel makes use of what could be called 'supporting evidence paragraphs'. In the original MS these paragraphs are indicated to put into small print. Following his practice in his 'Problems of Empiricism' we have put these paragraphs in the main text and have not turned them into footnotes. Eds.]

the learning of the theologians. Particularly in Italy where numerous monuments testified to the grandeur of classical antiquity the memory of the past was not dead. The Italian burghers looked up with envy to the achievements of ancient Rome when their own world appeared small and poor. It is strange that a youthful society, faced with the task of building up an intellectual culture of its own, looked back to the past. Yet this "renaissance"- process, one of the most impressive testimonies to the power of tradition in history, is susceptible to sociological explanation. Ancient civilization was but incompletely known to the Middle Ages. The fundamental differences between the nascent capitalistic society and the Roman republic of the Roman empire, therefore, could not be noticed. It was manifest, however, that classical civilization had been higher than the contemporary and, being a worldly civilization of city dwellers, it fitted the cultural desires of the trading noblemen and burghers better than the half military, half rural culture of the knights and the religious ideals of the monks. In the Flemish, French and German cities this congeniality was not able to produce the humanistic enthusiasm for antiquity; it was sufficient only later to make its adoption possible after it had developed in Italy. In the Italian cities, on the other hand, where the burghers considered themselves the direct descendants of the ancient Romans, the sociological congeniality was supplemented by patriotic pride in a past that was felt to be their own. Rienzi, the son of a tavern-keeper, thus could carry on his burgher insurrection against the Roman nobility with ancient slogans and by imitating political institutions of the Roman republic.

Rienzi was a political revolutionary and had no literary aspirations. Before he had himself proclaimed "tribune of the Roman people" in 1347 he had been a notary of the Roman municipality and later of the Papal See at Avignon. Other public notaries, enthusiasts or classical antiquity, combined burgher patriotism with literary activity. Half a century before Rienzi, Lovato des Lovati, a contemporary of Dante, called himself "judge and poet of Padua". His disciple, Albertino Mussato, proudly signed his letters as "poet and historiographer of Padua". He wrote a Latin tragedy composed in the style of Seneca with a patriotic-political purpose and several Latin works on contemporary history and moral philosophy. By profession Mussato too was a city official: he was a notary public, a member of the public council of Padua, and headed diplomatic legations of his native city to the Pope and the Emperor (1302 and 1311). From their legal education these political city clerks knew more of ancient Rome than the artisans and merchants. Hence they could express the contrast between the new burgher culture and the world of feudalism by ideals formed after ancient patterns and become, thus, the true initiators of the "revival of learning".

Usually Petrarch (1304-1374) is considered as the first humanist. Though a friend and admirer of Rienzi, he was more a literary man than a politician or office-holder. He too, however, was the son of a Florence notary, had studied law at Montpellier and Bologna, and was often employed as a political ambassador by the Pope and the Archbishop of Milan. Several times the position of an apostolic secretary, that is of a permanent official of the curia, was offered to him. He made his living as a protegé of wealthy noble families (the protection of the Colonna family, however, was lost by him, when he advocated the cause of Reinzi) and from numerous ecclesiastical sinecures. Powerful patrons - prelates, princes, and cities - competed for his services after he had become famous. Petrarch's friend Boccaccio (1313-1375) was like Petrarch primarily a literary man, but also in his life public offices play a certain part. He was the son of a merchant and, before he turned to literature, a merchant himself. As a young assistant to a merchant he made contact with the scholar officials at the court of Naples. After he had distinguished himself by his literary activity and classical scholarship he was frequently employed as ambassador by his native city, Florence. He lived on his modest wealth and from an annual stipend, allowed to him by the city of Florence for his public lectures on Dante. The lives of both Petrarch and Boccaccio show the close connection between office and scholarship in the period of early humanism. In the fourteenth century literary activity, if it did originate in governmental activity, was at least rewarded by political offices.

Up to the sixteenth century numerous Italian humanists were chancellors, secretaries, and officials of princes, cities, and the curia. The humanist office-holders chiefly had to conduct the foreign affairs of their employers. Their offices, however, tended to become sinecures. More and more humanists developed into court historians, court orators, and court poets or free literati dependent on princes, noblemen, and bankers as patrons. Others were engaged as tutors of the sons and, sometimes, the daughters of princes, or founded schools for children of noblemen. Or they travelled from city to city giving lecture-courses on classical authors to older students, the lectures being paid by the municipalities. Several humanists held academic chairs. At the universities, however, the spirit of scholasticism still predominated. The medieval and Renaissance universities were not devoted to research but to teaching: they were but institutes for the theoretical training of clerics, notaries and attorneys, and physicians; the seven liberal arts - grammar, rhetoric, dialectic, arithmetic, geometry, astronomy, and music - were at most universities not completely represented and everywhere regarded as merely preparatory subjects preceding the true, that is professional, training. The humanists in the Faculties of Arts - for them the first three of the liberal arts came into consideration - were, consequently, less esteemed and paid considerably less than the theologians, jurists, and medical doctors. Usually the humanistic teachers of "eloquence" were engaged for one year only and were more like travelling lecturers than permanent professors. Most of the fifteenth century humanists, however, took the occupations as professors, lecturers, and political secretaries alternately, and even the court poets and free literati were, at least occasionally, employed as political ambassadors by their patrons. Such court positions, connected with occasional official missions but without office and university routine, were best liked by the humanists.

A few humanists became bishops and cardinals and one - Enea Silvio Piccolomini - even pope, the secretarial activity was the start of their ecclesiastical careers. Before the invention of printing the copying of ancient manuscripts offered the possibility of a livelihood to many humanistic scholars. (In the university cities there had been professional copyists even in the scholastic period). After the establishment of the first printing press in Italy in 1465 many humanists worked with printers as assistants, editors, and proofreaders. The great printer of classical texts, Aldo Manuzio (1450-1514) employed over thirty classical scholars and was himself humanistically educated. He had been a tutor to the nephews of the count della Mirandola before he became a printer. Besides there were a few exceptional cases. Niccolo Niccoli (d. 1437), the scholarly collector of manuscripts, whose home was the center of the humanist circle in Florence, was the son of a trading nobleman and originally a merchant himself, he later lived without occupation on his modest wealth. Marsilio Ficino (1433-1499), the son of the physician to Cosimo Medici, was from his boyhood educated to become the head of the "Platonic Academy", lived as such in the house of Lorenzo Medici, and was ordained as priest in his old age. At the end of the fifteenth century the above-mentioned Giovanni Pico della Mirandola and his nephew Francesco were rich counts. Vespasiano da Bisticci was a humanist bookseller in the fifteenth, and so was the archaeologist Jacopo Mazochi in the sixteenth century; the former still had despised the printing press. Ciriaco de' Pizicolli (d. 1450), the collector of Roman inscriptions, was a travelling merchant. A few humanists were monks. In the sixteenth century the humanistic travelling lecturers, free literati, and political secretaries gradually disappeared. Since such positions were no longer available, in the later half of the century the humanist university professors regarded teaching as their permanent profession: the type of erudite and pedantic philology professor that flourished in 17th century France and Holland began to develop. Our survey refers to the Italian humanists only until the end of the sixteenth century; the humanists outside Italy will be discussed later.

The following list of occupations from Petrarch to 1600 is based chiefly on Georg Voigt: *Die Wiederbelebung des classischen Altertums*, 3rd ed. Berlin 1893; J.A. Symonds: *The Renaissance in Italy*, vol. 2: *The Revival of Learning*. The Modern Library, New York; J.E. Sandys: *A History of Classical Scholarship*, vol. 2, Cambridge 1904; and the *Enciclopedia Italiana*.

Political secretaries and officials: Giovanni di Conversino (1347-1406, secretary of Petrarch, chancellor of Ragusa and Carrara, professor at Florence, vagabond humanist), Zanobi da Strada (cf. below p. 11), Coluccio Salutati (1331-1406, assistant to an apostolic secretary, chancellor of Florence), Aurispa (1369-1459, prebends, temporarily apostolic secretary), Loschi (1365-1441, prebends, chancellor of Ferrera, apostolic secretary and protonotary), Lionardo Bruni (1369-1443, apostolic secretary, chancellor of Florence), Carlo Marsuppino (1398-1453, professor of eloquence, chancellor of Florence, title of an apostolic secretary), Gianozzo Manetti (1396-1459, Florentine nobleman, wealthy merchant, and ambassador; at the court of Naples; apostolic secretary), Flavio Biondo (1388-1463, apostolic secretary), Lorenzo Valla (1407-1457, professor of eloquence, secretary of King Alfonso of Naples, apostolic scriptor), Pier Candido Decembrio (1399-1477), apostolic secretary; at the Milan court), Poggio Bracciolini (1380-1459, copyist, apostolic secretary, chancellor of Florence), Bartolomeo Facio (1401-1457, teacher, later chancellor in Genoa, secretary and historiographer to the King of Naples), Benedetto Accolti (1415-1466, professor of civil law at Siena, chancellor of Florence), Platina (1421-1481, tutor, secretary to Cardinal Gonzaga, abbreviator apostolicus).

Lecturers and university professors: Manuel Chrysoloras (d. 1415, Greek ambassador to Venice, 1396 Greek chair at the University of Florence), Argyropulos (after 1456 lecturing on Greek and philosophy at Florence and Rome), Georgios Trapezuntios (after 1420 Greek lectures at various universities), Theodorus Gaza (1400-1448, Greek chairs at Ferrara and Rome), Poliziano (1454-1494, professor of eloquence, tutor to the son of Lorenzo Medici), Pomponio Leto (chair of eloquence Rome), Pomponazzi (1462-1524, professor of philosophy), Alciato (1492-1550, professor of civil law at various universities), Leonicus Thomaeus (after 1497 professor of philosophy at Padua), Mario Nizolio (1498-1576, professor of philosophy at Parma and Sabionetta), Robortelli (1516-1567, professor of eloquence at various universities), Sigonio (1524-1584, professor of eloquence at various universities).

Educators: Gasparino da Barzizza (d. 1431, professor of eloquence at Padua, court orator to Filippo Maria Visconti, the tyrant of Milan; principal of a school at Milan), Guarino (1370-1460, professor of eloquence at various universities, tutor to the son of the Duke of Ferrera and principal of a school), Vittorino da Feltre (1378-1446, professor of eloquence at Padua, principal of a school for the sons of the Marquess of Gonzaga and other children).

Free literati: Beccadelli-Panormita (1394-1471, lecturing in Bologna and Pavia; receiving presents; at the court of the King of Naples, tutor to the crown prince and often ambassador), Porcellio (born 1406, historiographer to the condottiere Malatesta and the King of Naples, vagabond humanist); Filelfo (1398-1481, professor of eloquence at the University of Padua, lecturing in Venice, secretary at the Constantinople imperial court and in a Byzantine legation to the sultan and the kings of Hungary and Poland, lecturing in Venice, professor of eloquence in Bologna and Florence, receiving presents, professor of eloquence in Rome, died impoverished in Florence), Pontano (1426-1503, at the court of Naples: secretary, ambassador, tutor).

Prelates: Bessarion (1403-1472, Greek archbishop of Niccea, converted to catholicism at the council of Florence, cardinal), Enea Silvio Piccolomini (1405-1464, secretary in the Vienna imperial chancery, bishop, cardinal, pope), Marco Musuro (d. 1517, professor of eloquence at Padua, assistant to the printer Manuzio, bishop of Malvasia), Paolo Giovio (1483-1552, physician in Milan, apostolic secretary, bishop of Nocera), Bembo (1470-1547, at the courts of Ferrara and Urbino, apostolic secretary, cardinal), Sadoleto (1477-1547, apostolic secretary, bishop, cardinal), Aleander (1480-1542, professor of eloquence in Paris, librarian of the Vatican, apostolic nuntius to Germany, archbishop, cardinal).

Monks: Luigi Marsili (1330-1394, Augustinian), Ambrogio Traversari (after 1431 General of the Camaldolentic order).

2. ERUDITION, FAME AND MASTERY OF STYLE - THE PROFESSIONAL IDEALS OF THE HUMANISTS

The professional group of the humanists arose from the diplomatic and administrative needs of the early capitalistic cities and principalities. Florence, whose municipal offices in the thirteenth century had been known as the best place for public notaries to acquire higher training, became in the fifteenth century the center of humanism. From 1375 to 1466 Florence had seven chancellors: five of them - Salutati, Lionardo Bruni, Carlo Marsuppini, Poggio, and Benedetto Accolti - were famous humanists. The fact, however, that everywhere in Italy the humanistic offices turned into sinecures and the officeholders into literati shows that the original needs were supplemented by others of a less vital character. With growing wealth analogous processes frequently occur in social evolution. In the case of the humanists it was the desire for prestige that came into play - prestige, incidentally, being in politics hardly less useful than efficiency of office work. The political secretary and ambassador was required to increase the prestige of his employer and this secondary function gradually became primary. Since the office-holder was working with his pen, his tongue, and his brain, the prestige he could give was based on his style, his eloquence, and his learning. Naturally, classical antiquity presented the literary models and the sources and contents of the erudition. Hence mastery of style, learning and prestige became the professional ideals of the humanists. The embryonic stage of these ideals is disclosed in a passage in Giovanni Villani's chronicle on a thirteenth century public official. It reads: Brunetto Latini, the chancellor of Florence, "was the first *to teach the Florentines the rudiments* and to make them skilled *in well speaking* and the knowledge of how to govern our republic according to the art of politics". Both Latini and Villani still belong more to the Middle Ages than the modern era. Latini was born half a century before Dante; his Latin still is entirely medieval and his cyclopedias epitomize the learning of scholasticism. Villani, the contemporary of Petrarch, is a merchant and, consequently, hardly touched by the spirit of humanism. Yet he praises the late medieval city clerk as the pioneer of worldly learning and eloquence: two of the three ideals which we are analyzing appear as early as in this voice from the very dawn of the modern era. The public officials turned into humanists a century later, when they put the tasks of public administration, still emphasized in Villani, in the background behind their function as givers of prestige. The humanists, who in the Renaissance crowded the office of the Papal See and the Italian princes and cities, wanted to have as little as possible to do with office routine. Most of them, of course not the Florentine chancellors, were employed primarily for display as official orators, ambassadors, and authors of polished diplomatic notes. Besides there were other clerks, trained in civil or canon law, who did the real office work. The humanists, however, despised the jurists because of their lack of eloquence and, at the same time, envied them their higher salaries and greater influence. When the problems of public administration multiplied under the pressure of growing capitalism the public officials for mere display disappeared and were again replaced by jurists in the late sixteenth century.

The sociological reasons why the officials chose classical authors as literary models have been indicated above pp. 19ff. We repeat the dates of the officials and authors mentioned: Brunetto Latini 1220-1294, Lovati 1236-1309, Dante 1265-1321, Mussato 1261-1329, Giovanni Villani's chronicle about 1345, Rienzi 1313-1354, Petrarch 1304-1374. On the reputation of the Florentine notaries, the chancellors of Florence 1375-1466, and Villani's judgement on Latini cf. Georg Voigt: *opere cit.*(1891) 1, 392.

When in 1329, Petrarch who often was temporarily employed as official ambassador once aspired to a permanent position as apostolic secretary he had to undergo an examination in the papal business style. Since his Ciceronian style was found too pompous he failed (cf. *Ibid*. II, 4). Petrarch's failure seems to indicate that the humanistic ideals have originated rather elsewhere than in the office and invaded it only later. Actually, office routine always and everywhere is very conservative. The new, specific humanistic style was, therefore, first used by men as Petrarch and Boccaccio who were not ordinary and permanent officials. Yet, not only did all of them have close contacts with offices, but also appreciation of literary skill and worldly learning first arose among office-holders in modern civilization. In 1358, six years after Petrarch's failure, a certain Zanobi da Strada, was as the first true humanist permanently employed in the Papal chancery. Zanobi had been a Latin teacher in Florence and a political secretary of the King of Naples before he became apostolic protonotary. Petrarch's admirer, Coluccio Salutati, entered the municipal office of Florence after a legal training in 1373.

The skills of the humanistic secretaries were required also for the humanists who had developed into free literati. In all periods in which authors or artists are not yet dependent on a large public but on individual patrons, the protegé has the sociological function of increasing the prestige of his protector. The writer humanist was maintained by a prince, a pope, a city tyrant, or banker. The more impressive his writings were and the more famous he became the more fame redounded to his patron. Viewed sociologically, the writer-humanists were primarily "dispensers of fame": in the dedications of their works they took care adequately to fulfill this task. The humanistic professors, lecturers, and schoolmen, finally, are but a necessary consequence of this development; when a special group of professional dispensers of fame has monopolized the intellectual leadership of the age and attracts gifted young people, teachers who prepare for this profession must develop, and the upper classes must feel the desire to make also their children familiar with the new spirit, too. Altogether mastery of style, erudition, and fame are the specific professional ideals of humanism in all its varieties and, manifestly, the secretarial office is the soil from which they have sprung.

Among the humanists there were followers of all kinds of philosophies. Although there were Platonists and Aristotelians, idealists and Epicurean materialists, orthodox Catholics, a few admirers of the Cabbalah, and many irreligious freethinkers, pornographers and highly moral family men: they all shared the three ideals of fame, perfection of style, and classical erudition. And they agreed only in these ideals - and, of course, in the veneration of classical antiquity. The three professional ideals of humanism, therefore, must be discussed in greater detail. Though mastery of style is the basic element in the triad, we shall start with the analysis of erudition, since this is comparatively nearest to the aims of science. The differences between phenomena that are nearest to one an other usually are most instructive.

3. USEFULNESS OF KNOWLEDGE AND PRIDE IN KNOWLEDGE. THE PSYCHOLOGICAL AND SOCIOLOGICAL ROOTS OF SCHOLARSHIP

Knowledge is esteemed by man for two different reasons. First, in numerous cases, knowledge is useful biologically. He who knows where food can be found is superior, biologically, to the man who is ignorant of this fact. Since one must know the causes to be able to produce desired effects this biological usefulness applies to knowledge chiefly of causes and physical laws, that is of recurrent associations of phenomena. Viewed more accurately, even the given illustration implies a regular connection between quite a number of facts. He who knows the habitat of an edible plant knows that at a place with certain properties always or frequently certain objects are found which, if eaten, appease hunger. It is not a single fact but a recurrent association of several phenomena that is known and proves useful to him. This kind of knowledge plays a decisive part in all economic acts, in technology and all crafts and, obviously, is the biological basis and economic root of science. A man who, faced with the task of lifting a load, studies the law of the lever may be taken as the archetype of the scientist. As has been proclaimed by Francis Bacon, knowledge is power: it enables man to control nature. Since control of processes is based on the ability to predict them and since all actions point to the future, scientific knowledge tends to refer to the future. At any rate it aims at general statements. Universal implications are its adequate logical expressions: always, if certain conditions, A, are realized, certain phenomena, B, occur.

Man is a social animal. It is a matter of course, therefore, that the man who possesses useful knowledge enjoys social esteem, just as the strong are more highly esteemed than the weak, the skilful more highly than the awkward. It is remarkable, however, that knowledge also of disconnected facts which are of no use at all can be the object of social esteem and considerable pride. The origin of this pride implies a problem. It is not the well known sublimation of values that we have here in mind. Rather often activities, originally esteemed only because of their usefulness, later become values per se. In this way, on a higher cultural level, scientific investigation of causes is esteemed for its own sake. This sublimation is not only understandable, psychologically, but also quite indispensable. Many abstract theories, developed without regard to any use, only later have met with practical application; satisfaction even of the practical needs of society, therefore, can not be safeguarded unless science, i.e. knowledge of causes and laws, is esteemed as a value per se. All this is well known and does not need further discussion. On the other hand, it implies a problem: how it happens that so many people are proud of knowing isolated facts unknown to others. Why are these polyhistors conceited? Whoever has observed quarrelling scholars and the ardor with which they endeavor to clear themselves of the suspicion of some very unimportant ignorance will not doubt that pride in knowledge can be a strangely strong motive of human behavior.

This motive is not restricted to the ranks of scholars. Otherwise it could not be explained why crossword puzzles, "quizzes", and similar opportunities to display one's knowledge of entirely useless details have met with such popularity. Certainly a social motive plays a part in this appeal. When one succeeds in such tests one is considered "educated" and the lower ranks of society are characterized by lack of education: nobody wishes to be counted among them. But behind this additional motive the original problem reappears. Why is it that in all civilized nations accumulation of knowledge, not referring to any practical needs, is a component of higher education and gives certain social privileges? Knowledge of isolated facts is obviously a luxury. Certainly any fact can, occasionally, become a stepping stone to later knowledge of causes; no detail is so unimportant that this possibility can ever be excluded in advance. The possibility, however, is too indirect to explain why erudition is considered a value.

What is, for example, the use of knowing the names of rare and remote objects? Do primeval ideas come into play here? Primitive societies believe in word magic and are convinced that things can be influenced by pronouncing their names. Before technology had separated from magic, knowledge of names could, therefore, be considered just as useful as knowledge of causes. The medicine man had to know even the most secret names: he had to undergo a specific training, his occupation was probably the first profession, and he and his colleagues formed the first privileged group in human society. The analogy between a modern "quiz" contestant who is proud of his knowing the names of the nine Muses, a Renaissance humanist, and a medicine man believing in word magic may appear artificial. Yet pride in knowledge is, sociologically, a primitive, psychologically, an infantile trait. It most probably originates in the fact that children are both weak and ignorant and look up to their father, who not only protects but also teaches them. To him they ascribe both surpassing strength and surpassing knowledge. All children want to be like the father: they want to be grown up and, certainly, showing knowledge is as good a proof of one's being grown up as proving one's strength. This motive becomes particularly manifest in children who are proud of knowing the facts about birth and procreation. There is no evidence that all pride in knowledge derives from the infantile pride in sexual knowledge. The infantile desire, however, to be grown up like the father offers the best if not the only explanation of the remarkable phenomenon that even quite useless knowledge is esteemed by men. All men have been children and once have looked up to their fathers.

In many cases the relation of man to his ideals and authorities mirrors the relation of the child to the grown up and, particularly, the father. Gods are always imagined superior to man both in power and knowledge. In the monotheistic religions, together with omnipotence, omniscience is attributed to the deity. The same attributes appear among the professional ideals of some of the most ancient professions. The medicine man, the priest, and after the invention of writing, the scribe (as in ancient Egypt) all laid claim to superior knowledge. Since in primitive civilisations every knowledge is still believed to be useable for magic, usefulness of and pride in knowledge can not yet be separated in the case of the medicine man. The wisdom of the priest, too, may have, originally, been used for magical purposes. The case is different with the scribe. The secular scribes in ancient Egypt had nothing to do with magic and proved useful in public administration and in administration of the big estates. Both the priests and the scribes, however, were privileged groups. They were in the position, consequently, to disregard practical utility and based their claims to social esteem more on the superiority than the usefulness of their knowledge: especially the Egyptian scribes were exceedingly proud of knowing more than the ignorant common people, as the precepts of the scribe Duauf to his son Pepi disclose. Together with the medicine man, the priest and the secular scribe are the most ancient scholars. In prescientific civilizations practically useful knowledge appears almost exclusively in the form of technological skill in the lower ranks of society, namely among the despised artisans. It is a remarkable phenomenon but it hardly can be doubted that social esteem of the mere volume of knowledge is older than esteem of its utility: scholarship is older than science.

Egyptian scribes, proud of their knowledge cf. A. Erman: *Ägypten und ägyptisches Leben im Altertum*, ed. H. Ranke, Tuebingen 1923, pp. 374f., pp. 443, 448; looking down upon artists and artisans, *Ibid*, pp. 504, 533.

Scholars are proud of knowing as many facts as possible. In prescientific theoretical literature there is, therefore, a tendency to the accumulation of unconnected details. Thus the compilations and cyclopedias are composed that are so characteristic of the medieval scholastics and all priestly scholars when they turn to secular subject matters. Scientific connection of facts by general implications is still unknown in periods of mere scholarship. Even today this prescientific form of theoretical activity prevails in works ordering their contents alphabetically. The *Nouveau Petit Larousse Illustre*, for example, the most popular French dictionary, gives definitions of all words. Under the heading "mer" it explains that the sea is "a vast accumulation of salt water covering the greater part of the globe". Under the heading "mère" it says that a mother is "a woman who has given birth to one or several children". Although such dictionary articles are familiar to us, it is worthwhile to question their ends. Manifestly the definitions given are practically useless, since every Frenchmen knows the meanings of both headings and non-Frenchmen who do not know them understand the French definitions even less. Where then do they originate? The addictedness to exact definitions would indicate a survival of the scholastic spirit. The sentence, however, given to illustrate the second definition, clearly points to humanism as the historical source of the method used: "Agrippinna", the dictionary says, "was the mother of Nero". In prescientific civilizations fledgling scholars learned in this way what a mother is. Even after the rise of science, however, the spirit of scholasticism and humanism survives in many fields, especially in education.

A French dictionary has been selected as example since the humanistic spirit is especially strong in French education. -Our remarks were not intended to advocate practical utility as the only aim of education. As mentioned above, not even in the merely intellectual training of a scientist would this goal be sufficient. In addition, education aims at development of emotional patterns always at least as much as the training of technicians and theorists. Particularly the advocates of humanistic education have always stressed emotional and esthetic values: classical culture is connected, by indissoluble links, with western religion, literature and art. We have, however, not to discuss goals of education but to describe sociological facts and to compare the intellectual procedures of science and scholarship.

A remarkable relation to time must be disregarded. Science originates in the needs of action and action points to the future. The scientist, being primarily interested in recurrent associations of events, endeavors to predict what will happen if certain conditions are realized: he tries to extrapolate the regularities observed in the past to the past. The scholar, on the other hand, is primarily interested in the past. Factual knowledge is based on experience and tradition, that is on recollection of past events. When the scholar is proud of knowing disconnected facts and of knowing as many of them as possible, he inevitably must turn to the great receptacle of facts: the past. History, archeology, philology, therefore, are the very fields of scholarship. Hence a noteworthy difference between science and scholarship results. Scientific zeal is frequently linked to progressive ideas. The more the scientist stresses action and the more he regards science, as Francis Bacon did, as a means of controlling and changing events the more science develops into a tool of progress. Sometimes science was, and more frequently it was considered, even a tool of revolution. Erudition and scholarship, on the other hand, are eminently conservative. This becomes manifest as early as in the most ancient representatives of the scholarly professions: the medicine man, the priest-scholar, the Egyptian and Babylonian scribe, all were custodians of tradition. Just as pride of knowledge, apparently, is older than esteem of its usefulness, so, in the social development of knowledge, the conservative tendency precedes the progressive one. Both phenomena originate in the division of society into subgroups. Learning and preservation of tradition are specific ideals of certain, numerically small, professional groups. To the whole of society action, utility of knowledge, and progress are more important. The historical development, however, is determined rather by interaction of social subgroups than by the interests of an abstract whole of society.

In the case of Renaissance humanism the retrospective tendency of the scholars met halfway with certain progressive desires of the rising middle classes. The burghers tended to detachment from the feudal past; they were, however, ignorant and still unable to settle their problems intellectually by their own means. The scholars, on the other hand, looked back to antiquity. Yet the retrospective learning of the scholars had something to offer to the burghers. The link between the two apparently opposite tendencies was formed, as we have pointed out before, by the urban and worldly character of both the classical and the early capitalistic civilization. Thus the birth of the new city culture, intellectually, took the shape of a "Renaissance" and a "revival". The revival of learning appears "reactionary" when we view only the words of the humanists; it points to the future in so far as it expresses cultural desires of the new middle classes - to which, after all, the humanists themselves belonged. However, the technical needs of manufacture and trade eventually proved stronger than the professional ideals of a few scholar-officials and literati: humanism declined and scholarship was superseded by science in the seventeenth century.

4. THE HUMANISTIC ERUDITION

After having discussed the characteristic features of scholarship in general it is easy to demonstrate them in Renaissance humanism in particular. Since we are interested in the sociological origins we shall restrict ourselves primarily to the fourteenth and fifteenth centuries. In the humanist, pride in knowledge is expressed by a deep contempt for the non-scholars. Petrarch would rather be not understood than extolled by the multitude; as he explains, it is a disgrace to the learned to be praised by the mob. Coluccio Salutati, the chancellor of Florence (1331-1406), points out that knowledge and eloquence distinguish man from beast and man from man; in this respect, however, the distance between man and man is even greater than between man and beast. The terms Salutati uses - wisdom, eloquence, intelligence - are obviously meant to characterize the humanist and manifestly, in his opinion, the scholar is further above the non-humanist than man above the beast.

Petrarch, *epist. famil.* XIV, 2 (ed. Fracassetti, vol. I, p. 279); cf. *Ibid.* I, 7 (vol. I, p. 63). Salutati, *Epistolario*, ed. F. Novati in *Instituto Storico Ital.*, Fonti 15-18, I, 77, 79, II, p. 204.

The contrast between the theoretical aims of humanism and of science stands out most distinctly in the tendency of the early humanists towards accumulating scraps of knowledge without theoretical connection. Thus Petrarch wrote "On things to be remembered" and "On famous men". Boccaccio on "The genealogies of the pagan gods", on "The vicissitudes of famous men", on "Famous women", on "Mountains, woods and rivers". In the merely enumerative method these compilations agree with the numerous "summae" of the scholastics, and, especially, with the late medieval cyclopedias for laymen, such as the Trésor of Brunetto Latini. The only difference is that the humanist compilations drew the facts from a considerably greater number of classical authors and replaced the medieval Latin with Ciceronian style. In the fifteenth century these half medieval compilations develop, on the one hand, into essays such as Poggio's *On the vicissitudes of Fortune* and *On the calamities of princes*, on the other, to learned archaeological encyclopedias such as Biondo's *Roma Instaurata, Roma Triumphans*, and *Italia Illustrata* that were the forerunners of the modern handbooks of classical archeology. Even as late as in 1506 Rafael Volaterranus wrote a encyclopedia, *Commentarii Urbani*, which tried to comprehend the whole of humanistic knowledge in three volumes, entitled geography, anthropology and philology. Volaterranus` work differs from a modern encyclopedia by the absence of scientific criticism and alphabetic order, from the medieval cyclopedias for laymen by the fact that it abounds with veneration of classical antiquity. Also, collections of biographies and lists of celebrities are very numerous in Renaissance literature. Among them are several collections of famous women and such a strange work as Manetti's six books *On famous long-lived persons* (c. 1450) in which the lifes of "all" celebrities who reached the age of sixty years, from Adam to the humanist Niccoli, are described. A tendency to collection of curiosities is rather manifest in humanist literature from Petrarch up to the beginning of the sixteenth century.

The encyclopedia of Volaterranus was widely read. The *Catalogue of Printed Books* of the British Museum gives seven Latin editions and one Italian translation between 1506 and 1603. On collections of biographies cf. Edgar Zilsel: *Die Entstehung des Geniebegriffes*, Tuebingen 1926, pp. 159-175. Collections of curiosities: Domenico di Bandino d'Arezzo: *Fons memorabilium universi* (c. 1370); Gulilelmus Pastrengo (a friend of Petrarch): *De originibus rerum* (printed Venice 1547; deals with inventors, founders of cities, ancient offices and names etc.); Polydorus Vergilius: *De rerum inventoribus*, Venice 1499 (deals with comparatively few technological inventors, but reports on the genesis of writing, marriage, prostitution, various sects, etc. For the greater part it gives fabulous stories and myths collected without any criticism. It appeared from 1499 to 1680 in at least 20 Latin editions and 12 translations); Sabellicus: *De rerum et artium inventoribus* (a poem printed in 1560); Alexander Sardus: *De rerum inventoribus* (intends to fill the gaps in the work of Polydorus Vergilius); Pierio Valeriano (d. 1558): *De infelicitate literatorum* (printed 1620; a collection of biographies of "unhappy" literati).

Reference works, encyclopedias, and tables are required also in modern science. They are not regarded, however, as the true achievements of science but contain only the material from which the scientific structures are built. A considerable part of the humanist literature, on the other hand, resembles such collections of material, rather inadequately arranged, while scientific elaborations are absent. The deficiencies of humanism are best illustrated by a comparison with a contemporary pioneer of the really scientific spirit. In 1554 Niccolo Tartaglia mentions the rule for the solution of equations of the third degree, discovered by him. He tells how he, fortunately, found the method after a rival in a mathematical competition had set him several problems leading to cubic equations. If he had not discovered the rule he would have been blamed "by the ignorant crowd but certainly not by intelligent people". For, as he adds, "one particular secret does not make a man a scientist, because science deals with general rather than with particular subjects; the number of the particular subjects is infinite and it is not possible, consequently, to know every one of them". This is the voice of a representative of the true modern era. As self evident as Tartaglia's remark sounds to us, one will meet with a similar remark in none of all the humanists. Much too proud of their erudition to regard the slightest scrap of knowledge as unimportant, they knew as little of the difference between fruitful and sterile insights as the medieval scholastics. This is a decisive difference between humanism and science. Humanism, and all prescientific scholarship, esteems the mere volume of learning; scientists appreciate knowledge only if it results in further knowledge. In the case of Tartaglia the question is of a general mathematical rule but in the same desire to make knowledge work and bear fruit Galileo's general physical laws also have their origin. Tartaglia, by the way, was also one of the forerunners of Galileo in the investigation of mechanical laws. He was, of course, not a humanist but wrote in the Venetian vernacular and belonged rather with the artisans. This remarkable man was a self educated mathematics teacher who sold mathematical advice to gunners and architects, ten pennies one question, and had to litigate with his customers when they gave him a worn out cloak for his lectures on Euclid instead of the payment agreed upon.

Tartaglia on the scientific insignificance of particular subjects *Quesiti et inventioni diversi* IX, 25 (Venice 1554, fol. 106 v.); "ten pennies (scudi) a question" *Ibid*. III, 10 (fol. 42 r.); the worn out cloak *Travagliata inventione*, Venice 1551, appendix terzo ragionamento (sig. F ij v.).

The merely accumulative and enumerative method of humanism manifests itself also in other traits. Even those works that are not just compilations are always interwoven with unnecessary references to classical authors. Manifestly, the Renaissance scholars used every occasion to display their classical reading. By citation reputation as a scholar was acquired. In a contemporary report on the first lecture at the University of Florence of Carlo Marsuppini the later chancellor is expressly praised in that "there was no Greek nor Roman author from whom he did not quote". By this, the biographer adds, "he gave a great proof of his memory". The emphasis upon memory is significant. The official speeches which the political secretaries were required to deliver on the occasion of princely weddings, coronations, and diplomatic missions had to be made from memory. Otherwise they would not have befitted the festive occasions and would not have gained credit for the employers of the speakers. A good memory, therefore, belonged to the professional requirements of a humanist. From the fourteenth century, when Petrarch dedicated a chapter of his book *On the Remedies for the Vicissitudes of Fortune* to the praise of memory; good memory was mentioned time and again when the eminent gifts of a famous author were enumerated. To scientists too a good memory is useful; yet it would never be counted among the characteristics of a good scientist in a scientific age. As early as in the fifteenth century Leonardo da Vinci, who was not a humanist but on artist-engineer, that is a superior craftsman, had the scientific attitude towards memory: "who ever appeals to authority, he says, applies not his intellect but his memory". Galileo and Descartes also scoffed at the humanistic esteem for memory and, in contrast to it, stressed causal reasoning and mathematical demonstration.

The report on Marsuppini in Vespasiano da Bisticci: *Vite;* Petrarch's chapter on memory in *De remediis utriusque fortunae* I, 8; Leonardo on the contrast intellect-memory. Analogous passages in Galileo and Descartes cf. below p. 42; qualities of eminent authors were frequently enumerated. Instances: Boccaccio, *Eulogy of Petrarch* (printed in Petrarch, *De remediis*, Rotterdam 1649, at the beginning): Petrarch was distinguished by his "innate gifts (ingenium) and his *memory*"; Boccaccio, *Opere volg*, Firenze 1833, XV, 49: Dante was distinguished by his "capacity, *memory*, intellect, innate gifts (ingegno), and invention"; Alberti, *Opusc. mor.*, ed. Bartoli, Venice 1568, p. 160: authors of eminent *"memory,* mind, and innate gifts (ingegno)" are very rare; Erasmus, *Ciceronianus* (1528) in *Opera*, Basel 1540, I, 829: eminent authors are distinguished by "invention, arrangement of ideas, imagination, emotion, charm, *memory*, learning, spirit and genius"; Trissino, *Poetics* (1563), in *Opera*, Verona 1729, II, 120: Dante was distinguished by his "*memory*, his innate gifts (ingenium), his marvellous nature, and his learning".

To people who appreciate memory so highly, the past means more than the future. The merchants, artisans, and navigators of the Renaissance must have been conscious of the newness of their achievements and their age; otherwise they could not have accomplished the complete transformation of feudal technology and economy. From artists - who belonged with the artisans -we have a few remarks expressing such sense of youth. The literati, on the other hand, felt aged and tired even at the very beginnings of humanism. As early as in the 14th century Petrarch complains of the lack of eminent men and contrasts "the misery of his century" with the grandeur of classical antiquity. His friend, Salutati, points out that the authors of the period do not produce anything new; we are, he says, "botchers only, patching together garments from pieces of classical cloth". Similar expressions of resignation recur frequently in humanist writings. We are but diminutive men (homunculi) exclaims Lionardo Bruni (c. 1400) and the same term in Greek translation (anthropiskoi) is repeated in Bessarion (1462). Only the ancients, particularly the ancient authors, are in the opinion of the humanists real men. This senile attitude of the literati is among the strangest phenomena in the rise of the new society. How it derives from the professional ideals of the scholars has been explained before. Probably it is a somewhat artificial product and may be compared to a flourish by which the scribe attests his professional dignity.

Feeling of decay in Petrarch, *Epist, famil.*, ed. Fracassetti VI, 4, vol. I, 336 ff., cf. *ibid,* I, 1); Salutati's remark quoted in Karl Vossler: *Poetische Theorien der Fruehrenaissance,* p. 54; Bruni's remark *ibid.* 82; Bessario's remark in a Greek letter to Apostolios (Migne, *Patrologia,* Patres Graeci, CLXI, 688 ff.).

The sense of inferiority to antiquity is the emotional background of both the literary and the philosophical method of humanism. In their ideas of literary style the humanists virtually never got beyond the ideal of imitation. They occasionally disagreed on the question as to whether the perfect writer has to imitate one author - Cicero - or better imitates several classical models alternately. The idea, however, that writers could have their own personal style, though a familiar notion to a plebeian author such as Pietro Aretino, who wrote in the vernacular, occurred extremely seldom to learned humanists. And in their philosophical quarrels they always uncritically refer to their ancient authorities and attack the authorities of their opponents. Even if once a scholar tries to conciliate, as Bessarion did in the quarrel between the Platonists and Aristotelians of his period, he pleads for eclecticism and without discrimination proclaims all philosophers of antiquity as authorities that, by a modern, must be followed, not attacked. Just in this context the modern thinkers were called "diminutive men" by Bessarion, as mentioned above. The notion of autonomous investigation of truth, manifestly, was foreign to humanism. Laurentius Valla only (about 1450) occasionally advocated philosophical originality, Johannes and his nephew Franciscus Pico originality of literary style (about 1500).

Imitation of Cicero advocated in Paolo Cortese's letter to Poliziano (the letter before the last in Politianus, *Opera,* Basel 1553); likewise Bembo in his letter to Joh. Franc. Pico (Bembo, *Opera*: Basel without date III, 17 ff.); eclectic imitation of several authors advocated in Joh. Franc. Pico's letter to Bembo (in Pico, *Opera* II, 123 f., contained also in Bembo, *loc. cit.* III, 3 ff.); likewise Erasmus of Rotterdam, *Ciceronianus* (*Opera,* Basle 1540, I, 820 ff.). Personal style advocated in Pietro Aretino, *Lettere* I, 123; cf. I, 82 and 114; III, 176. Though a literary celebrity, Aretino was proud of his lack of education; he always scoffs at humanistic erudition. -Laurentius Valla advocating philosophical originality: philosophers always had the freedom of saying what they think not only against heads of other schools but also their own school head; this applies the more to philosophers who have joined no school at all (in *Dialectica,* preface). Valla's sense of originality, however, must not be overestimated, though he has a certain tendency to criticize established authorities. In the quoted work he is opposing Quintilianus, who at this time was not generally recognized, against the authority of Cicero, pointing out that Quintilian can be surpassed only by a god (*op. cit.* chap. 40, 1509 edition, fol. 29 v.). On both Pico's cf. below p. 44 f.

The ideal of imitation, though interesting to historiographers of literary style, need not be discussed here. The humanist belief in authority, on the other hand, directly concerns our problems. It hardly differs from the prescientific attitude of the medieval scholastics. These believed in the authority of the Scripture, the church fathers, and Aristotle, the Renaissance scholars in the authority of the secular writers of classical antiquity: this is the only difference. The humanist belief in authorities is entirely unscientific. It is in accord rather with the traditionalism and collective mindedness of the Middle Ages than the individualistic spirit of early capitalism. The merchants, many artisans and artists of the Renaissance were already used to relying on their inventive spirit and their individual abilities. In these ranks appreciation of novelty and, among a few artists of the fifteenth and sixteenth centuries, even ideals of originality had developed. It is remarkable how little this individualistic spirit influenced the scholars of the early modern era. Obviously it was not humanism that has produced modern thinking. Viewed sociologically humanism was the ideology of a caste of literati-officials that, in their time, monopolized the intellectual leadership of the age but since then has become extinct. Viewed historically humanism was a by-path rather than the main road of the advance of the scientific spirit, and the modern spirit in general. Certainly the humanists have helped to replace ecclesiastical with secular thinking. They have, moreover, rediscovered classical philosophy, literature, and art and, thus, transmitted the intellectual achievements and aesthetic ideals of antiquity to the subsequent centuries. In intellectual developments, however, methods are more important than material contents. The spirit of science, which sets off the modern era from all other periods, originated in social groups which, as artisans and navigators, were not touched by or were opposed to humanism. From these ranks came Francis Bacon's enthusiasm for progress and the seventeenth century insurrection against belief in authorities. In Galileo, Descartes, and Bacon this revolt was directed against scholasticism and humanism alike. Science arose in open opposition to humanism.

A considerable portion of the spirit of prescientific scholarship, however, survives in the age of science. Primarily our present philology and historiography, directly descend from Renaissance humanism. Both are more interested in single facts than in general laws, more in the past than in prediction, and they are not always free from certain implications of the pride in erudition. On the other hand our historians and philologists have not adopted the uncritical belief in authority and the passion for imitation from their Renaissance ancestors. In the modern world of natural science and machines even the "humanistic studies" could not remain unaffected by essential elements of the scientific spirit.

5. THE HUMANISTS AS DISPENSERS OF FAME

We have at length discussed the humanist ideal of erudition. The ideal of fame, having less relations to the scientific spirit, may be treated more briefly. The desire for fame is very strong in man. It is particularly powerful in the upper classes and can, hence, even become the economic basis of special professions. In many civilizations with a warlike nobility professional bards make their living by spreading the fame of members of the upper ranks. In the tribes of the North American Indians the warriors themselves sang of their deeds. In Greece of the Homeric period and among the ancient Norsemen the primeval tribal equality existed no longer; there a nobility had developed and professional rhapsodists and scalds had taken over the task of singing the deeds of the heroes. The prestige of the upper class in early capitalistic Europe was based more on wealth, political power, and display of luxury than on warlike deeds. Especially in Italy, however, the desire for fame reached a degree unknown in any another culture except classical antiquity. Probably it was the division in numerous small states, rivalling with one another, that in Italy produced the unusual intensity of the desire for fame. Apparently the same sociological cause produced the same effect in ancient Greece. Also a certain fading of religious faith in the early modern era contributed to the increase of the passion for fame, for the glory ideal cannot fully develop as long as the spiritual interest is directed towards non-worldly objects.

At any rate the municipal governments, princes, and popes, city tyrants, bankers and noblemen of Renaissance Italy competed with each other for fame. To this end they used a special professional group, the humanists. It has been mentioned how the political secretaries not only had to conduct foreign affairs but also to increase the prestige of their employers, by their literary activity. This function was discharged in two ways. Primarily the humanists were required to insert rather immoderate glorifications of their protectors in their writings or, at least, in their dedications. This was the direct method, used to excess by the literati. On the other hand the prestige of a prince or prelate was increased more indirectly by the mere fact that he was able to maintain outstanding writers or scholars. In this respect painters, sculptors, and architects could render the same services as literati. Famous authors or artists at the court of a prince discharged the same sociological function as his palace, his suite, and his luxurious garments and jewels. Almost in all civilisations princes and noblemen increase their prestige by display of costly luxuries that sometimes are very important for the development of civilization. Viewed sociologically, the writers and artists of the Renaissance belong with these luxury goods of the upper class. In addition most of the dynasties and all city tyrants in Renaissance Italy could not rely on the prestige of ancestors. Usurpers need dispensers of fame much more than old dynasties with inherited prestige and the same is true for popes who come to power by election. Hence in many cases the Renaissance patrons even hunted after celebrities with offers of donations and positions, endeavoring to win them over if they were in the service of a competitor.

To the writer the profession of a dispenser of prestige offered the financial basis for his literacy activity. This is a phenomenon common to all periods in which a large and educated public has not yet developed and in which, consequently, authors, and artists are dependent on individual patrons. In the Renaissance, particularly, there existed a kind of symbiosis between the humanist and his patron. The author was supported by his patron and, in return, made him famous. Sometimes the authors were fully conscious of this reciprocity. The more rationally the "give and take" was handled by the dispenser of glory and the more frequently he changed his patrons the more his activity degenerated into adulation and blackmail. The low of this development of an in itself morally neutral sociological relationship is represented by the humanist Filelfo in the fifteenth century and the vernacular writer Pietro Aretino in the sixteenth. Both were extremely gifted authors who procured themselves patrons according to mere rational business principles without any moral inhibitions. In a period in which authors do not live on donations from individual patrons but on the sale to an anonymous public of their books, writers like Filelfo and Aretino would probably have made use of publicity agents. In the Renaissance they used shameless extortion. It is significant that those humanists who, by their occupation, were more remote from the fame business lived up to the moral standards of the contemporary middle class. The bookseller Vespasiano da Bisticci, the educators Vittorino da Feltre and Guarino, the printer Aldo Manuzio were exemplary business, family, and professional men. In the world of the free literati and their upper-class protectors, on the other hand, glory displaced virtually all other ideals. The greed of fame, probably, was increased also by the specific reciprocity of the patron - protegé relationship. It was certainly more honorable to be protected by a famous than by an unknown patron and more glorious to be praised by a celebrity than by an unknown scribbler. Both sides, therefore, were interested in the increase of fame.

Ever since humanism had come into existence the ideas of the humanists were dominated by glory. As early as in the fourteenth century, Petrarch dedicated quite a series of chapters to the problems of fame in his book *Remedies for the Vicissitudes of Fortune*. There we find chapters "on glory" and "on the hope of fame", "on fame hoped from buildings" and "on fame hoped from intercourse", "on infamy", "on contempt", and "on posthumous fame". The book, however, is a dialogue and of the protagonists only one praises fame, whereas the other, advocating the vanity of all worldly goods, always opposes eternal bliss to desire for fame. Manifestly, in Petrarch the medieval ideal of Christian humility still combats the humanistic ideal of fame. Over the same conflict Petrarch drudges his life away in his dialogue *On Contempt of the World*. There St. Augustine advocates Christian humility whereas Petrarch, who himself appears as the other speaker of the dialogue, reproaches himself with his vanity. The dialogue ends with the promise of the humanist to collect all his strength against the allurements of fame: "may God assist me". Still, thirty years later, the same conflict of the two ideals - characteristic of a period of transition - appears in the letters of the chancellor Salutati. When, on the other hand, later humanists -

Erasmus, Lorenzo Valla, Francesco Pico - occasionally object to fame, the Christian arguments are, for the most part, replaced by Stoic ones. Such humanists attacks against greed for fame, however, must not be taken too seriously. Just as the analogous attacks in ancient Stoics and Sceptics, they only confirm the strength of the adversary. Before business people preachers declaim against the treasures that are eaten by rust and moths; before an audience of literati, and protectors of literati, other literati declaim against fame.

Petrarch *De remediis utriusque fortunae* (together with *De contemptu mundi*) Rotterdam 1649. *De rem*, I, 117, 122; II, 25. *De cont*. III, 808 ff., 812, 820, 823. *Epistolario di Coluccio Salutati* in *Instituto Storico Italiano*, Fonti no. 15-18: for fame: I, 10, 89, 105, 110, 198; II, 182, 204; III, 86; against fame: III, 349, 425, 471.

The idea that dispensing of fame is the chief function of the literati emerges very early. It is implied in an odd theory of Boccaccio on the sociological descent of the poets. In ancient times, Boccaccio explains, the kings had used priests to achieve veneration of themselves and their ancestors; from these priests, in his opinion, the writers descend. The economic background of the idea clearly stands out in the Latin letters of Filelfo. In 1433 Filelfo, one of the most influential and unscrupulous of the humanists, begins a letter to Cosimo Medici by mentioning the gracious reception given to him by the addressee and stresses that, in return, he has in his writings commended Cosimo's name to immortality. This is but the introduction to an attempt to set his protector against two humanists rivals. The main part of the letter explains that Niccoli and Carlo Marsuppini are worse than pestilence and accuses them of having called another competitor, the old Chrysoloras, a lousy beard. Curiously enough the writer protests a few lines later that he has not learned to flatter and to adulate. In another letter to a certain Simoneta of 1451, Filelfo first assures the addressee of his love. After mentioning that others prove grateful for benefits by gold and gems he continues: "I, however, make gods out of men and give them the immortality which is implied in the eternity of glory. Certainly you can see what you may expect from me". The real project of the letter is a petition for a donation. Filelfo's Latin letters were published and later printed. It is significant that in his Greek correspondence, which was not intended to be published, the "dispenser of fame" ideology occurs only once, in a letter to the Sultan. There Filelfo asks for the release of a few female relations who had been captured by the Turks. Beginning with the affirmation that he has already heard of the glorious deeds of the sultan, he introduces himself as follows: "I am among those who make mortals immortal by the glory that the word dispenses". The request follows. In Filelfo the give and take in the glory business is quite manifest.

Family tree of the poets: Boccaccio, *Opera volga*, Florence 1832, XV, 53 f. (*Vita di Dante*); Filelfo's Latin letters: *Epistolarum familiarum libri* 37, Venice 1502, fol. 12 r, and fol. 54 v.; letter to the sultan: Emile Legrand: *Cent-dix lettres Grecques de Filelfe*, Paris 1892, p. 63. Further evidence of the "dispenser of glory" ideology: G. Voigt, *op. cit.* I, 334 (Poggio), 446 (Petrarch, Beccadelli-Panormita), 527 (Filelfo).

Sometimes strange ideas result from the "dispenser of fame" ideology. In the middle of the 15th century Benedotto Accolti, the chancellor of Florence, states the dark ages had actually achieved as much as classical antiquity; only the historical writers had been deprived of their remunerations and had, for this reason, hushed up all eminent achievements. Or Porcellio, court humanist to the King of Naples, about the same time considers the dispensers of glory more important that the glorified persons; he begins his exposition of the deeds of the condottiere Piccinino with the praise of the writers "by whose documents the praiseworthy men live in eternal memory of mankind and, miraculously, become immortals from mortals". About half a century later an Italian court humanist to the emperor Maximilian I, Sbrullius, who had been portrayed by Dürer and, in return, had dedicated a poem to the painter, affirms with a strange reversal that Dürer will become immortal by the poem and he himself by the picture.

Accoltus: *De praestantia virorum sui aevi*. Parma 1697, p. 60 (the same opinion expressed in Poggio: *De varietate fortunae*, cf. Voigt, *op. cit.* II, 492); Porcellio in Muratori: *Rerum Italicarum scriptores*, Milan 1761, XXV, 2A; Sbrullius in J. von Schlosser: *Materialien zur Quellenkunde der Kunstgeschichte* III (*Wiener Akademie Berichte*, philo.-hist. Klasse vol. 180 (1916), 72) On Sbrullius or Sbrollius cf. C.G. Jöcher: *Allgemeines Gelehrtenlexicon*, Leipzig (1751).

The passion for fame of the Renaissance literati results in a remarkable phenomenon that essentially distinguishes humanism from modern science. Since the time of Francis Bacon scientists usually give control of nature, progress and furthering of human civilization as aims of their activity. If they ever mention fame as a motive they speak, at the highest, of the prestige of their scientific school, their university, or their fatherland, but even this is done only in somewhat backward countries. No modern scientist would admit that he does his research in order to become famous. Just this is plainly stated by the Renaissance humanists. In doing so, they only follow, however, classical models. To classical literature not only the "dispenser of fame" ideology is familiar but also the desire for fame is very often given by ancient authors as the decisive incentive of cultural activities. Thus Cicero declares in his *Tusculaneans* that "it is honor that nourishes the arts and men are impelled to the studies by fame". This idea was with enthusiasm adopted by the Renaissance. As early as in 1386 Coluccio Salutati quoted the saying of Cicero and expressed his agreement. Salutati, however, rejected the same saying as heathenish seventeen years later since, as mentioned before, he still wavered between the humanist ideals and Christian humility. In the following centuries fame is very often used even as an argument in theoretical controversies. Over and over again it is pointed out that the behavior, the literary style, or the doctrines of some adversary are not likely to make him famous: obviously this argument is considered to be a valid refutation. The fame ideology is a typical product of humanism. Originally it was foreign to the artists who in the fourteenth century still adhered to the guild ideals of the artisans and thereby, in some respect, were nearer to the modern spirit than the literati. When in the early Renaissance the artist rejected love of gain as incentive he demanded love of his art from the painter without even mentioning fame. In the middle of the fifteenth century, however, the fame ideology spread from the literati to the architects, painters, and sculptors who began to be ashamed of their descent from artisans. The sixteenth century artists gave desire for fame as a motive of their activity just as the humanistic dispensers of glory.

Fame as motive in classical antiquity: Cicero, *Tusc.* I, 2, 4 and *pro Arch.* 6 and 11. Similar passages: Plato, *conviv.* 208 C ff., Horace, *ep.* II, 3, 324. On the "dispenser of fame" ideology in classical antiquity cf. Alexander's complaint of having no equal herald of his deeds as Achilles had in Homer; furthermore Theognis 237 ff., Pindar, *Nem.* 4, 6 ff.; 7, 13; *Pyth.* 1, 90 ff.; 3, 114 ff.; *Ol.*, 9, 27; Cicero, *ad. fam.* V, 12, 13; *pro. Arch.* 6, 9 and 12; Horace, *carm.* IV, 8 and 9; *ep.* II, 1, 229 ff.; Vergil, *Aen.* IX, 440 f.; Pliny, *nat. hist.* pref. 25 (on Apion); Seneca *ep.* 21, 3 ff.; Claudianus, *de cons. Stilich.* 3 pref.

Renaissance: fame as motive: Salutati, *Epistolario loc. cit.* I, 70 and III. 86 (Cicero quotations). Bessario (*In calumniatorem Platonis*, Venice 1516, I, 1) starts his attack on George of Trapezunt with the remark that he had expected George's work to have been written "in order to gain posthumous fame". Franciscus Pico, letter to Bembo (1512) in *Opera* II, 123 f.: literary imitation must be avoided since it is detrimental to fame. - Likewise Erasmus, *Ciceronianus* (1528) in *Opera*, Basle 1540, I, 840 ff. -Cardano, *De vita propria* (1542) in *Opera*, Lugduni 1663, I fol. 7 r: "immortalization of name" praised as "glorious invention". Desire for fame given as motive for the composition of his book by the French humanist Bachet (1621), full quotation below, p. 48. Michelangelo Biondo, *Treatise on Painting* (1549) in *Quellenschriften für Kunst-Geschichte*, Vienna 1888, vol. 5, 32: all artists should strive for fame. The treatise begins with the wish for "immortal fame to all excellent artists of Europe" (Biondo is a medical doctor and not strictly a humanist; his treatise is written in the vernacular).

Guild ideals in artists: Cennini, *Treatise on Painting* (c. 1390), chap. 2 (in *Quellenschriften für Kunstgeschichte*, vol. 1, new ed. Vienna 1888 p. 4). Glory ideals in artists: Leone Battista Alberti, *Treatise on Painting* (1435) in *Quellenschriften für Kunstgeschichte*. Vienna 1877, vol. 11, pref. and pp. 49 and 99. Leonardo, *Treatise on Painting* (c. 1500): I am addressing myself rather than to painters greedy for money, to artists "who want to gain fame and honour through their art" (ed. Ludwig, *Quellenschr.*, vol. 15-17, Vienna 1882, I § 81, cf. I § 65; Leonardo is otherwise very little influenced by humanism). Vasari, *Biographies of Artists*, in *Opera*, ed. Milanesi, Florence 1887, I, 91: eminent artists produce perfect works "inflamed by desire for fame"; *ibid* I, 11: Michelangelo created his works "in order to leave posthumous fame like the ancients"; cf. *ibid.* VIII, 163. Francesco d'Ollanda (a friend of Michelangelo), *Da pintura antiga* (in *Quellenschr.* N.F., vol. 9, Vienna 1899) p. 27: immortal name is the only valuable aim in human life.

It can hardly be assumed that the writers, and artists, of the Renaissance were essentially vainer than their modern colleagues. Many modern scholars too may be motivated in their research by the desire for prestige. The fact that Renaissance authors openly admit personal fame to be their aim whereas modern scholars, as far as such questions are discussed at all, put fame in the background behind impersonal ideals makes the real difference between the two periods. Of course, many humanists, more or less sincerely, professed religious ideals. Impersonal intellectual ideals, however, were far less developed in the period of the Renaissance than of modern science. This, apparently, is correlated to the absence of any co-operative organization of intellectual activities before the seventeenth century. Scholars, by their very nature, seem to tend towards personal rivalries. In the Middle Ages the strength of group tradition and the common membership of the church were counterweights to such individualistic impulses. Yet even at the late medieval universities, where institutions corresponding to modern laboratories and research institutes were unknown, the practice of disputation produced a considerable quarrelsomeness among the schoolmen. In the Renaissance rivalry among the scholars greatly increased. The disintegration of feudalism and the rise of economic competition had, sociologically and economically, prepared the soil for literary individualism. The professional conditions of the literati produced it. In their belief in authorities the humanists still were medieval; in their quarrelsomeness hyperindividualistic. Literary polemics became as frequent and vehement as never before and personal quarrels and intrigues were regarded as an unavoidable part of the life of a literary man. In this general struggle of all against all, every kind of humanist took part: political officials as Lionardi Bruni, Carlo Marsuppini, and Poggio, free literati as Filelfo, university professors as Robortelli and Sigonio are among the most quarrelsome of the scholars. Only those humanists who were more remote from the "fame business" - the booksellers, printers, and educators - abstained from the general passion for polemics.

The idea that scholars have to promote knowledge by mutual co-operation was as yet unknown to all of these individualistic advocates of the fame ideal. Even at the end of the sixteenth century Henri Estienne (the younger), because of scholarly rivalry, did not allow his son-in-law Isaac Casaubonus to use his library for philological studies. Both Estienne and Casaubonus were Geneva Huguenots and very remote from the amoral literati of the Filelfo and Panormita period, both were outstanding classical scholars, but their scientific ideals were entirely individualistic. It is not mere coincidence that Francis Bacon, who first proclaimed the objective ideals of progress of science and control of nature, at the same time rejected personal fame and advocated foundation of research institutes based on co-operation of scientists. Not before the middle of the seventeenth century were Bacon's ideas realized, the Royal Society was founded and the first scientific periodicals were published. At any rate the hypertrophy of fame in the Renaissance is but the reverse of the absence of any co-operative scientific institutions. Only the scientific age views science as a great building, rising stone by stone through co-operation of scientists, each of whom uses the results of his fellow workers and predecessors. The complete lack of this idea in the Renaissance is among the most characteristic differences between science and humanism. It must not be overlooked, however, that even in the scientific age, and even after the rise of learned periodicals and research institutes both for the sciences and the humanistic studies, a certain liking for learned polemics might occur more frequently among philologists than among natural scientists. This, certainly, is a survival of the humanistic spirit.

On Henri Estienne vs. Casaubonus cf. Sandys, *loc. cit.* II, 205. Francis Bacon against personal ambition as scientific motive *Novum Organum* I, 129 (Fowler, p. 337); for research institutes, co-operation, and division of labor in scientific research *Nova Atlantis*.

6. MASTERY OF STYLE

Mastery of style is the third of the professional ideals of the humanists. We discuss it last not because it plays a smaller part than the ideals of erudition and of fame but because it is remotest from the scopes of science. It has grown out from the tasks of the political office and refers both to the written and the spoken word since the political secretaries had frequently to act as official orators. Because of this connection with speech-making, mastery of style was usually called "eloquence" in the Renaissance. As far as humanism was represented at the European universities the chairs of eloquence were its seats. How far valuation of eloquence went may be shown by a few remarks of an early humanist. Eloquence, Coluccio Salutati says: "is the greatest of all humanistic studies, the most beautiful of all sciences". In a letter on the death of his master and friend Petrarch Salutati expresses his conviction that in heaven the deceased "with his eloquent breast" will succeed in persuading God early to reunite Petrarch's admirers with their master. He, Salutati, is looking forward to meeting his friend again and will regale himself in the other world of the "nectarean suavity of Petrarch's eloquence". The quotation at once gives a good instance of the rhetorical exaggerations that by the humanists were considered eloquent style. The emphasis upon style lasted up to the end of the Renaissance. At the beginning of the sixteenth century Bembo, papal secretary and later cardinal of the Roman church, cautioned authors against reading the epistles of St. Paul; St. Paul's Greek would have spoiled the style of the readers. Not before the decline of the literati and the rise of the professors in the later half of the sixteenth century did classical erudition become more important in humanism than imitation of classical eloquence. Even the most learned and pedantic professors, however, attached great value to the linguistic purity of their Latin expositions.

Salutati on eloquence *loc. cit.* I, 179 21 f., I, 230 1. 9 f., II. 295; on Petrarch in heaven, *ibid*. I, 199. Bembo on St. Paul cf. J.A. Symonds op. cit. 511.

Science is interested in factual and logical content, mastery of style is a formal and aesthetic ideal. A great portion of the humanist aims, therefore, has nothing to do with scientific knowledge. Even from the merely stylistic point of view, however, the prose writings of the Renaissance humanists differ considerably from modern scientific works. Before the rise of the professors at the end of the Renaissance these writing are virtually always pathetic, often satirical, and abound with declamations and rhetoric repetitions even when theoretical arguments are brought forth. It is not here our task to give value judgements but to investigate humanism causally and to compare it with science. Theoretical knowledge certainly does not exhaust the totality of human activities; in the scientific age, too, there are in addition to science fine arts and letters. From the aesthetic and the educational point of view the emphasis upon style and the refinement of literary language, due to humanism, are of considerable historical importance. In classical antiquity rhetoric had played a great part in higher education. The sense of the aesthetic values of language was widely spread and also considered a prerequisite also of historiography and every philosophical and scholarly activity. Among the monks of the Middle Ages this sense had been lost: they were too much interested in the other world and subtle theological arguments to care much for language. Literary style and language were rediscovered as objects of interest by the political secretaries of the early Renaissance and this discovery has left its mark on the civilization and the education of the Western world.

It is another question whether the specific style favored by the humanistic still appeals to the modern mind. The language of modern science is exact, factual, and concise. Since also the non-scientific literature has not been left untouched by this new style the literary taste of our age may take offense at the overheated declamations and the verbosity of the humanists. The merely factual writings on technological topics of Renaissance artisans (their contents will be analyzed later) and the witty letters of the completely uneducated Pietro Aretino make a much more "modern" impression on readers of our time than the works of the humanists. They are nearer to the modern sense of style just because their authors were not touched by humanism. Of all sixteenth century Italian prose writers with humanistic education, only Macchiavelli is virtually free of the Renaissance rhetoric that appears so unmodern to us. Macchiavelli, however, wrote in the vernacular just as the artisans.

The specific style of the humanists has more to do with the absence of the scientific spirit than historians of thought who disregard the literary form of the publications possibly assume. Actually the great advocates of experimentation at the beginning of the seventeenth century scoffed in their writings not only at the sterile subtleties of the scholastics but also at the rhetoric of the humanists. Such antihumanistic attacks occur frequently in the works of William Gilbert, Galileo and Francis Bacon. It is remarkable, however, that Gilbert and Bacon themselves are still strongly influenced in their style by humanistic verbosity; Galileo only writes an unsophisticated, though witty, Italian which stems from the plain language of the plain people. Descartes too is entirely free of humanistic "eloquence". Even from the stylistic point of view modern science arose in manifest opposition to humanism.

Attacks against humanism: William Gilbert, *De Magnete*, London 1600, preface to the reader: the contemporary writers are "destroyers of the good arts, literary idiots, *grammarians*, sophists" etc. He, Gilbert, will not "refer to *the ancients and Greek auxiliaries*, since neither Greek arguments nor Greek words are able better to prove or to illustrate the truth... And we have not used *the paint of eloquence or the adornment of words* in this work but have restricted ourselves to discussing difficult things in such a style and by such words that are necessary to understand them". -Galileo, *Dialogo sopra i due massimi sistemi del mondo*, Edizione nazionale VII, p. 87, line 20 ff. : if the argument were on "human studies where there is neither truth nor falsity... skill of speaking" would be instrumental, but "in the natural sciences oratory is inefficient"; *ibid* 135 I, 1 ff.: by mere combining and interpreting everything could be proved from Virgil and Ovid and even better from the alphabet; *ibid*. 139 I. 4 f.: opponents who refer to authorities in their argumentation would better call themselves "historians or doctors of memory" than philosophers; subject of the argument is "the world of the senses not the world of paper"; *ibid*. 293 I, 7: "rhetorical flowers" do not fit in with scientific arguments; they belong to "orators and poets". -Francis Bacon, *Advancement of Learning* I, 4, 2: the humanists prefer words to the matter and believe in authorities; -Descartes, *Récherche de la vérité, Oeuvres*, ed. Cousin XI, 341: Latin and Greek are of no more importance than the Swiss and Breton dialects; rational argumentation, not memorized knowledge is the point that matters.

In an argument Pico della Mirandola, the younger, says that things are more important than words. Pico's remark is a noteworthy exception in humanist literature for, in general, the interest of the humanists is primarily directed towards language. They were engrossed in "eloquence" and the classical purity of the used phrases, in products of classical literature and Latin and Greek grammar. Objects of nature and theoretical problems interested them, in general, in so far as they had been treated before in classical literature. A humanist once pointed out that Caesar had made himself immortal through his description of the conquest of Gaule - not by the conquest itself. Another humanist dispenser of glory, in a work on famous men, mentions the invention of printing among the great achievements of the age. He does not give, however, technological details of the invention or appreciate its import on the spread of material knowledge, but sees its merit in "the destruction of linguistic barbarism". The odd report on printing ends with the praise of a humanist colleague, well deserved, for his pure Latinity - the name of Gutenberg is not mentioned. The numerous "philosophical" writings of the humanists are primarily exercises in eloquence. They are more literary collections of Platonic, Pythagorean, Aristotelian, and Stoic quotations than original efforts to solve philosophical problems. Only the neo-Platonic philosophy of Ficinus seems sincerely interested in material questions, his interest, however, originating rather in his Christian faith than in a zeal for scientific explanation of the world. Not until the decline of humanism at the end of the sixteenth century did a more original natural philosophy arise - Telesio, Giordano Bruno - and this was, in Bruno, combined with vehement attacks against humanistic overestimation of language and linguistic pedantism. The analogous attacks of the first modern physicists and scientific philosophers have just been mentioned. In their combats with rising humanism the scholastics too had brought forth the same arguments against their victorious rivals. They too reproached the humanists with disregard of the material problems and with the preference given to words. In this respect, after the humanistic interlude, science has returned to the standpoint of scholasticism - though the scientist and the scholastic might considerably disagree on the question as to which problems belong to the "material" ones and by which methods they have to be investigated.

On Pico cf. the end of this note. -Caesar and the conquest of Gaule in Muratori, *Rer. Ital. Script.* XX, 448, 453. -Printing and Latinity in Egnatius (Cipelli) *De exemplis illustrium virorum*, Paris 1554, fol. 299 f. Polydorus Vergilius too in his well known *De rerum inventoribus*, Venice 1499, only points out that printing has diminished the price of books and made preservation of classical authors possible (book 2, chap. 7). -Bruno against humanism, *della causa*, dial. 3 (at the beginning), dial. 4 (beginning). -Scholastic opposition to humanist overestimation of words: the Cologne professor of theology Hochstraten (1521) in his reply to Hutton's *Letters of Obscure Men* and a scholastic magister in a protest against the new humanistic curriculum at the University of Leipzig (1519), cf. Friedrich Paulsen: *Geschichte des gelehrten Unterrichtes*, 3rd ed. Leipzig 1919, I, 52 n. and 109; the Vienna professor of theology Saeldner against followers of Enea Silvio Piccolomini (about 1450) cf. G. Voigt, *loc. cit.* II, 292.

"Things more important than words", Johannes Franciscus Pico in his second letter to Bembo on the imitation of Cicero (*Works* of both Pico's, Basle 1601, II, 145). Pico himself, however, ends his first letter on the same subject with a long apology for the lack of stylistic polish. Yet few parallels to Pico's attack on overestimation of words might be found in humanistic literature. Altogether the younger Pico (1470-1533) and his uncle Giovanni Pico (1463-1494), who is so often quoted as a representative of the Renaissance spirit, are rather exceptions among the humanists. The elder Pico, proclaiming love of truth as the only motive of a true philosopher, disregards fame (loc. cit. I, 212, *de hominis dignitate*); he argues against overestimation of classical antiquity, underestimation of the own period, and demands study of the scholastics, Arabs, and Chaldeans (ibid. I, 79, *apologia*). The younger Pico argues against fame as aim: philosophy and science must be studied only for the sake of God and truth (*ibid.* II, 24 *de studio philosophiae*); he attacks imitation (II, 123 ff. letter to Bembo) and underestimation of the own period (*ibid.* II, 125 f.). The special position of the elder Pico possibly is partly explained by his training in scholastic philosophy at the Paris university; except for the few humanistic monks, he is, probably, the only humanist who had such a training. In addition both Pici, being wealthy counts, were farther remote from the usual fame business of the literati.

7. THE HUMANISTS AND THE SCIENTIFIC LITERATURE OF CLASSICAL ANTIQUITY

In the humanist literature, the almost complete silence on the contemporary technological inventions and geographical discoveries is striking. The period of the Renaissance was a period of technological revolution and an unprecedented expansion of the geographical horizon. Of these great historical events in which the artisans, manufacturers, and merchants were naturally highly interested virtually nothing is to be noticed in the writings of the Italian humanists. This silence is closely linked with the separation of liberal and mechanical arts or, what is the same, the disdain of manual labor. Technological inventors and navigators were "mechanics". Humanists, on the other hand, were proud of being representatives of the "liberal arts" and, particularly, representatives of their more distinguished division. This superior division (the "trivium" of grammar, dialectic, and rhetoric) comprehended those arts in which relationship to speech is especially manifest. To the inferior "quadrivium" (arithmetic, geometry, astronomy, and music) the Italian humanists paid virtually no attention.

The treatment of the discoverers and inventors in contemporary literature will be discussed later more extensively.

Certainly several humanists worked also on ancient mathematics and natural science. They edited Euclid, Archimedes, Apollonius, and Diophantus, translated them into Latin and, by these editions and translations, considerably influenced the rising modern physics and mathematics. Yet it would be erroneous to ascribe understanding of mechanical and mathematical problems to the editors. All of these editions and translations are the work of philologists who edited the ancient scientists because they were ancients rather than because they were scientists. The texts and translations are not only full of material mistakes but the prefaces reveal the merely literary attitude of the editors and, sometimes, a remarkable lack of mathematical understanding. The first Greek edition of Archimedes (Basle 1544), the work of a certain Thomas Gechauff, a German humanist, is the most superficial. Archimedes was by far the most eminent mathematician of antiquity and virtually the only ancient physicist to investigate quantitative laws by means of experimentation and mathematics. Compared to Aristotelian and medieval "physics" his work represents an entirely new approach to the problems. Actually it contributed a great deal to the development of modern scientific mechanics after it had come into the right hands. Of all this the humanist editor is not aware. His preface is manifestly intended to hide his mathematical ignorance behind the usual laudatory phrases of the professional dispensers of fame; as scientific achievements of Archimedes he gives statements that were known both to every ancient and every sixteenth century schoolboy; and the value of mathematics is proved by means of a classical quotation from Quintilianus stating that mathematical knowledge is necessary for the orator. The application to technology of mathematics, with which the contemporary architects and engineers were intensely occupied, obviously is considered as too inferior by the humanist. The verbiage of Gechauff's prefaces is in a remarkable contrast to the first Latin edition of Archimedes that had appeared one year earlier (1543). Its editor was Tartaglia, one of the forerunners of Galileo, who really understood the great ancient scientist and, in practice and treatises, applied Archimedean methods and results to technological problems. Tartaglia, however, was, as mentioned above, not a humanist but a self-educated mathematics teacher and mathematical adviser to gunners, architects, and merchants. Of course, Tartaglia, who admitted that he knew little of the ancient languages, had not himself translated the Greek text but edited a thirteenth century Latin translation. All of his own words are written in the vernacular.

The preface to the first Greek edition of Diophantus (Paris 1621) - the editor was Bachet, a French humanist - is sounder than Gechauff's preface to Archimedes. Bachet compares his tract with a previous Latin translation and is concerned with textual correctness and the biography of Diophantus. In a lengthy and "eloquent" explanation, however, he gives as incentive of his publication emulation with another scholar and his desire to become famous. He is an adherent of the humanistic glory ideology and a conscientious philologist but not a mathematician. The preface to the first Greek edition of Euclid (Basle 1533) is typical of humanism. It too, however, does not contain any evidence of real mathematical knowledge.

The prefaces to the first editions of ancient scientists are reprinted in Beriah Botfield *Prefaces to the First Editions of the Greek and Roman Classics*, London 1861. The following quotations refer to this work. The editor of the first Greek Archimedes edition, Thomas Gechauff, called Venatorius, was a Nuremberg preacher who wrote several theological works and translated also Aristophanes. His Archimedes edition also contains the Latin Archimedes translation of Jacob of Cremona. It has three prefaces. In the first the scientific importance of Archimedes is proved as follows (loc. cit. 416 f.): Gechauff points out that the circle has no beginning and no end: "who of the ancients, I ask, has more clearly written on this fact than Archimedes?" The cube stays stable however it falls: "who, I adjure you, has more eruditely, more accurately, more diligently explained these facts than our Archimedes?" There follows an enumeration of ten contemporary eminent "mathematicians"; all of them are classical scholars, only three - Regiomontanus, Schöner and Rheticus - actually were mathematicians. Several classical anecdotes on Archimedes conclude the first preface. The second preface (420-425) begins with long references to ancient philosophers on mathematics and quotations from Homer and Vulcanus. The value of mathematics is proved (423 f.) by means of Quintilianus' statement that the orator requires mathematical knowledge. As an example Gechauff at length discusses the size of an ancient acre that must be known to the orator and gives no fewer than five diagrams to illustrate the concept of the area of a rectangle.

The editor of the first Greek Diophantus, Claude Gaspard Bachet, sieur de Meziriac, was a humanist and lawyer, author of Latin, French, and Italian poems and a French translation of Ovid. His Greek Diophantus contains also the editor's Latin translation of the text. The first preface is addressed to a lawyer. Just as Themistocles competed with Miltiades "in honourable emulation for fame", Bachet begins, I am thinking day and night of how I might become as famous as you are. You have surpassed all jurists and even Papinianus. To which field am I to turn? After having examined all other disciplines, I decided to choose mathematics "since it wonderfully delights the minds and since in mathematics the subtlety of the intellect specially comes to light. This Diophantus will give evidence of my achievements and show whether I have deserved fame beyond the ordinary mathematician". The preface to the reader (658-665) is less personal. It gives learned biographical notes on Diophantus, mentions an earlier Diophantus translation of Xilander (1575), "that is much worse than ours", and states that in the following edition the text has been purified for the first time and that additions of the editor have been enclosed in brackets. A French translation of the first books of Diophantus (from Xilander's Latin translation) had been published thirty-six years earlier by Simon Stevin (*L'arithmetique*, Leyden 1585) who was not a humanist but, originally, a bookkeeper and cashier of the municipalities of Bruge and Antwerp, later a military engineer and Quartermaster general of Holland, and who really understood mathematics.

-The editor of the first Greek Euclid (Basel 1533) was Simon Grynaeus, a Basle professor of Greek, a friend of Thomas More, and editor of works of Aristotle, Plato, Plutarch, and Ptolemy. The preface *loc. cit.*, 381 ff. -The German humanists Peuerbach and Regiomontanus, however, in contrast to their colleagues, were eminent mathematicians.

The editions of Euclid and Archimedes exerted their influence rather in circles linked to the mechanical arts and opposed to humanism than among the colleagues of their humanist editors. And only a small minority of the humanists were engaged in work on classical scientific literature. Scientists were edited scarcely before the sixteenth century, that is about two centuries later than the ancient orators, poets, and philosophers. Interest in ancient science did not belong to the original aims of humanism. Apparently the humanists began to occupy themselves with classical scientists only when other circles became interested in mathematics and natural science. In his *History of Classical Scholarship*, Sir John Erwin Sandys gives two lists of first prints of Latin and Greek authors. The Latin list contains seventy-one first editions from 1464 to 1596, starting with the Mainz print of Cicero's *de officiis* of 1464. Among them there are only three books - Pliny the Elder (1469), Lucretius (1473), and Vitruvius (1486) - that can be called works on topics of the natural sciences or technology. The list of Greek first editions begins with Aesop's fables of 1478 and contains one hundred and eight prints. Seven works - ancient astronomers, Galen, Hippocrates, Ptolemy, Euclid, Archimedes, Diophantus - deal with astronomical, medical, geographical, mathematical, and mechanical subjects. The first Greek Euclid was printed in 1533 - sixty-nine years after Cicero - the first Greek Archimedes in 1544, the first Greek Diophantus not until 1621. Neither Euclid nor Archimedes appeared in Italy but in Basel, Diophantus in Paris. Certainly, these printed first editions do not exhaust the ancient scientific literature that was at the disposal of Renaissance readers. There were handwritten copies before the invention of printing and there were Latin, and later vernacular, translations of Greek authors. And, certainly, it was not the fault of the humanists that the great majority of preserved ancient writers were orators, philosophers, poets, and historiographers. But just this fact discloses the part played by science in the humanist "revival of learning". Somewhat less than 6.5% of the first prints of ancient authors dealt with scientific problems. It is historically understandable that Galileo, scoffing at the humanistic way of thinking, would remark that his was the world of the senses not the world of paper.

The two lists of first prints in Sandys *op. cit.*, vol. 2, Cambridge 1908, 102 ff. Galileo on the world of papers, cf. above p. 42.
The rise of interest in zoology and botany too is somewhat linked with humanism. The first print of Pliny's *Natural History*, Rome 1469, was edited by the humanist Theodorus Gaza, a Greek refugee who also translated the biological works of Aristotle into Latin. Among the earliest zoologists of the modern era is Aldovrandi (c.1522-c.1605) who published thirteen folio volumes on natural history. Aldovrandi was a professor of philosophy at Bologna, an Aristotelian who had turned to zoology upon stimulation of the

Montpellier physician Rondelet. Humanist influence is distinctly noticeable in his work. Of his two books on the eagle the first treats its subject philologically and archaeologically and only the second describes the biological facts. The other eminent zoologists of the sixteenth century - Rondelet (1577-1666), Salviani (1514-1572), Gesner (1516-1565), and Belon (1517-1564) - were medical doctors with few relations to humanism (cf. E.W. Gudger: *The five great naturalists of the sixteenth century*, Isis 22, 1934, 21 ff.). Between 1469, the year of Gaza's first edition, and 1600 thirty reprints of Pliny's *Natural History* were published, but before 1469, there was no humanistic literature dealing with biology. Obviously also the biological interest of the late Renaissance originated outside humanism.

It cannot be overlooked that preoccupation with language and related problems is characteristic primarily of the Italian humanists. In Germany, England, and France a not negligible fraction of the humanists took part in the religious struggle of the period, advocating the cause of the Reformation. Other non-philological questions too engaged the non-Italian humanists. A humanist like the English Chancellor Thomas More considerably differs, in his problems, his style, his life, and his death, from his Italian literary and office colleagues. The humanist alchemist to Queen Elisabeth, John Dee, wrote a preface in the vernacular to an English Euclid translation (1570) and showed interest in, and a limited understanding of, the mechanical arts, cartography, and navigation. His fellow countrymen and contemporaries, the mathematicians Recorde, Leonard and Thomas Digges, furthered commercial arithmetic, surveying, and gunnery in spite of their humanistic education. The two most eminent astronomers before Copernicus, Peurbach and Regiomontanus, lectured in the middle of the fifteenth century at German universities alternately on astronomy and Latin poets. And after all, Copernicus himself had received a humanistic education at the University of Cracow. Obviously, outside Italy, eloquence and language had absorbed less of the interest of the humanistic scholars, a fact that should require a sociological explanation.

8. RATIONAL METHODS IN HUMANISM

The professional ideals of eloquence, fame, and erudition basically distinguish humanism from science and its aims. Yet humanistic scholarship is not less rational than science and in some way the Renaissance scholars were even more intellectualistic than modern scientists. This intellectualism appears in their idea of poetry. Though the ancient idea of poetical "enthusiasm" - the idea of divine frenzy - was, at least as a metaphor, very familiar to the Renaissance, poetry was considered a learned and often a learnable activity. Even when the innate gifts of the eminent poet or writer were stressed, Minerva, the goddess of wisdom, was considered a fitting allegory for the donors of these gifts. Only the uneducated and anti-humanist Pietro Aretino has an anti-intellectualistic conception of poetry that is much more modern than everything written on this subject by humanists. The Renaissance conception of painting is not less intellectualistic.

When the painters were rising from handicraft they attached great value to their lack of relationship to the mechanical arts. After the second half of the fifteenth century they therefore emphasized that painting, since it requires geometrical knowledge, is a science. This argumentation is humanistic as is disclosed by the frequent reference to Pamphilus, a Greek painter who about 400 B.C. had used the same argument for the same social reasons. Great painters, moreover, are frequently called "great intellects" in the numerous Renaissance treatises on painting.

Minerva as allegory for innate poetical talent in Erasmus, *Ciceronianus* (*Works*, Basel 1540, I, 830 f.) and in Trissino, *Poetics* (*Works*, Verona 1729, II, 116). On the idea that painting is a science and the reference to Pamphilus cf. Edgar Zilsel: *Die Entstehung des Geniebegriffes*, Tuebingen 1926, p. 147. Great painters as great "intellects": Alberti, *On painting* (about 1440) in *Quellenschriften für Kunstgeschichte* XI, 47; Giovanni Santi (*The Father of Rafael*, about 1480) in *Federigo de Montefeltroe*, ed. Holtzinger, Stuttgart 1893, XXII, 16 vers 90a; Michelangelo Buonarotti, *Rime e Lettere*, Firenze 1903 p. 432; cf. Zilsel *loc. cit.*, p. 266.

The intellectualism of the Renaissance is nearer to the Middle Ages than to the spirit of our century. In prescientific periods the majority of the population is strictly bound to tradition whereas the small "learned" minority which is able to read and write overestimates reason. The medieval theologians were extremely intellectualistic and there is a continuity of the same attitude in Renaissance humanism. Even in the scientific era the importance of the irrational elements of the human mind was discovered, by Rousseau and the German Romanticists, not before the end of the eighteenth century. In the evolution of human civilization reason is younger than irrational tribal instincts, irrational custom, irrational tradition. But this is true only if humanity is viewed as a whole. The majority of mankind is mute and does not leave written documents. As soon as literary men appear they are so proud of their exceptional position as scholars that they stress the characteristics by which their profession stands out from the rest of the population. For this reason everywhere written literature starts with intellectualism and only very late do the men who produce and leave written documents discover the irrational elements by which the behavior of mankind, and of themselves, is still dominated. The intellectualism of Renaissance literature, therefore, is not a scientific but a primitive trait and one must be very careful not to mistake the rational procedure of the humanists for a scientific one.

In humanism the systematic method that is characteristic of science was but slightly developed. This is true even of the quintessential field of humanist activity, philology. Up to the middle of the fifteenth century imitation of classical "eloquence" was the chief aim of the humanists. The classical scholars felt enthusiastic for ancient manuscripts, collected them, and emended the text if passages were not understandable. They replaced the mistakes of the medieval copyists, however, rather arbitrarily with phrases conforming with their opinion on classical "eloquence". A sense of historical exactness was foreign to them.

Though as early as in the fourteenth century Salutati had detected the spuriousness of the pseudo-Ciceronian *On Differences*, Lorenzo Valla's proof of the spuriousness of the donation of Constantine is the methodically most eminent achievement of the fifteenth century and, probably, all Renaissance philology. This document, on which the pope based his claims to worldly domination was shown by Valla to be a medieval falsification. His treatise was written in 1440 but, for political reasons, could be printed only seventy-seven years later by Ulrich von Hutten. Valla's analysis of the document uses both linguistic and historical considerations and is as rational as that of a modern philologist. He worked by order of the King of Naples, a political adversary of the curia with an aversion to the church - Valla was an Epicurean and, probably, a free thinker - which might have sharpened his criticism. It is significant both of the scholars and the church in the period of humanism, that Valla a few years later made his peace with the pope and became an apostolic writer. Valla also knew that the correspondence between St. Paul and the philosopher Seneca was a medieval falsification.

Valla's treatise on the Donation of Constantine reprinted and translated by C.B. Coleman, New Haven, 1922. On his philological achievement cf. George Voigt, *op. cit.* I, 69 and II, 475 f., 496 f.; J.E. Sandys *op. cit.* II, 66 f., and Ulrich von Wilamowitz-Moellendorf *Geschichte der Philologie* (in *Einleitung in die Altertumswissenschaft*, ed. A. Gercke and E. Norden, 3rd. ed. Leipzig-Berlin 1927) I, 11 f.

In the fifteenth century there was some more exact philological analysis. Poliziano who in his *Miscellanies* (1489) wrote on the chronology of Cicero's letters and the use of the tenses in Greek inscriptions was, in the judgement of U. von Wilamowitz-Moellendorf, "a real philologist though not a textual critic". Many of the humanists who lectured on or edited Latin and Greek authors were eminent classical scholars. It certainly was not easy to read the manuscripts, to correct the mistakes of the copyists, to understand and to interpret the often very difficult texts without the help of the numerous reference works that are at the disposal of modern scholars. All these activities presupposed a considerable amount not only of learning but also of rational thinking. After the invention of printing philological exactness also increased. The humanists who in the late fifteenth and early sixteenth century edited classical authors in the printing offices in Venice, Paris and especially in Basel, and did a great deal of textual criticism and textual emendation. They proceeded with much greater care than the early humanists and habitually compared various manuscripts of the same text. Yet, all of this rational philological work was lacking in systematic method. To some extent this is true even for the eminent classical scholars among the sixteenth century professors, Robortelli and Sigonio in Italy, both Scaligers, Henricus Stephanus and Casaubonus in France, Lipsius in Holland. True, imitation of classical eloquence was no more their chief aim. But they did not systematically use the results of their colleagues; they had no method for determining and comparing the age and reliability of the codices; and even the sixteenth century scholars more frequently edited, emended, and interpreted single texts than investigated general questions.

It is remarkable how rarely the humanists gave an account of their methods. There are a few humanistic expositions of logical problems, composed by Laurentius Valla and the German Rudolphus Agricola in the 15th century, by the Italian Nizolius and the Frenchman Petrus Ramus in the sixteenth. All of these humanistic logicians attack the Aristotelian logic of the scholastics, reproach it for artificiality, and want to replace Aristotle with Quintilianus. All of them conceive logic as a branch of rhetoric, an approach that is typically humanist - and unscientific. The methods of humanist philology, however, were not discussed in these treatises. Methodological writings were rarely composed by the classical scholars. Even for elementary instruction in Latin grammar the medieval memorial verses of Alexander of Villadei were used up to the end of the fifteenth century. The first modern Latin grammar was composed by the learned bishop Perotti, one century after Petrarch, in 1468, the first successful modern Latin prosody by the same author in 1453. The *elegantiae* of Lorenzo Valla (c. 1440) deals more with "eloquence" than with proper grammar and is an extremely learned juxtaposition of details without any systematical arrangement.

The humanists occupied themselves with the correct spelling of Latin words as early as in the fourteenth century. The first learned Latin dictionary was published by Robert Estienne in 1532, two centuries after Petrarch, its Greek counterpart by Henri Estienne in 1572. The first treatise on the method of textual emendation was written in 1557 by Robortelli (*On the Art and Method of Emending Ancient Books*), seven years later a treatise *On the Methods of Emending Greek Authors* by the Dutch professor, Willem Canter followed. The codices were always used without exact methods of determining their age. The first treatise on paleography in which such methods were given was composed by Mabillon in 1681, almost one century after the period which is the subject of our analysis. Not until 1697 did Bentley publish his treatise on the spuriousness of the Epistles of Phalaris - the first contribution to historic-philological criticism that equals and surpasses the achievement of Lorenzo Valla of 1440. Investigation of the "genealogical tree" of the codices was first demanded by protestant theologians in the middle of the eighteenth century and the exact method of textual criticism was accomplished by Lachmann in the middle of the nineteenth century. Thus one of the outstanding classical scholars of our time, Ulrich von Wilamowitz-Moellendorf, could say that it was "the seventeenth and eighteenth centuries that elevated humanism to the level of a science". Of course, this development was a gradual one and, certainly, the professor-humanist in the sixteenth century proceeded more critically and thoroughly than the political secretaries and literati in the fourteenth and fifteenth centuries.

Humanistic treatises on "logic": Valla: *Dialecticae disputationes contra Aristotelicos*, printed posthumously 1499; Agricola: *De inventione dialectica*, 1480; Ramus: *Dialecticae partitiones*, Paris 1543; Nizolius: *De veris principiis et vera ratione philosophandi contra pseudophilosophos*, 1553. On the development of the humanistic methods cf. Voigt, *op. cit.* II, 373 (*Latin grammar*), 376 ff. (*Latin grammar*, Perotti), 378 (Valla's elegantiae), 379 ff. (prosody and Greek Grammar), 381 ff. (textual emendation); Sandys *op. cit.* II, 202 f. (Scaliger the elder), 141 (Robortelli), 216 (Canter); Wilamowitz-Moellendorf *op. cit.*, 22 f. (Scaliger the elder), 28 f. (humanism rising to the level of science only after the Renaissance); Giorgio Pasquali: *Storia della critica del testo*, Firenze 1934, p. 3 (Lachmann), 9 (protestant theologians), 90 and 93 (16th century humanists guessing on the age of codices).

On the whole the sixteenth century humanists investigated their problems by relying only on their intelligence and without giving an account of their methods. One might object to this characteristization that it fits the nascent natural sciences as well. Yet this objection is erroneous. The first representatives of modern natural science were so well conscious of the novelty of their aims that they proceeded much more methodically than the sixteenth century, let alone the early humanists. Galilei very often discussed his new scientific method in interspersed remarks, most extensively in his God-Weigher (1632); Francis Bacon opened his combat for the new scientific approach to nature with a most extensive exposition of the method of induction (1620); and Descartes' first publication, his *Discourse on the Method for well directing one's Reason and investigating Truth in the Sciences*, is, as the title indicates, a program not only of the new philosophy but also, in spite of the disregard of experience, of the new scientific procedure. In humanism analogous methodological expositions are absent, since it had started from stylistic ideals and had turned to theoretical aims only gradually. If one disregards this difference one may say that more critical and more exact methods arose in humanism in about the same century (1590-1690) as in the natural sciences. Humanism, however, was two and a half centuries older than science.

This synchronism is remarkable. Since direct influence between the two competitors can hardly be assumed, both phenomena may be considered as two effects of one common cause: the increase of individual thinking and rationality in sixteenth century society. Of both critical philology and natural science there are certain beginnings in classical antiquity. Both are absent in the oriental cultures, though in China philological activities of literati-officials, comparable to the early humanists, were richly developed. The emergence of critical and systematical methods in philology is also a characteristic peculiarity of modern Western civilisation as the rise of science, the development of machine technology, and modern capitalism. In the Renaissance, however, only the first beginnings of the exact philological methods appeared.

What distinguished the humanist from the scholastic method? The humanists themselves were well conscious of the difference. The early humanists derided the "barbarous" Latin of the scholastics and their ignorance of classical authors. The attacks on the scholastics in Ulrich von Hutten's *Letters of Obscure Men* (1515) were pointed in the same direction. Still about 1600 Casaubonus, after having attended a disputation at the Sorbonne, is said to have remarked: I have never heard so much Latin spoken without understanding it. A similar Casaubonus anecdote, however, already points to a difference that passes beyond linguistic and stylistic ideals. To a friend, explaining that in this auditorium of the Sorbonne scholars have been disputing for four hundred years, the philologist is said to have replied: what have they decided? This reply of a humanist could as well have been made by a scientist. The method of disputation at the late medieval and early modern universities is scholastic. Casaubonus' answer shows that humanism rejects this method and, at the end of the fifteenth century, had developed a concept of exactness that differs from the exactness aimed at by the scholastics in logical distinctions and syllogisms. Certainly, Casaubonus did not miss experimentation and mathematics in the scholastic disputations. What else he did miss is, unfortunately, not pointed out.

Casaubonus' aversion to the method of disputation is typically humanist. When, at the beginning of the sixteenth century, humanism conquered many German universities, the obligatory disputations were replaced in the curriculum by "declamations", i.e. exercises in public speaking. Cf. Friedrich Paulsen: *Geschichte des gelehrten Unterrichts auf den deutschen Schulen*, 3rd. ed., Leipzig 1919, I, 120 f.

We, heirs of a scientific evolution of three hundred years, are in a better position than Casaubonus to see the methodological problems. There is not only one kind of rationality. Compared to the methods of knowledge and action in every day life in a precapitalistic society bound to tradition, the methods of scholasticism, humanism and science are equally rational. These three varieties of rational procedure, however, are substantially different. In the eyes of the scholastics logical distinctions, syllogisms, and criticism of the opponent, based on the doctrines of some authority, represented the peak of rationality. The humanists proceeded rationally even when their chief endeavor still pointed to imitation of the style of the classical authors; this endeavor, though not a theoretical one, was an entirely intellectual and learned affair. Later the humanists gradually developed the rational methods of historical criticism and textual emendation, methods which, unfortunately, up to now have been much less analyzed than those of physics and mathematics. Neither the scholastics nor the humanists, however, used the methods of science. As far as humanism is concerned, this fact is partly a consequence of the difference of subject matters: experiments can not be performed in the study of literary products of the past. It can not be explained by the peculiarity of the objects of their studies, however, that the humanists virtually never investigated causes. This is a difference of mental attitude that basically distinguishes humanism from science. The humanist, proud of his erudition, gathers single facts, the scientist wants to explain and to predict. And even methodologists of our time might disagree on the question as to whether the humanists did not use quantitative methods and never investigated general laws, because their field of research does not admit these methods, or because their interests lie in a different direction. Even today these methods are almost exclusively reserved to the natural sciences; they are used in the social sciences more rarely and extremely seldom in the investigation of literary and historical objects. In our opinion this fact has less to do with intrinsic differences between natural objects and human activities than with the descent from Renaissance humanism of our humanistic studies. Investigation of causes and general laws, quantitative methods (and experimentation) do not fit, sociologically and intellectually, into scholasticism and humanism. These methods, the very characteristics of science, do not go back to the "liberal arts" of the Middle Ages and the Renaissance; they rather ascended to the scholars and professors from the ranks of the artisans. But this did not happen before the end of the sixteenth century and will be discussed in the following sections.

9. SOCIOLOGY OF EXTRA-ITALIAN HUMANISM

A few words are necessary on the sociology of the extra-Italian humanists. The Italian origin of European humanism has hardly ever been doubted. Numerous German, French, English, and Spanish humanists had studied in or visited Italy. The Councils of Constance (1414-1418) and Basle (1431-1449), at which prelates and princes with their secretaries met from all parts of Europe, contributed much to the spread of humanism. Just as in Italy, in Germany, France, and England political secretaries were the first representatives of the humanistic spirit. Even before the Council of Constance, in the time of Petrarch, there were humanists in the imperial offices of Charles IV at Prague. Jean de Montreuil, the first French humanist, was secretary to the curia at Avignon, to the Dauphin, and later chancellor to Charles VI (1380-1422) of France. A few decades later Adam de Molyneux, the first English humanist, was secretary of state to Henry VI of England. Still in the sixteenth century there were such humanistic officials outside Italy: the chancellor Thomas More (1480-1535) in England and the secretary to Louis XII, Guillaume Budé (1467-1540) in France are the most famous examples.

In extra-Italian humanism, however, political secretaries played a smaller part because the Papal See and the great number of Italian princes, city republics, and city tyrants were lacking in France and England. In Germany there were a considerable number of princes, free city republics, and prelates subject to the Emperor alone and, consequently, a considerable number of humanist office holders and city clerks. Yet even in Germany as early as in the first half of the sixteenth century virtually all the more eminent humanists were professors. This predominance of the teaching profession can be accounted for by the specific development of higher instruction in Central Europe. In Germany between 1456 and 1544 no fewer than ten universities were founded. These, as new institutes, were less open to medieval traditions than the old Italian universities and some of them owed their foundation even to the direct intention of the reigning princes to promote humanism. After the appearance of Luther, in addition, many protestant princes and cities founded secondary schools, all of them with humanistic curricula. Most of the old Latin schools had given up the medieval curriculum even before Luther. In Germany, for all these reasons, classical scholars with theoretical interests had considerably more opportunities to teach as university professors or as rectors and masters of secondary schools than in Italy.

In western Europe too most of the fifteenth and sixteenth century humanists were professors. Many of them taught eloquence at the universities of Paris, Montpellier, and Bourges, Oxford and Cambridge, Louvain and Leyden. Since outside Italy the universities, usually, offered resistance to the intrusion of humanism, at two of them special colleges for the studies of ancient languages were established, the collegium trilingue at Louvain in Belgium (1518) and the collège de France at Paris (1531). Everywhere the humanists were members of the faculties of arts which, however, were considered merely preparatory for the other faculties and afforded less pay to their professors. Only at Paris were there also humanist professors of law, the great role played by the newly introduced Roman Law in France accounting for the considerable number of jurist-philologists. In England after 1512 several new "public schools" came into existence where also many humanists were headmasters and masters. In Switzerland the University of Basel (founded in 1459) was a center of humanism. In the late sixteenth century the French professors, in the early seventeenth the Dutch were the leading classical scholars in Europe.

Paris jurist-philologists: cf. *The Cambridge Modern History*, vol. I, New York 1903, p. 577; English Public School, *ibid*. 1, p. 582.

The early appearance and the great number of professors must not be interpreted as evidence of the origin of extra-Italian humanism from academic instruction. It was, on the contrary, rather want of skilled political secretaries that produced the new educational establishments. In Prussia, before the foundation of the university of Königsberg, the elector of Brandenburg applied to Melanchthon, then the leading classical scholar in Germany, for an expert Latinist. The elector, as he expressly wrote, needed good Latinists, then lacking in Prussia, for his diplomatic correspondence with Poland. To satisfy this want the university was established (1544) and Melachthon's son-in-law, professor Sabinus, became its first rector. A few years earlier the same Sabinus had introduced the humanistic reform at the University of Frankfurt-on-the-Oder with a speech on the importance for the statesman of a polished Latin style. The foundation of the University of Königsberg is not an isolated case. Outside Italy the humanistic reform of the universities was everywhere carried through under pressure from the princes against the resistance of the Scholastic minded professors. The collegium trilingue in Louvain was established by the bequest of the royal counselor Busleiden; the collège de France by King Frances I, upon the instigation of his political secretary Budé: in both cases humanist statesmen were the real founders. The charters of several German Renaissance universities disclose the influence of Enea Silvio Piccolimini, then an official in the imperial chancery. As far as humanism spread to eastern Europe the royal chanceries at Prague, Ofen, and Cracow were its first seats. Everywhere in Germany the princes, reigning prelates, municipalities, and, primarily, the emperor were the real protectors of humanism. Their motives, probably, were identical to those which had led to the establishment of the University of Königsberg. In the last analysis, probably, most of the numerous new universities and secondary schools in all parts of Europe were founded to promote education of skilled political officials: "skilled", in the Renaissance, meaning "able to write polished Latin".

Greek and Hebrew were not used in the diplomatic correspondence and yet were almost always included in the new humanistic curricula. Educational aims, however, and the development of public instruction must not be interpreted too narrowly. Ideas are not separated by impenetrable walls. The ruling ranks of the Renaissance sought after spiritual values, apt to embellish their lives and to increase their prestige. The monastic ideals of the Middle Ages contradicted their love of luxuries and the university professors were pedantic scholastics. In this period of transition the humanistic officials were the only intellectuals able to present the required values. Since the officials considered eloquence and philology to be the very keys to the new world of humanism, the princes promoted the studies of the ancient languages even beyond the direct diplomatic requirements.

Some features of extra-Italian humanism, however, exceed the ideology of the literati-officials as it had developed on the other side of the Alps. The studies of Hebrew and Greek belong to them. That is the language of the Old Testament, this of the New. Very few Italian humanists were interested in Hebrew and the enthusiasm for Greek did, in Italy, not at all refer to the New Testament. Manifestly, in central and western Europe, the philological interest in language and words was much more frequently combined with a religious interest in the word of God than in the country of the Papal See. After Luther and Calvin this combination resulted in the well known alliance between humanism and Protestantism. Both partners hardly had more in common than certain individualistic tendencies and hostility to catholic scholasticism. A few Renaissance universities, however, were founded at least as much to further protestant theology as to promote humanistic eloquence. All these relations need not be analyzed here. The Lutheran and Calvinistic varieties of humanism have nothing to do with our problems, since their relationship to the science of philology is independent of the question as to whether the philologist deals with Tacitus or the text of the Bible. Sixteenth century protestant theology is certainly not nearer to science than the contemporary secular philology. At the German universities Lutheran theology rather soon returned to methods not very different from those of the scholastics. Even disputations, which had first been eliminated from the artistic faculties by the humanists, were reintroduced a few decades later. Historical criticism in the manner of Laurentius Valla was first applied to the Bible by the heretic Jew Spinoza in the late seventeenth century and by Jean Astruc, the catholic physician, in the early eighteenth century. Protestant theology did not contribute, substantially, to the development of the philological methods until the eighteenth century.

On the development of the German universities in the period of humanism cf. Friedrich Paulsen: *Geschichte des gelehrten Unterrichts auf den deutschen Schulen*, 3rd. ed., Leipzig 1919, vol. I, book 1, chap. 4 and book 2, chap. 1-4; on the German secondary schools, *ibid* 1, 5 and II, 4-7. A comprehensive exposition and sociological analysis of the intrusion in the non-German universities of humanism would be desirable. A few data in Stephen d'Irsy: *Histoire des Universités*, vol. I, Paris 1933, chapter 10 and 11. On the foundation of the University of Königsberg cf. Paulsen *op. cit.* I, 241. Foundation of the collegium trilingue *ibid.* 129; Aeneas Sylvius and German university charters *ibid.* 138; the German princes, prelates, and municipalities and the humanistic university reforms *ibid.* 172 and 112 (Wittenberg), 121 (Rostock, Greifswald), 123 (Mainz), 128 (Cologne), 131 (Vienna), 135 ff. (Heidelberg), 139 (Basel, 142 (Tuebingen), 153 f. (Nuernberg); foundation of protestant universities for theological reasons, p. 252 f. (Jena and Helmstedt) (1558 and 1576). Elimination of the disputations *ibid.*, 120 f., their reintroduction, 271 ff.

In spite of their great number the professors were not the only humanists in western and central Europe. There were in Germany numerous court-humanists to prelates and princes and wandering poets making their living as dispensers of fame by selling laudatory lines to more or less munificent municipalities. Culturally, however, fifteenth century Germany had not yet caught up with the native country of the Renaissance. Compared to their Italian colleagues the German literati-humanists, therefore, were rather poor fellows, both financially and intellectually. Ulrich von Hutten (1488-1523), the merciless antagonist of the medieval spirit, is the most brilliant of them. In his adventurous life - he was successively a student, soldier, courtier with the archbishop of Mainz, poet laureate, and a persecuted fugitive - he resembles more the wandering scholars of the late Middle Ages than the successful Italian literati such as Filelfo and Panormita. The first German humanist professors too really belonged with the wandering poets. "Professors" Peter Luder (1415-1476) and Conrad Celtes (1459-1508) moved from university to university, giving lectures on poetry, sought to sell their eulogies to cities and prelates, occasionally worked as political secretaries, and lead a rather loose life, more similar to the Italian literati than to pedantic university men. Only after 1520, in the period of Melanchthon, when humanism had gained a firm footing at the faculties of arts, did the German humanist professors adopt the mode of life of their more respectable colleagues. In France Jean Dolet (1509-1546) differed considerably from the professors. He was successively secretary of the French embassy in Venice, poet, orator, printer, and eventually was executed as a heretic. Furthermore there were, as in Italy, humanist printers in Basle (Amerbach, Froben, Cratander), in Paris (Robertus Stephanus), in Antwerp and Leyden (Plantin, Elzevir); many other humanists were employed by these printers. There were quite a number of humanist catholic canons and protestant pastors in Germany, humanist physicians to prelates, and humanists living with aristocratic patrons in France.

One humanist, the most famous of all of them, Erasmus of Rotterdam (1467-1536), resembles sociologically, even the modern literary celebrities. Erasmus originally had been a priest. He entered the service of the bishop of Cambrai and went with him to Paris where he lived as a Latin teacher. As a tutor to an English nobleman he came to England, then gave lectures at Oxford and Cambridge and became a professor of divinity at Cambridge. Later he went to Basel and lived in the house of the humanist printer Froben. Froben not only paid him a salary for his activity as a literary adviser and proof reader but, as a novelty, also royalties for his books. Though Erasmus also received pensions from patrons, he is the first author in history to live, to a substantial degree, from the sale of his publications to an anonymous public. Of his *Adagia* (translations included) thirty thousand copies are said to have been sold in Europe during the author's lifetime. Only three hundred years later when, with the rise of the middle classes, a large educated public had come into existence, did professional writers, living on the return of their publications, become a common phenomenon. In his period Erasmus is an exceptional case, accounted for by his unusual fame. He does, therefore, not essentially differ in his intellectual attitude from the other humanists as, in general, ideas are influenced by sociological changes only if considerable groups of individuals are affected. Not until the nineteenth century did the rise of professional writers leave noticeable marks on modern ideology, as, for example, on the modern ideas on genius, posterity, and misunderstood persons.

On Erasmus as a professional writer, cf. John Clyd Oswald: *A History of Printing*, New York 1928, p. 135 ff., and G.H. Putnam: *Books and their Makers During the Middle Ages*, New York 1897, II, 214 ff.

On the whole the sociological bases of extra-Italian and Italian humanism do not substantially differ. Beyond the Alps too the humanistic style and the humanistic spirit arose first in the political offices. Everywhere the political secretaries underwent the same social development as in Italy. Everywhere they turned into literati, dependent on patrons, living as dispensers of fame on the one hand, as professors on the other. Everywhere, therefore, eloquence, erudition, and fame were the professional aims of humanism proper. Outside Italy the humanists took a considerably greater part in the religious struggles of the period although, sociologically, the alliance between humanism and protestantism was a rather extrinsic affair, however seriously it may have been taken by many humanists. Apart from this more religious attitude the greater percentage of professors is, sociologically, the greatest difference between European and Italian humanism. The ideals of eloquence and fame, therefore, had fewer and less brilliant advocates beyond the Alps. Literary dispensers of fame were virtually absent especially in England. One more sociological phenomenon that seems to be peculiar to England would require further analysis. In the later half of the sixteenth century, apparently, more English scholars with academic training published works in the vernacular on problems of mathematics and the mechanical arts than in any other European country. England, furthermore, is the only country in Europe in which the first printed book - Caxton's *The Dictes or Sayings of the Philosophers* - appeared in the vernacular. It dealt, of course, with classical philosophers and was the translation of a French book. These two facts are possibly connected. English scholars, apparently, looked down on the mechanical arts less than their continental colleagues; and a public with theoretical interests though without university affiliation was in England possibly more numerous than abroad. If both facts are correct their historico-sociological explanation would be of importance for the problem of the genesis of modern science.

The following list of occupations of extra-Italian humanists before 1600 is based, primarily, on Sir John Edwin Sandys *op. cit.*, The *Cambridge Modern History*, vol. 1, chap. 16, the *Allgemeine Deutsche Biographie*, the *Grande Encyclopedie*, the *Biographie Nationale de Belgique*, and the *Oxford Dictionary of National Biography*. The list contains 72 persons; 45 of them (62,5%) are professors, nine (12.5%) political office holders. The corresponding list of Italian humanists (above pp. 6 ff.; 43 persons) contains 15 (34.9%) professors and educators and 15 (34.9%) political secretaries. Although both lists are not at all complete, they are composed according to analogous principles and may be compared. The comparison shows the much greater percentage of professors among the extra-Italian humanists. Both lists contain only better known authors and, in the period of early humanism, also humanists noteworthy for sociological reasons. Authors remote from classical scholarship proper have not been listed, even if they play a leading part in the Renaissance literature of their countries. Of the seven leading French Renaissance poets ("la Pleiade") one (Dorat) was a professor, one (Belleau) secretary to a marquis, and five (Ronsard, Du Bellay, Jodelle, Baif, Pontus de Thyard) were noblemen.

secretaries and political officials:
France: Jean de Montreuil (1361-1418, chancellor to Charles V, friend of Lionardi Bruni, historiographer); Jean Lemaire (1473-1525, royal financial clerk, secretary to the count of Ligny, historiographer and poet); Guillaume Budé (1467-1540, secretary to Louis XII, maître de requêtes, diplomatic missions; pioneer in the studies of Roman Law and Roman coinage); Jacques de Thou (1553-1617, councillor of state). **Germany**: Johann of Neumarkt (d. 1380, notary, bishop and chancellor to Charles IV, friend of Petrarch); Willibald Pirckheimer (1440-1530, counselor and ambassador of Nuremberg, historiographer, translator); Sebastian Brant (1457-1521, professor of law, Basle; city clerk Strassburg); Johannes Cuspinianus (1473-1529, poet and statesman). **England**: Adam de Molyneux (d. 1450, keeper of the privy seal to Henri VI), Thomas More (1480-1535). **Holland**: Busleiden (d. 1518, royal counselor).

Professors, master of secondary schools:
France: Nicolas de Clemanges (1360-1440, professor of eloquence, Paris); Faber Stapulensis (1455-1537, professor, Paris); Alciati (1492-1550, professor of Roman law at Avignon, Bourges and Italian universities); Grouchy (1520-1572, professor of philosophy at Bordeaux and Paris); Pierre Ramus (1515-1572, professor of philosophy and eloquence, Paris); Cujas (1522-1590, prof. of law at Toulouse, Geneva and German universities), Hotman (1524-1590, professor of law); Doneau (1527-1591, jurist); Brisson (1531-1591, jurist); Godefroy (1549-1621, jurist); Casaubonus (1559-1614, professor at Geneva and Montpellier, lectuer du roi and librarian at Paris); Passerat (1534-1602, professor of eloquence, Paris); Turnebus, Dorat, Lambin (sixteenth century, royal readers). **Germany**: Peuerbach (1423-1461), magister, lectures as the first at the University of Vienna on Latin poets, visits Italy, astronomer-humanist); Regiomontanus

(1436-1476, visits Italy, magister Vienna, librarian to Mathias Corvinus, Budapest; lecturer Nuremberg, bishop, astronomer-humanist); Peter Luder (1431-1474, wandering poet and professor; M.D. Padua, political secretary to Sigismund of Austria); Hegius (1433-1498, master at secondary schools); Rudolphus Agricola (1440-1485, studies in Italy; town clerk Groningen, prof. Heidelberg, often diplomatic missions); Wimpeling (1450-1528, professor); Reuchlin (1455-1522; visits Italy, counselor to the count of Würtemberg, judge of the Swabian Confederation, professor); Conrad Celtes (1459-1506, poet laureate, wandering professor); Von den Busch (1468-1534); Heinrich Bebel (1471-1528) and Helius Hessen (1488-1540): Wandering poets and professors; Melanchthon (1497-1565, prof.); Simon Grynaeus (1493-1541, head master, professor); Joachim Camerarius (1500-1570, prof.); Johannes Sturm (1507-1589, prof.); H. Wolf (1516-1580, secretary to J.J. Fugger, headmaster); Neander and Basilius Faber (16th cent., headmasters); Crusius, Frischlin, and Xilander (16th century, professors); Justus Lipsius (1547-1606, secretary to Cardinal Granvella, prof. at Jena, Leyden, and Louvain). **England and Scotland**: William Lily (1468-1522, highmaster of St. Paul's); Richard Croke (1522, public orator, Cambridge); John Cheke (1540 regius professor, Cambridge); George Buchanan (1506-1582, professor, public official); Roger Ascham (1515-1568, Cambridge); Thomas Wilson (1525-1584); Andrew Melville (1545-1622); John Owen (1560-1622, master).

Physicians:
Julius Caesar Scaliger (1484-1558, in Italy soldier, physician to French bishop); Rabelais (1490-1553, proofreader, physician); Linacre (1460-1524, Greek studies in Italy, M.D. Padua, physician to Henry VIII of England, later priest); Hartmann Schedel (1440-1514, M.D. Padua, physician).

Theologians, monks, clergymen:
Jean Heynlin and Guillaume Fichet (professors of theology at the Sorbonne, introduce first printing press in Paris in 1470, first printed book: the letters of Gasparino Barzizza); Amyot (1513-1593, professor Bourges; bishop of Auxerre); William Selling and William Hadley (benedictines, 1460-170 Greek studies in Italy); William Grocyn (1446-1519, professor of theology, Oxford prebends); Colet (1466-1519 dean of St. Paul, London, prebends); Latimer (1485-1555, bishop of Worcester); Rudolf von Langen (1438-1519, canon Münster); Conrad Muth (1417-1526, canon Gotha); Thomas Gechauff (1510-1551, German pastor).

Living with patrons:
Joseph Justus Scaliger (1540-1609, living with French nobleman).

4

PAUL O. KRISTELLER
REMARKS ON ZILSEL's 'THE METHODS OF HUMANISM'*

(P. 22) Representatives of worldly learning before the humanists were the notaries, jurists, grammarians, and physicians of the Middle Ages. Besides the secretaries of the cities, you might as well mention those of the princes. Most famous example Pier delle Vigne.

(P. 23) I would not say that Rienzi had no literary aspirations. He was a notary, and his bombastic letters certainly have a literary ambition. There is even reason to assume that he was used as a kind of secretary at the court of Charles IV at Prague. Why do you not say something about the notarial tradition of the late Middle Ages and its connection with grammar and poetry?

(P. 23) The notaries had not a legal education. The Ars notorae was different from the study of law and rather connected with rhetoric. In this connection, you might say something about the revival of the Roman law, and about the teachers of grammar and ars dictaminis.

(P. 24) The teachers of grammar were an important section of the early humanists and of Petrarch's correspondents. Dialectics had no connection with humanism. Only grammar and rhetoric had. In the sixteenth and fifteenth centuries the humanists were well paid at the universities. I remember that Antonio Ricconi had a higher salary at Padua than Galilei. You might study the available documents for the salaries paid to professors. The humanists in Italy very often were permanent professors if they cared to stay. The most famous ones passed from one place to the other because they had so many offers. But so did the famous jurists, physicians, etc. Among the duties of the court humanists, you might mention the public orations which played a great role in that time.

(P. 25) The remark on Ficino's education is incorrect in this form, although close to his own words. He actually studied philosophy and medicine. Beginning

---

* [Kristeller's notes on Zilsel's 'Methods of Humanism' clarify several philological aspects of Zilsel's essay and offer a glimpse into the uphill battle Zilsel faced: the refusal of scholars to view intellectual history 'as determined by economic and sociological factors'. Kristeller and Zilsel were in regular contact while they were both in New York. Kristeller was affiliated to Columbia University.
  The original references in Kristeller's typescript were of course to the pages of the Zilsel typescript. We have changed the page numbers in such a way that they now refer to the pages of this edition. Eds.]

with the fifteenth century, you should mention the numerous humanists who were theologians, jurists, physicians, or noblemen. At that time, humanism spread through secondary education and influenced even those who were not to be professional humanists. Ficino did not live in Lorenzo's house, but Cosimo gave him a house and a villa to live in. He was ordained in 1473, at the age of forty. The bookseller Vespasiano was not a humanist. He probably knew no Latin. Ciriaco of Ancona collected also Greek inscriptions.

(P. 25) Giovanni da Ravenna was not chancellor of the city of Carrara, but of the Carrara family which at that time ruled Padua. Antonio Loschi was chancellor of Milan, not of Ferrara. Decembrio lived 1392-1477, according to Rossi. Poggio died in 1459. Benedetto Accolti taught at Bologna, not at Siena, if I am not mistaken. Pomponazzi certainly was not a humanist. Nizolio was not a professor of philosophy, but of Greek, if I remember well.

(P. 26) Did Filelfo ever teach at Padua? I do not think he died impoverished. I confess that I do not understand what your criteria were in compiling this list and other similar lists. Since the lists are neither complete nor selected from complete sources, I cannot see how you can base on them any general inference.

(P. 27) In your list of Florentine chancellors you might include Bart. Scala. The ideal of style is already medieval. Examples the papal chancery and Pier delle Vigne. Is Giovanni Villani not earlier than Petrarch?

(P. 27/28) You know that later humanism influenced also the jurists: Giasone del Maino, Lod. Bolognini, Andrea Alciati, Cujas, Bodin, Hotman, Zasius, etc.

(P. 28) I do not agree with you that the secretarial office was the only source of humanism. How about Petrarch? What do you mean by idealism in the Renaissance? You better mention the Sceptics and Stoics.

(P. 29-33) Are these remarks on knowledge pertinent?

(P. 32) I do not think that encyclopedias are characteristic of scholasticism. And I do not see that modern science does without them.

(P. 31/32) Why do you ridicule Larousse? It is needed. Do you never use Webster which is also English and English? Such dictionaries are used by foreigners, and also by teachers and writers who want to check on the precise use of certain words.

(P. 32) Does society need only progress and no conservatism?

(P. 34) The scholastic method is not purely enumerative.

(P. 34) Sabellicus died in 1506. Modern reference works are by no means containing the materials from which the scientific structure is built. They are rather summaries of results obtained so far by the sciences. You know that the priority of Tartaglia is debated. Cf. the studies of Bortolotti. When Tartaglia talks of the ignorant crowd, he seems to follow the aristocratic attitude of the humanists. In talking of the works of the humanists, you should consider the numerous commentaries, and such works as Poliziano's Miscellanea which became increasingly frequent in the sixteenth century.

(P. 34) Erasmus, long before Bacon, ridiculed both the scholastics and the humanists (cf. his Praise of Folly). What is your criterion for fruitful knowledge? The same as Bacon's?

(P. 36) Leonardo's notes probably belong to the early sixteenth century.

(P. 36) Memory is always listed among numerous other gifts of the mind, even in modern handbooks of psychology. Behind it is a definite scheme of traditional psychology. Cf. Augustine.

(P. 36) What is feudal technology? Resignation is not characteristic of the humanists. The passages you quote belong to the reaction against the Middle Ages as being inferior to antiquity.

(P. 37) The question of imitation was disputed. You quote Cortese's letter, but not Poliziano's answer which takes the opposite point. Also Erasmus in his Ciceronianus takes a stand against pure imitation. Cf. Sabbadini, Storia del Ciceronianismo.

(P. 37) Valla is not an advocate of philosophical originality. The passage belongs to his criticism of the blind adherence to Aristotle. Pico upholds content against eloquence in his correspondence with Ermolao Barbaro. The change of authorities effected by the humanists was a first step toward the full liberation from authorities.

(P. 38) Your emphasis on methods over materials is doubtful in this form. Do you think that the artists and artisans were not affected by humanism? Are there no scholastic or humanistic elements in Bacon, Descartes, Galileo? Alberti is a good example for the combination of art and humanism. I disagree with your remarks about the methods and task of modern historiography and philology.

(P. 40) Here you bring Aretino together with the humanists.

(P. 40) The "protagonists" in Petrarch's De remediis are Ratio, Spes etc. Please revise your bibliographical references.

(P. 41) What is the first name of Simonetta? If it is Cicco, he is not a certain Simonetta, but the famous secretary of Francesco Sforza. Why do you say Filelfo's Greek letters were not written for publication? In the mss. they appear among the Latin letters, and Legrand picked them out because he was only interested in Greek philology.

(P. 43) Do you think modern scientists and learned societies have no sense of prestige?

(P. 46) I should make a distinction between medieval theology and the "monks".

(P. 46/47) The emphasis on style appears also among the "dictatores".

(P. 47) Also Galileo was influenced by humanism, and so were Descartes, Montaigne ...... and the plain people.

(P. 47) Your remarks on things being more important than words seem to refer rather to the older Pico, Giovanni, cf. his letter to Ermolao Barbaro.

(P. 47) Ficino and Pico were not pure humanists.

(P. 48) Also Ficino had scholastic training, and so had many other humanists. It is largely a matter of the division of subjects of learning. There was some humanistic interest for Columbua, cf. Peter Martyr of Anghiari and Nicolaus Scyllacius.

(P. 51) I should add the scientific translations of Giorgio Valla, Zamberti and others. There were new translations of Galen, Dioscorides, etc., many of them belonging to the fifteenth century. Your survey is incomplete because it is based only on printed editions of Greek and Latin originals. Aside from ms. materials, you should include editions of Latin and vernacular translations. The Latin translations often were printed long before the Greek texts and had a wider influence. I think even the list of ancient scientists is not complete enough. The source for such statements should be Hain, not Sandys.

(P. 51) I should qualify your judgment on humanism and science. How about Alberti?

(P. 54) The humanists sometimes were good textual critics.

(P. 54) Are you sure Valla's Donation of Constantine was not printed before Hutten? Please check Hain. Valla's epicureanism and his attitude toward the Church are highly complex matters.

(P. 55) Is Quintilian the authority for Ramus, Agricola, Nizolius too?

(P. 55) I think the grammatical works of Guarino and Barzizza were earlier than Perotti. Valla's Elegantiae have the purpose of teaching a correct Latin style and for this purpose they have been used ever since by philologians who had to write in Latin. There was a Latin dictionary by Varinus Camers, a Greek one by Johannes Crestonus, both long before the Estienne. Numerous scholastic grammars were composed by Mancinelli, Sulpitius Verulanus and others.

(P. 55) Lachmann's method is now antiquated.

(P. 56) The anecdote of Casaubonus does not prove very much.

(P. 57/58) You should specify in what sense you believe in an application of scientific methods to history.

(P. 59) There are numerous schoolmasters and professors among the Italian humanists. Louvain is not in Holland. Alciati and Cujas did not teach at Paris, and Alciati was an Italian.

(P. 59) Was the University of Königsberg founded after the union of Prussia with Brandenburg? In connection with the College of France etc., you should say Francis I, not Louis XII (who died before 1515). In connection with the spreading of humanism, you might mention the Italian humanists abroad and the foreign students in Italy.

(P. 60) Greek and Hebrew were not equally diffused in the Renaissance. There were studies of the New Testament in Italy (Valla, Manetti). You should also remember the numerous translations from the Greek Fathers made by Italian humanists, beginning with Traversari.

(P. 60) Are you sure the disputations had ever disappeared from the universities? How about Erasmus" philological study of the Bible and the Fathers?

(P. 61) You should not forget the printers of Lyon. It was one of the most important printing centers of the sixteenth century. You might also mention Badius Ascensius, a famous humanist printer.

(P. 62) I do not agree with what you say about ideas and sociological changes.

(P. 62) There are numerous Catholic humanists in the sixteenth century, both in Italy and elsewhere. You cannot identify humanism with protestantism. I do not think there was a greater percentage of professors among the non-Italian humanists.

(**P. 63/64**) The criteria for compiling this list are not satisfactory, see above. Statistics can prove something only when they are based on consistent criteria of selection. The list is actually incomplete and arbitrary. What are the sources?

(**P. 63**) Alciati does not belong to France. Does Germany include Switzerland and the Low countries?

POK

5

THE ORIGINS OF WILLIAM GILBERT'S SCIENTIFIC METHOD

William Gilbert's *De Magnete* appeared in 1600, six years before Galileo's first publication, five years before Bacon's *Advancement of Learning*; it is the first printed book, written by an academically trained scholar and dealing with a topic of natural science, which is based almost entirely on actual observation and experiment. In the learned literature of the period, among the writing of both contemporary university-scholars and the humanistic literacy, it is an isolated case. An analysis of the origins of its scientific method, therefore, is not only interesting in itself but is likely to throw some light on the origins of modern natural science in general. The results of Gilbert's investigation of magnetism and electricity being generally known, we shall consider first a few characteristics of his method and shall then try to trace its sources. Unfortunately very little is known of Gilbert's life and nothing at all of his way of working. The investigation, therefore, must be based entirely on his two printed books.[1]

---

1 *De Magnete Magneticisque Corporibus et de Magno Magnete Tellure, Physiologia Nova plurimis et argumentis et experimentis demonstrata,* Londini, 1600. If no other source is given all quotations in the following paper refer to this work and this edition. An English page-for-page version by Silvanus P. Thompson has been edited by the William Gilbert Society, Chiswick Press, London 1900. It contains valuable notes. Gilbert's second work is quoted from the only edition, *De Mundo nostro sublunari Philosophia nova, Opus posthumum. Ab Authoris fratre collectum pridem et dispositum ......* Amstelodami, 1651. - *De Mundo* does not shed much light on the origin of Gilbert's ideas. We are not even sure whether it was composed before or after *De Magnete*. In the margin of page 139 of *De Mundo* a reference to *De Magnete* VI, 4 is given and a similar remark is added at the end of the chapter. But since the author's brother who edited *De Mundo* declares himself in the preface not to know which of the books was composed earlier, obviously both remarks have been added by the editor later on. On the other hand in *De Mundo* (pp. 118 and 151) two statements of Patrizzi are criticized. These quotations can refer only to Patrizzi's *Nova de Universis Philosophia*, part Pancosmia, book 26 and book 12 respectively (in the second edition, Venice 1593, fol. 132 col. 2 and fol. 91 col. 3). The first edition of Patrizzi's work was printed in 1591. *De Mundo*, therefore, must have been composed after 1591 (Gilbert died in 1603). Altogether *De Mundo* gives the impression of greater immaturity; it is more pedantic and contains more remnants of Scholastic terminology than *De Magnete*. The first book of *De Mundo* combats the doctrine of the four elements, the second deals with astronomy, books 3 to 5 discuss "meteorological" problems, beginning with comets, the milky way, and clouds, and ending with the sea and the air. Very few experiments are given. *De Mundo* contains some modern-looking results - *e.g.,* space above the terrestrial atmosphere is thought to be empty and cold - but the methods and arguments are in no way outstanding.

I

1. Gilbert's scientific method combines essentially modern with metaphysical, scholastic, and animistic elements. Several of his experimental devices are still in use today. He dresses the poles of his spherical loadstones with sheet-iron and thus invents the armature of magnets (II, chap. 17). In order to examine weak magnetic forces he fixes small iron pieces on cork floating on water or suspends them on threads (I, 12 and 13; III, 8; V, 9). He even uses a few physical instruments. One of them is of his own invention and is the first of its kind in the history of physics. It is a - still somewhat imperfect - electroscope which obviously is constructed after the pattern of a magnetic needle (II, 2 p. 49).

Besides Gilbert describes at length and illustrates by woodcuts four magnetic measuring instruments, two declinometers and two inclinometers (IV, 12; V, 1; V, 3). They had, however, been neither invented nor essentially improved by him, though Gilbert omits that point.[2]

It is significant with respect to the origin of Gilbert's interest in scientific accuracy that all of his physical instruments are actually nautical instruments or are at least nearly related to the mariner's compass. On the whole he performs measurements practically only when he deals with quantities which are important in navigation, such as magnetic declination and inclination, altitudes of stars, and geographical latitudes (e.g., IV, 4 p. 160; IV, 12 p. 176; V, 8; VI, 1 p. 214). In other fields he usually restricts himself to qualitative observations and experiments. His best quantitative experiment verifies the hypothesis that magnetism is imponderable by weighing pieces of iron "on most exact gold scales" before and after magnetization (III, 3). It is taken over, however, from the compass-maker Robert Norman without the source being given. The few quantitative investigations which are original with him are not very outstanding.[3] Altogether, quantitative investigation appears considerably developed in *De Magnete* if compared with physics in the Middle Ages; it cannot compare, however, with the use of scientific measurements in the works of Galileo and his followers. Calculations are lacking entirely.

---

2 Gilbert's electroscope consists of a light horizontal metal needle, which is put on a point so that it can be turned easily. In *De Magnete* it is called by the same name *versorium* that is employed for magnetic needles. The description of Gilbert's four magnetic measuring instruments must be omitted here. The declinometer was invented in 1525 by Felipe Guillen. It was improved before Gilbert by Francisco Falero (*Tratado del Esphera*, Sevilla 1535), Pedro Nunes (*Tratado da Sphera*, Lisbon 1537), William Borough (*A Discourse of the Variation of the Compass*, London 1581), and Simon Stevin (*De Havenvinding*, Leyden 1599). The inclinometer had been invented by Robert Norman (*The Newe Attractive*, London 1581). These works are reprinted in G. Hellmann: *Rara Magnetica, 1269-1599*. (*Neudrucke von Schriften ... über Erdmagnetismus No. 10*) Berlin 1898. As quotations at other places of *De Magnete* show, the cited works of Nunes, Borough, Norman, and Stevin were known to Gilbert. Gilbert also invented and constructed two nomograms. The first (IV, 12, p. 176) simplifies determination of the astronomical meridian by means of graphic calculus. The second (V,8) - which, however, is based on incorrect assumptions - is meant to determine geographic latitude graphically.
3 II, 17 p. 86; II, 25 p. 92; II, 29 p. 97; II, 32 p. 99; III, 15 p. 145; III, 17 p. 150.

Mechanics also plays a very small part in *De Magnete*. Twice Gilbert shows some mechanical insight. Once (II, 35) he vehemently attacks medieval attempts to construct a perpetual motion engine. At another time (II, 24 p. 92) he knows that unstable equilibrium cannot persist for a long time and that, therefore, Fracastoro's story of a piece of iron suspended in the air between the earth and a magnet is "absurd". These two passages, however, are the only ones in his book dealing with mechanical questions. Both the interest in mechanics and the mechanical interpretation of all natural phenomena which dominated physics from Galileo to the nineteenth century are still lacking in Gilbert.

2. It is not easy to draw the picture of Gilbert's scientific attitude correctly. He is usually as critical-minded as a modern experimentalist, does not rely on any authority, and always tests reports of others by his own experiments. Superstitious ideas are emphatically rejected by him. He derides the ancient and medieval stories of diamonds and garlic destroying magnetism, the stories of magnets detecting faithlessness of women and unlocking locks (pp. 2f. and 6f.). He vehemently attacks alchemists and their obscure language (pref. fol. iij; I, 3 pp. 19 f. and 24). He rejects the explanation of electric and magnetic attraction by means of sympathy and, on that account, scoffs at Fracastoro (II, 2 p. 50; II, 3 p. 63 f.; II, 4 p. 65; II, 39 p. 113). On the other hand he believes in horoscopes, like most of his contemporaries: the magnetizing effect of the earth on pieces of iron being forged in the smithy is compared by him to the influence of the stars on a child during its birth (p. 142).[4]

Aristotelian and Scholastic concepts play a major part in his theoretical conceptions. Gilbert believes in the two basic principles matter and form, "out of which all bodies are produced" (II, 2 p. 52). In his opinion electric effects get their strength (*invalvescunt*) from matter, magnetic effects from a "distinguished" (*praecipua*) form (p. 53), for he thinks that the spherical form of the stars and especially the earth, being "primary and powerful" (I, 17 p. 42), is "the true magnetic potency" (II, 4 p. 65). Obviously his explanation of magnetism is based on the Scholastic metaphysics of active forms. In all his experiments he uses spherical loadstones, although he himself knows (II, 15 p. 83; III, 31 p. 99) that bar-like magnets are more effective. He calls them "little earths" (*terrellae* I, 3) and presumably clings to the medieval shape of his magnets because he believes in a metaphysical connection of spherical form and magnetism.

Cardanus's story that "the magnet lives and feeds on iron" is derided by Gilbert as old women's talk (I, 16 p. 37; II, 3 p. 63). He refutes it, using experimental methods, by ascertaining that the weight of the iron filings in which a magnet is kept does not diminish. Again he proves himself an empiricist, but he is opposed to vitalistic explanations only in so far as they

---

[4] The astrological theory of correspondence between metals and planets, however, is called "insane" (p. 20). In Gilbert's opinion metals, especially iron, are the very essence of the earth and, therefore, do not depend on the stars.

contradict single empirical facts. His own "philosophy" of magnetism, so far as it can neither be confirmed nor disproved by observation, is as animistic as the theory of Cardanus. A chapter of his book (V, 12) is entitled: "The magnetic force is animated or is similar to soul; it by far surpasses the human soul as long as that is bound to an organic body". The chapter refers to ancient philosophers from Thales to the Neoplatonists, who taught the existence of a soul of the universe, and adds the Egyptians, Chaldeans and (p. 209) even authorities on occult science, such as Hermes, Zoroaster and Orpheus. It explains (p. 209 f.) that the earth and the stars have souls, although they have no sense-organs, and that God himself is soul;[5] and, quoting Thales, he calls the magnet "an animated stone that is a part and beloved offspring of the animated mother, Earth".

The last quotation shows that Gilbert's theory of magnetism is embedded in a vitalistic philosophy of the terrestrial globe. To him the earth is "the common mother" of all things. Again and again in *De Magnete* this term is repeated, whenever the earth is mentioned.[6] We can therefore scarcely doubt the strongly emotional background of the idea of the maternal earth. The power of the magnet derives directly from the earth in Gilbert's opinion. For nothing but the magnet has preserved (I, 17 p. 42) "this distinguished substance which is homogenous to the internal nature of the earth and most akin to its marrow itself". Iron and magnets are (I, 16 p. 37) "the true and most intimate parts of the earth", because "they retain the first faculties in nature, the faculties of attracting each other, of moving, and of adjusting by the position of the world and the terrestrial globe".

Gilbert was the first to conceive the earth as a large magnet (I, 17; VI, 1). He was the first to teach that the interior of the earth consists of pure iron and that its surface and rim only are "soiled by other impurity" (I, 16 p. 39). Thus he has anticipated important empirical results of modern geophysics. But the resemblance of his magnetic philosophy to modern science is merely a matter of chance. Gilbert's terms "interior" and "intimate" combine spatial and metaphysical meaning and are always used as concepts of value. How near his "magnetic philosophy" still is to medieval vitalism is revealed by the fact that he believes in a metaphysical correlation of magnetism and rotation. He speaks of the "magnetic rotation" of the terrestrial globe (VI, 3 p. 214), and would like to accept the statement of Pierre de Maricourt that a spherical magnet rotates continuously by itself, were it not for his conscience as a cautious experimentalist. He reproduces Pierre's statement and adds (VI, 4 p. 223): "until now we have not succeeded in seeing this. We even doubt this movement because of the stone's weight and because the whole earth moves by itself, as it

---

5 Gilbert's religious belief obviously is rather Neoplatonic than Protestant. The whole chapter is strongly influenced by Patrizzi. *Cf.* below § 4, footnote 13.

6 *E.g.* pref. at the beginning and pp. 12, 26, 38 (twice), 41, 117, 152, 210. Moreover Gilbert likes to compare the interior of the earth with the mother's womb. In his opinion all metals originate from exhalations of the innermost part of the earth that are condensed and congeal nearer the surface in warm cavities "as the sperm or embryo congeals in the warm uterus" (I, 7 p. 20). *De Mundo* advocates the doctrine (p. 39) that all kinds of matter originate in earth and that earth, therefore, is the only element.

is moved by the other stars also. That does not hold proportionally of some part [the *terrella*]". Everyone who remembers how vehemently Gilbert attacks the reports on perpetual motion machines must notice the difference in emphasis.[7]

II

3. The material thus far presented may serve for a general indication of Gilbert's way of thinking. Animistic and Neoplatonic ideas are abundant in his book; the traces of Scholasticism and astrology are scarcer. But it is not these pre-scientific features that are conspicuous, for his work shares them with the whole learned literature of his period. What really counts is that his animistic metaphysics is nothing but the emotional background of his thinking and does not affect the empirical content of his science. The writings both of the Scholastics and the Renaissance philosophers abound with superstitious stories and magic. Gilbert rejects all that with unswerving criticism and bases his findings on experience and experiment only. This attitude is so exceptional in his period that the questions arises where it originates. Since critical minded experimentalists appear more and more frequently among the scholars a few decades after Gilbert, a satisfactory answer would at the same time contribute to the solution of the problem of the origin of modern science in general.

Even in a period in which quoting was more favored by scholars than nowadays, Gilbert is remarkable for the number of his references and his wide reading. He stresses, nevertheless, the novelty of his ideas. His attitude to contemporary literature is explained in the preface of *De Magnete*. There Gilbert says:

> What business have I in that vast ocean of books? ... By the more silly ones among them the crowd and most impudent people get intoxicated, insane and haughty ... They declare themselves to be philosophers, physicians, mathematicians, and astronomers and neglect and despise the learned men. Why should I add any thing to this disturbed literary republic? Or am I to offer this eminent philosophy that because of its unknown contents, as it were, is new and unbelievable to people who blindly trust authorities, to most absurd destroyers of the good arts, to literary idiots, grammarians, sophists, pettifoggers, and perverse mediocrities? ... No! I have presented these principles of magnetism that belong to a new kind of philosophy, to you true philosophers ... who look for knowledge not in books only but in things themselves.

Continuing, he announces that he will not call upon ancient writers for help, "because neither Greek arguments nor Greek words" can assist in finding truth. He promises that he will avoid "the ornament of eloquence" and will not darken things by words "as the Alchemists are wont to do". He plans to write with the same "liberty of mind" (*licentia*) as the ancient Egyptians, Greeks, and Romans.

---

[7] The story of the rotating spherical magnet is mentioned in *De Mundo* also (p. 138). There Gilbert gives the same reasons why the *terrella* does not rotate "although it is fit and inclined by nature to rotation." -In order to understand Gilbert's argument we have to realize that he was among the earliest adherents of Copernicus in England and was already convinced of the rotation of the earth.

The "sciolists" of present times still keep the errors of the ancients, but Aristotle, Theophrastus, Ptolemy, Hippocrates, and Galen themselves are sources of wisdom. "Yet our own period has discovered and brought to light very many things which those men too would be glad to accept if they were alive".

These vehement attacks on believers in authority and words, and the emphasis on the novelty of his ideas, are characteristic of the period of the expiring Renaissance, and anticipate Francis Bacon, and in some degree Galileo also. As the mention of grammarians, Greek words and eloquence shows, Gilbert's attack is aimed at declining humanism. Similar attacks are repeated several times in *De Magnete*.[8] Gilbert's other book, *De Mundo*, contains less polemics and is written more dispassionately. But it also opposes belief in authority: the slogan "he himself has said so, Aristotle has said so, Galen had said so" is considered a nuisance (*De Mundo* I, 3 p. 5).[9]

4. We shall therefore not expect to meet with much agreement with other authors in Gilbert's book. In fact most of the numerous references he gives are critical and negative, whereas the real sources of his ideas are chiefly to be sought elsewhere.

Ancient authors are often quoted. Comparatively favorable judgments are pronounced on philosophers who believe in universal animation, such as Plato and most of the Pre-Socratics. Atomists and mechanists are rejected. The Stoics are not mentioned. Although Gilbert is still greatly influenced by the concept of substantial form, he is opposed to Aristotle. In *De Magnete* (p. 116 and 209) Aristotle's astronomical doctrines are chiefly attacked, in *De Mundo* (I, 3) his doctrine of the four elements. The first book of *De Mundo* is even entitled "New Physiology against Aristotle", the third, "New Meteorology against Aristotle".[10]

References to medieval authors are rarer. Thomas Aquinas is twice quoted (I, 1 p. 3 and II, 3 p. 64) and his ingenuity and scholarship are highly praised. Yet Gilbert adds that Thomas did not experiment and consequently committed

---

8 Gilbert scoffs (I, 1 p. 2) at "precocious sciolists and copyists" who add fictitious stories to ancient authors. He accuses "the modern philosophers" (I, 10 p. 28) of having drawn their knowledge from books rather than from things. He derides (II, 2 p. 48) the books "cramming the bookshops" that deal with mysterious stories instead of experiments, and are as fond of Greek words as barbers who try to impress people by using scraps of Latin. He charges Fracastoro (II, 39 p. 113) with his predilection for Greek words and reproaches "the crowd of philosophers and copyists" (II, 38 p. 109) with repeating old opinions and errors.
9 As is generally known, the *ipse dixit* (α τός α) was the slogan of the Pythagorean school by which they referred to their master.
10 Thales, Empedocles, Anaxagoras, Pythagoreans, Plato praised V, 1; Plato attacked p 61; Aristotle: his importance admitted (pref. about the end), his (and Galen"s) opinions on iron approved, p. 39; Hippocrates praised because he did not advocate the doctrine of the four elements *De Mundo*, p. 5, attacked *De Magnete* p. 35; Galen criticized, p. 35 and 62, his importance admitted, pref. about the end; Strabo, Ptolemy, Tacticus, and Pliny the Elder quoted on iron mines p. 25; Pliny the Elder (on glass-making) attacked, p. 112.

errors. A few Arabian authors are mentioned, but for the most part their opinions are attacked.[11]

Almost the same holds of the authors of the modern era. Gilbert does not seem to have known the humanists very well. Among modern scholars cited most frequently are the philosopher-physician Fracastoro, the mathematician and physician Cardanus, the philologist and physician Scaliger, and the learned compiler of curiosities, Giambattista Porta. The first three authors were among the most famous scholars of the late Renaissance. Nearly always Gilbert derides all four of them, Facastoro because of his belief in "sympathy", the others because of their credulity and superstition. Gilbert - he was physician in ordinary to Queen Elizabeth - wrote two chapters (I, 14 and 15) on the medical effects of iron. There he proves to be familiar with modern medical literature, but practically all authors cited are refuted. He vehemently attacks Paracelsus, who among the physicians was the first to rebel against the authority of Aristotle and Galen, and he twice mentions (pp. 34 f.) the eminent and empirical-minded anatomist Fallopius without bringing him into any prominence. Gilbert's personal medical opinions are remarkably sound and free of superstition. Contemporary astronomical literature is well known to him (*De Mundo* II, 10 and 20) and Copernicus is highly praised in *De Magnete* (VI, 3). In the preface to *De Magnete*, written by Gilbert's friend Wright, the heliocentric theory is defended at length against scientific and religious objections.[12]

More may be learned of the origin of Gilbert's ideas from the references lacking than from those he gives. Among ancient authors three are conspicuous by their absence in *De Magnete*: Euclid, who is most important for the development of geometrical knowledge in the fifteenth and sixteenth centuries; Archimedes, who greatly influenced mechanics in the same period; and Vitruvius, who is the main source of knowledge in the field of ancient engineering. The three omissions show that Gilbert was not concerned with the mathematical literature of the period, that he was not interested in mechanics, and that he had connections neither with the humanists nor the architects of the Renaissance, who often quoted Vitruvius. With artists, presumably, Gilbert did not have any contacts at all. He could have found real experiments in the papers of the Italian artist-engineers (Brunelleschi, Ghiberti, Leonardo), which, however, were not yet printed. He never mentions Biringuccio either, who belonged with the architects of the Renaissance. Biringuccio's work *Della*

---

11 Avicenna is quoted on meteorites, p. 26; the medical opinions of Avicenna, Razes (= Abu Bekr al Rasi), and unnamed Arabian physicians attacked, p. 34f.; the alchemists Geber and Gilgil Mauretanus attacked, p. 19.

12 Nicolaus Cusanus ("not to be despised"), p. 64; Marsilius Ficinus, p. 3 ("ruminates ancient opinions") and p. 16; Fracastoro *De Sympathia* (1554), mentioned, pp. 5, 9, 110, 113; his theory of planetary movements (given in his *Homocentricorium seu de Stellis Liber*) discussed in *De Mundo* II, 10; Cardanus's *De Subtilitate* (1552) attacked, pp. 5, 27, 37, 42, 63, 107, 110, 169; Scaliger's *Etotericarum Exercitationum* (1557) attacked, pp. 5, 27, 37, 42, 63, 107, 110, 169; Porta's *Magia Naturalis* (1589) quoted, pp. 6, 24, 63, 137f., 166ff.; Paracelsus's "shameless charlatanry" attacked, p. 93, his merits admitted but Paracelsists attacked, *De Mundo* p. 7; the Antiparacelsist *Thomas Erastus* quoted, pp. 3 and 23. Tycho Brahe (on the coordinates of the Polaris) referred to, p. 174.

Pirotechnia, printed in 1540, treats metallurgy quite empirically and by experiments, but still discusses the magnet in a rather superstitious way.

The omission of one more group of authors is instructive. Gilbert's opposition to belief in books and authorities and his pride in the novelty of his ideas, are greatly reminiscent of Bernardino Telesio. Telesio was the first among the scholars of the Renaissance to oppose his "own principles" to Aristotelian natural philosophy (*De Rerum Natura Iuxta Propria Principia*, 1565 and 1570). Actually the influence of Telesio appears a few years after *De Magnete* in the works of Bacon, in which the anti-Aristotelian rebellion is carried on with even greater impetus. Gilbert, however, neither mentions Telesio nor seems to have known his work. The case of Telesio's pupil Patrizzi is somewhat different. Patrizzi always attacks Aristotle but is not much of a champion of originality: he likes quoting Plato and the authorities of occult science too well. He was known to Gilbert and is twice quoted in *De Mundo* (II, 2, p. 118 and II, 10, p. 151). Both times, however, statements of Patrizzi - on the shape of the globe and on the cause of the motions of the stars - are rejected. In *De Magnete* also both content and wording of the Neoplatonic chapter on universal animation (V, 12) obviously are influenced by Patrizzi, although he is not even mentioned.[13] Campanella and Giordano Bruno are also intellectually related to Telesio. Both attacked Aristotle and rejected the humanistic veneration of books with the same vehemence. Yet they are never mentioned in Gilbert. Bruno lived in England from 1583 to 1585; it would have been easy, therefore, for our author to make contact with him.

Gilbert's ideas - he describes, as we have seen, parts of *De Mundo* as *Physiologia nova contra Aristotelem, Nova Meteorologia contra Aristotelem* - belong to the same intellectual current as those of Telesio, Patrizzi, Campanella, and Bruno. Modern technology and modern economy had changed civilization too thoroughly for the Scholastic belief in Aristotle or the humanistic veneration of antiquity to endure. Telesio, Patrizzi, Campanella, and Bruno, however, were metaphysicians, not experimentalists, though Telesio and Campanella, theoretically at least, emphasized the importance of experience. It is rather instructive to realize that three of these philosophers exerted no influence at all on Gilbert and only Patrizzi contributed a few Neoplatonic ideas to his philosophy. In a sociological analysis the young experimental science of the early seventeenth century and the antidogmatic but fantastic metaphysics of the late Renaissance might prove to be connected: in both the same rebellion of the nascent modern society against the antiquated erudition and authorities of the past manifests itself. Yet the natural philosophy of the late Renaissance was the

---

13 Patrizzi's main work *Nova de Universis Philosophia* appeared in Venice, 1591. We quote, however, from the second edition, Venice, 1593. The part *Panpsychia*, book 4 refers to the Presocratics, Plato, the Neoplatonists, the Egyptians and Chaldeans, and to Zoroaster, Hermes, and Orpheus; it stresses the fact that stars do not need organs, though they have souls; it three times (fol. 55 col. 2 and 3) calls Aristotle's philosophy a "monstrum", because in his doctrine the whole universe is animated except for the earth. Quite the same theses and references are repeated in *De Magnete* V, 12 and even the term "monstrum" appears there (p. 209).

older brother of experimental science, not its father. The experimental method did not and could not have descended from the metaphysical ideas of the natural philosophers. We have to look elsewhere and in other social ranks for its immediate predecessors.

Among all the scholars quoted by Gilbert there is one who really did influence his investigation and method a great deal, although he does not at all emphasize indebtedness. This is the medieval nobleman Pierre de Maricourt, who in 1269 wrote a short but remarkable account of his magnetic experiments. About his life almost nothing is known. Written copies of his letter on magnetism were circulated until the sixteenth century, when it was printed under the title *Petri Peregrini Maricurtenis De Magnete, seu Rota perpetui motus libellus*, Augsburgi, 1558. Gilbert mentions Petrus Peregrinus five times in *De Magnete* and once in *De Mundo*.[14]

The first reference is in the first chapter of *De Magnete* which compiles the opinions on magnetism of the authors of the past. There (p. 5) Gilbert says: "About 200 years before Fracastoro there is a short work, sufficiently learned considering the period, under the name of a certain Petrus Peregrinus, which many think to have originated in the opinions of the Englishman Roger Bacon of Oxford. From that Johannes Taysner of Hainolt excerpted a booklet and published it as a new one".[15] Twice (III, 1 p. 116 and IV, 1 p. 153) Petrus is mentioned among the advocates of the erroneous opinion that "the magnetic needle is attracted by the celestial pole". In a short chapter (II, 35) Gilbert vehemently rejects the perpetual motion engines of Cardanus, Antonius de Fantis, Petrus Peregrinus and Johannes Taysner. And, finally, in *De Magnete* VI, 4 (p. 223) and *De Mundo* II, 7 (p. 13) he criticizes Pierre's story of the always rotating *terrella* (*cf.* §2, above). Except for the first passage, which, however, is rather general and rather tepid, Gilbert always differs with and criticizes the opinions of Pierre de Maricourt.

But in fact he owes more to Pierre than his words indicate. Pierre already knew (Chap. 6) that unlike poles attract, like ones repel one another. He knew (Chap. 9) that, when a magnet is divided, the pieces become new magnets with new poles. But Gilbert's knowledge of these facts need to have been taken over directly from the medieval experimentalist. The case is different with the spherical shape of the magnets. This shape is not a matter of course, but is, from the modern point of view, rather inexpedient. Pierre uses spherical loadstones, and the complicated way of determining the magnetic poles of the sphere - short pieces of iron wire are put on them and meridians are drawn with chalk until they intersect - is so identical in both authors (Pierre, Chap. 4, Gilbert I, 3, p. 12

---

14 On Pierre and his letter *cf.* Silvanus P. Thompson: *Petrus Peregrinus de Maricourt and his Epistola de Magnete*, Proc. Brit. Acad. vol. 2 (1905/6), pp. 337-408, and Erhard Schlund: *Archivum Franciscanum Historicum* vol. 4 (1911) and vol. 5 (1912). The letter on magnetism is reprinted in G. Hellmann: *Rara Magnetica (Neudrucke etc.)* Berlin, 1898. On the origin of Pierre's scientific method *cf.* below § 8.

15 As a matter of fact Roger Bacon depends more on Pierre than Pierre on Bacon. -Taysner's plagiarism was printed Coloniae 1562.

f.) that literary influence cannot be doubted. Gilbert is indebted to the outstanding medieval experimentalist as well for one of his experimental devices. Pierre (Chap. 5-7) had already made his loadstones float on water by means of wooden vessels. The cork pieces which are used by Gilbert of course were not yet known to him.

5. Up to this point we have not been able to give many positive contributions in answer to our main question. We have traced numerous authors to whom Gilbert was not indebted for his scientific method and only one - Pierre de Maricourt - to whom he was. The origins of his experimental technique and his scientific criticism are almost as enigmatic as they were before we started collecting his quotations. But we may have proceeded incorrectly. It was wrong, in fact, to look for his intellectual predecessors among scholars and philosophers. One has but to turn over the leaves of *De Magnete* in order to realize that he was interested in unscholarlike people and non-scholastic subjects too. Of the 240 pages of the book only 97 (40%) explain physical experiments. On the other hand 60 pages (25%) deal with nautical instruments and navigation, 25 pages (10%) with mining, melting, and fashioning of iron. The rest discusses astronomical questions (25 pp.), the opinions of numerous authors (18 pp.), the terrestrial globe as a magnet (11 pp.), and the medical effects of iron (4 pp.). Obviously *De Magnete* differs a great deal from a modern textbook on magnetism. The very first printed book on experimental physics deals so extensively with practical problems, that in some respects it is nearer to a technological than to a physical work of our time. And this gives the clue to the solution of our problem.

We may discuss first Gilbert's interest in mining and metallurgy. The literature on the subject is well known to him. George Agricola, the best known sixteenth century author in this field, is quoted most frequently. Gilbert esteems him highly but corrects errors uncritically taken over by Agricola from antiquity. Not less than three chapters of *De Magnete* (I, 2, 7, and 8) give extensive accounts of the distribution of iron in the world, describe the various ores, and quote ancient, Arabian, and modern authors on the subject.[16] Iron-manufacturing

---

16 The books of Agricola (1490-1556) on mining and metallurgy are still the best source of knowledge on this branch of technology in the 16th century. Gilbert (I, 1 p. 2) calls him "most outstanding in science", but regrets that he took over the ancient stories of antimagnetic effects of garlic and diamond. He rejects (I, 38 p. 110) Agricola's statement that the magnet is useful in glass manufacturing and reproaches Agricola for being influenced on this point by the "ignorant philosophy" of Pliny the Elder. Of course Gilbert knows that glass is not attracted by magnets. He approves (I, 7 p. 19) Agricola's chemical opinion that iron is composed of earth and water. Agricola and other - unnamed - "learned metallurgists" are referred to (I, 2 p. 10) on occurrences of iron-ore in Germany and Bohemia. On a special kind of iron-ore the opponent to Paracelsus, Thomas Erastus, is given as literary informant (I, 7 p. 23). *De Magnete* I, 8 quotes Strabo, Ptolemy, Tacitus, and Pliny on iron-mines in various parts of Europe and emphasizes that iron is the most frequently occurring mineral, as "every expert on metallurgy and chemistry" can confirm. Again Agricola is given as a reference for the occurrence and working of meadow-ore (p. 26). "As some authors write" (obviously Spanish cosmographers or mariners), there is iron in the West Indies too, "but Spaniards are looking for gold only." The chapter ends with a report on iron meteorites and quotes on that subject Avicenna, Scaliger, and Cardanus.

THE ORIGINS OF WILLIAM GILBERT'S SCIENTIFIC METHOD 81

also is discussed at length (I, 7). Gilbert reports (p. 23) on the manufacturing of cast iron, wrought iron, and steel in Styria and Spain, he refers to the description of iron- foundries in Porta's *Magia Naturalis*, and gives (p. 24) a list, eleven lines long, of iron devices. It contains among other things various kinds of guns, "the plague of mankind", and ends with a hint at other "numerous devices unknown to Latins". His reports on England are most interesting, as they are obviously based on personal experience. He tells (I, 2p. 11) that "newly" in an English mine, owned by the gentleman Adrian Gilbert, magnetic iron ore was found.[17] He reports (I, 7 p. 23) on the handling of iron in English gun foundries. And he knows (I, 8 p. 26) that English clay always contains iron and that, if bricks are baked in open kilns, "which are called *clampa* with us", the bricks next to the fire show "ferruginous vitrification".

Gilbert is also familiar with forging. In a chapter dealing with magnetic experiments (I, 11 p. 29) he describes how he himself manufactures the wrought iron he needs for experiments, and adds: "out of that the hammersmiths (*fabri*) form quadrangular pieces but mostly ingots (*bacillas*) which are bought by merchants and blacksmiths (*ferrarii*) and out of which various devices are manufactured in the workshops (*officinis*)". In a chapter (III, 12) which explains how iron is magnetized by the magnetic field of the earth he even gives a large woodcut of a smithy with furnace, bellows, anvil, and tools.

That very woodcut, which would be impossible in a modern textbook on magnetism, illustrates the intimate connection of Gilbert's theoretical investigation with practical metallurgy. Moreover, we must not forget that Gilbert did not live in the period of tradition bound medieval handicraft. The mining and metallurgy he is interested in is the mining and metallurgy of rapidly advancing early capitalism. As we know from Agricola, hauling engines, stamping mills, ventilators, and tracks for the dogs came into use in mining during the sixteenth century. In the same period the introduction of the blast furnace revolutionized the whole technique of iron manufacture. English mining and English metallurgy participated in that development.[18] Since the miners and foundrymen of the period belonged to the lower ranks of society and were uneducated we know neither their names nor their ideas. Yet we cannot doubt that many of them, stimulated to improvements by economic competition, were wont to try new techniques and to observe natural processes. Technology could not have progressed so rapidly if the laborers in the manner of the medieval guilds had simply clung to the traditional working processes of the past. Obviously, among such manual laborers there were experimentalists, though experimentalists with practical aims only and without theoretical knowledge. With their ranks Gilbert must have had many contacts. By a lucky accident we are even able to prove that he must have himself descended into an iron mine. Once (III, 2 p. 119 f.) he tells

---

17 The owner was no relation of the author. *Cf.* the family-tree in Silvanus P. Thompson: *The family and Arms of Gilbert of Colchester*, Trans. Essex Archaeol. Soc., vol. 9, new series (1906) p. 211.

18 *Cf.* Ludwig Beck: *Geschichte des Eisens*, Braunschweig 1893-95, vol. 2, pp. 879-97.

how he verified the hypothesis that the direction of magnetism in magnetic iron is induced by the earth. He says:

We had a twenty pounds" heavy loadstone dug and hauled out after having first observed and marked its ends in its vein. Then we put the stone in a wooden tub on water, so that it could turn freely. Immediately the surface which had looked to the North in the mine turned itself to the North on the water.

It is almost symbolic that Gilbert performed a laboratory experiment just after having left a pit and talked to miners. Of course Gilbert's experiments were not plain copies of the trials of the miners and foundrymen. But his spirit of observing and experimenting was taken over not from scholars but from manual workers. Sometimes, however, even his experiments simply repeated the working processes of contemporary iron manufacture. In three chapters of *De Magnete* (I, 9-11) he describes magnetic experiments with iron ore and wrought iron: he makes pieces of ore and iron float on water, he suspends them by threads, and has them attracted by magnets; but first he heats the ore for hours in a furnace and melts it; then he hammers the product, puts it into a second furnace and so on. All this is described, not as a mere preparation, but as a part of the experiments themselves. At least a part of his laboratory must have looked like a smithy.

6. Navigation and nautical instruments play an even greater part in *De Magnete* than mining and metallurgy. About 32 pages (13%) of the book are dedicated to nautical instruments, about 28 (12%) to general navigation. Already at the very beginning of *De Magnete*, in Wright's preface, geographic discoveries and circumnavigations of the globe are mentioned. In his survey of previous writers on magnetism (I, 1 p. 4) Gilbert reports (erroneously) the history of the invention of the compass and remarks that "no invention of human arts has ever been of greater use to mankind". He mentions Sebastian Cabot as the discoverer of magnetic declination and gives (p. 7) the names of four men "who have observed the variety of magnetic declination on long voyages": Thomas Hariot, Robert Hues, Edward Wright, and Abraham Kendall.[19] Gilbert proves to be

---

[19] Since Gilbert's authorities on navigation are characteristic of the social soil from which modern natural science has sprung, their activities and occupations are important. The mathematician and astronomer Hariot or Harriot (1560-1621) who was mathematical tutor to Sir Walter Raleigh as a young man, was sent by him as a surveyor to Virginia, and came back to England later. He published a report on Virginia, and mathematical works. The mathematician Robert Hues (1553-1632) accompanied Thomas Cavendish on his circumnavigation of the globe and published a *Tractatus de Globis et eorum Usu*, London 1594, dedicated to Sir Walter Raleigh. The mathematician Edward Wright (1558-1615) accompanied the Earl of Cumberland on his voyage to the Azores. He was lecturer on navigation to the East India Company. In his book *Certain Errors in Navigation*, London 1599, he introduced the cartographic projection that usually is ascribed to Mercator. Abraham Kendal or Kendall is the only non-scholar among the four men. He was sailing-master of Sir Robert Dudley's ship the Bear and later joined Drake's last expedition (cf. *The Oxford Dictionary of National Biography* and the Chiswick Press translation of *De Magnete*, London 1900, notes p. 19). Wright wrote the second preface to *De Magnete*. Most probably the three other men also were personal friends of Gilbert (cf. footnote 22 below).

familiar with mariners also in a chapter on the terrestrial globe. There (I, 17 p. 39) he gives numerical statements on the depth of the ocean according to the soundings of the mariners. He must have been told of their results by personal friends.[20]

The full extent of his nautical knowledge appears in the fourth book of *De Magnete* which deals with magnetic declination. Gilbert knows (IV, 1 p. 152) that declination differs at different places and gives its amount for places dispersed over all oceans and continents.[21] The remarkably wide range of his statements proves his familiarity with the reports of the English, Spanish, Portuguese, and Dutch navigators and the books of the learned cosmographers of the period. Moreover he mentions (IV, 5 and 10) that declination is great in high latitudes and that it is not influenced by the iron mines of the island of Elba in the Mediterranean. He knows (IV, 8 p. 165 f.) that the Portuguese royal cosmographer Pedro Nuñes (*Tratado da Sphera*, Lisboa, 1537) disregards declination entirely and that the Spanish historian Pedro de Medina (*Arte de Navegar*, Valladolid 1545) is wrong on it. He complains of the inexactness of most mariners in determining declination and warns especially of the reports of Portuguese navigators on their voyages to the East Indies. He knows that the Portuguese mariner Roderigos de Lagos, the Spanish mariner Diego de Alfonso, the Dutchmen, and "the experienced Englishman" Abraham Kendall contradict each other in their numerical statements (IV, 13 p. 177 f).[22] Since determination of geographic longitude was a difficult and, consequently, an often discussed problem at that period, he tries to solve it by means of the declination of the magnetic needle. He mentions (IV, 9 p. 167) that the learned compiler of curiosities Giambattista Porta (*Magia Naturalis,* 1589), the Venetian geographer Livio Sanuto (*Geografia*, 1588) and the mathematician Giambattista Benedetto give wrong solutions of the problem, since declination does not vary proportionally with the distance on the surface of the earth, as had been assumed by them. In the end he quotes the correct solution of Simon Stevin, the eminent Dutch expert in military engineering, navigation and book-keeping.[23]

---

20 He states that the depth of the ocean reaches one mile at a few places only and generally is no more than 50 to 100 fathoms. As the greatest depth of mines he gives 400 to 500 fathoms, as the diameter of the earth 6,872 miles.
21 P. 153f. East coast of the Atlantic from Guinea to Norway, West coast from Florida to Cape Race in New Foundland; p. 161 Azores; p. 163f. London; p. 167 North Cape in Norway, Corvo in the Azores, Plymouth; p. 178f. on the equator and in the South Atlantic (St. Helena); pp. 179-182 Nova Zembla (from Dutch observations), South Pacific, Mediterranean, Indian Ocean.
22 The sailing-master Kendall (*cf.* footnote 19) did not publish any book. Since Gilbert is familiar with his experiences, he must have known him personally.
23 The (antiquated) solution is: the declination at the various places of the surface of the earth has to be listed at first and then the geographic position of the ship can be determined by comparing observed declination with the list. Stevin's paper (*De Havenvinding*, Leyden 1599) is reprinted in G. Hellmann, *Rara Magnetica*, Berlin 1899. Gilbert does not quote the original paper but its Latin translation by Hugo Grotius (the elder) *Portuum Investigandorum Ratio*, 1599. It was in the same year also translated into English by Gilbert's friend, Edward Wright (*The Havenfinding Art*, London 1599) and into French (*Le trouve Port*, Leyden 1599). The four publications in one year, three vernacular, one Latin, illustrate rather well the rapid development in scientific navigation at this period and the kind of people Gilbert

Gilbert is familiar with the astronomical aids to navigation too. He knows how geographic latitude is determined astronomically, even takes into account atmospheric refraction, and gives a long list of bright stars with their declinations and right ascensions for the practical use of navigators (IV, 12 p. 174f.).

Gilbert got his nautical knowledge not from reading only. Again, as with the miners, an occasional mention in *De Magnete* reveals the personal contacts of the author. Once (III, 1 p. 117f.) Gilbert explains that the compass works under all latitudes from the equator up to the 70th and 80th degree N.L., and adds: "This the most famous captains and also very many of the more intelligent sailors confirm to us. This our most famous Neptunus Francis Drake, and the other circumnavigator of the globe, Thomas Cavendish, have told and confirmed to me". Obviously he is proud of the friendship of the two great circumnavigators who by their naval victories over the Spaniards - and by their successful privateering - had access to the court of Queen Elizabeth. Cavendish was a gentleman by birth, Sir Francis Drake was knighted because of his naval success: the names of the ordinary master mariners and helmsmen Gilbert had contact with are not given by him.[24]

At the end of the passage just quoted (III, 1 p. 118) Gilbert states that the compass works badly only when the needle has rusted or when the point on which it turns has got blunt. This leads us to his interest in nautical instruments. The measuring instruments described at length in *De Magnete* have already been discussed, and it has been mentioned that they are less new than the reader of Gilbert's description would assume.[25] After the publication of *De Magnete* Gilbert was still engaged in improving his instruments and making propaganda for them. One year before Gilbert's death a certain M. Blundeville published a booklet *Theorique of the Seven Planets*, London, 1602. It is written in English and contains as an appendix "the making description and use of the two most ingenious and necessarie Instruments for Seamen ... First invented by my good friend, Master Doctor Gilbert ". Obviously Gilbert had suggested the publication in English. The two instruments are the nomogram of *De Magnete*, which is supposed to make possible the determination of geographic latitude, and a somewhat more simplified inclinometer than the one in *De Magnete*.[26] As this

---

was in touch with. He quotes Stevin just one year after his paper had appeared. In this period this is remarkable.

24 Gilbert himself in the quotation just given distinguishes "illustrissimi naucleri" and "nautae etiam sagaciores plurimi" among his authorities. The sailing master Abraham Kendall (*cf.* footnotes 19 and 22) was personally acquainted with him. Edward Wright was his friend and so probably were Thomas Harriot and Robert Hues (*cf.* footnote 19). These three men, however, were academically trained mathematicians who had intimate relations with navigators and navigation.

25 § 1, footnote 2.

26 Blundeville is one more of the friends of Gilbert. He wrote popular scientific books in English for gentlemen. Besides the quoted work he published treatises on horsemanship, on Aristotelian logic, on map-making, on morals, and on counsellors of princes. The subtitle of his *Theoriques of the Seven Planets* illustrates rather well which social ranks outside the universities were interested in astronomy at Gilbert's time. It reads: *A Booke most necessarie for all Gentlemen that are desirous to be skillful in*

improvement shows, Gilbert does not deal with instruments as a mere theorist, but is familiar with the practical demands master-mariners make. He realizes (*De Magnete* IV, 12 p. p. 172) that in navigation simply built instruments are necessary which can be handled in spite of the rolling of the ship, and he invents and draws nomograms because he feels complicated calculations and "the exercises of mathematical genius" to be out of place on shipboard. On the method of preparing, magnetizing, and balancing the needle of the compass he gives a few practical hints (III, 17 p. 147 f.). He discusses at length (IV, 8 p. 165 f.) the various types of compasses that are used by the sailors of the various European nations. This chapter, however, is based on statements of Robert Norman without mentioning his name.

7. Norman's influence on Gilbert's investigation is so important that it must be discussed in greater detail. Gilbert himself does not emphasize it at all, but rather hides it. In the first chapter, after mentioning Wright and his friends, Gilbert goes on (p. 7 f.): "Others invented and made public magnetic instruments and expedient methods of observation, necessary to navigators and long-distance travellers, e.g., William Borough in his booklet on the Declination of the Compass, William Barlow in his Supplement,[27] and Robert Norman in his New Attractive". He adds that Norman, "an expert mariner and ingenious artificer", discovered the dip of the needle. A second time also Norman is quoted with approval. There (IV, 6, p. 161f.) Gilbert explains that the adjusting of the magnetic needle with the meridian is not effected by attraction but by some "disposing and turning faculty" of the earth, and adds that this was stressed by Norman as the first. Then he describes at length and illustrates by a woodcut an experiment which is supposed to prove the explanation given.[28] The experiment in every detail (and its incorrect interpretation) is borrowed from Norman's book. Twice more (I, 1 p. 5 and IV, 1 p. 153) Norman is mentioned in three words as the author who suggested the name "point respective" for the place that all magnetic needles point to, instead of "point attractive". Strangely enough, three of the four quotations refer to an opinion in which Norman is wrong. If we

---

Astronomie and for all Pilots and Sea-men or any others that love to serve the Prince on the Sea or by the Sea to travell into forraine Countries. This means that astronomical papers - if they were written in English - were of interest to overseas-traders and ship-owners, their master-mariners and helmsmen, and the gentlemen in the Royal Navy. Blundeville's booklet is based not only on Ptolemy but also on the ephemerides of Peurbach, Copernicus, and his followers Reinhold and Mestlin.

27 Barlow was the son of a bishop and himself a clergyman, and was interested in navigation, though he had never gone to sea. He published among other papers *The Navigators Supply*, London, 1597. He was a personal friend of Gilbert. Cf. Gilbert's letter to him, published in Barlow, *A Briefe Discovery of the Idle Animadversions of Mark Ridley*, London, 1618.

28 He makes a magnetic needle float in water by means of a piece of cork and carefully sees to it that it is completely submerged; from the fact that the needle adjusts itself with the direction of earth-magnetism but is not drawn to the rim of the vessel he concludes that there is no attraction. He (and Norman) forget that the needle has two opposite poles which are drawn in opposite directions.

wish to learn what Gilbert actually owes to him, we have to examine Norman's treatise.[29]

Norman was a retired mariner who had turned to compass-making. That can be referred from his booklet, which is the only source available on his life. The booklet itself begins with a few mineralogical remarks on magnetic iron ore, and reproduces a story of Paracelsus on loadstones which can be strengthened to such a degree by making them red-hot so that they can draw nails out of a wall. It is the same story which incites Gilbert to abuse Paracelsus as a shameless charlatan (*De Magnete*, p. 93). Norman, however, believes it. It is more important that Norman's very first chapter describes experiments in which magnets are suspended by threads and made to float on water.[30] The second chapter discusses earth-magnetism. Norman does not believe that it can be explained by loadstones at the North Pole of the earth, because he knows that the iron mines at Elba do not deflect the magnetic needle - a statement simply taken over by Gilbert. Then Norman discusses (chap. 3) the dip of the magnetic needle "not before having heard nor read of any such matter", and describes (chap. 4) and illustrates by a wood-cut the very first inclinometer.[31] The descriptions of two outstanding and most carefully performed experiments follow, both taken over by Gilbert. The first (chap. 5) proves by means of a gold balance that magnetism is imponderable; this is experimentally and theoretically entirely correct. The second (chap. 6) has been mentioned above (footnote 28); it is meant to prove that the earth does not attract but only turns the magnetic needle. It is illustrated by a woodcut, is even more carefully performed than in Gilbert - Norman stresses that any current of air must be avoided - but its theoretical interpretation is wrong, just as it is with Gilbert. The same chapter (6) introduces the term "point respective" which we have already mentioned.

The rest of the book does not contain experiments. Norman discusses (chap. 7) how the "point respective" might be determined by comparing magnetic needles at different places on the earth. As a simple mariner and instrument maker he is unable so he confesses (chap. 8), to explain the cause of terrestrial

---

29 *The Newe attractive, Containing a short discourse of the Magnet or Lodestone and amongst other his virtues, of a Newe discovered secret and subtil propertied concerning the Declining of the Needle touched therewith under the plains of the Horizon. Now first finds by Robert Norman Hydrographer. Hereunto are annexed Certaine necessarie rules for the art of Navigation by the same R.N.*, London, 1581. The book, reprinted in 1585, 1592, 1596, 1614, and 1720, has become a bibliographical rarity. G. Hellmann, *Rara Magnetica*, Berlin 1898 gives a reprint of the 1720 edition. The preface, the introductory poems, and the astronomical tables are not reproduced by him. We quote from the extremely rare second edition, London, 1592.

30 Norman did this by means of small pieces of cork. It is to be remembered that these new experimental devices were simply taken over by Gilbert in *De Magnete*. [In the original this note was numbered 29a, all subsequent notes are accordingly renumbered; eds.]

31 As a matter of fact the dip had been observed before, though less exactly, by the German physician Georg Hartmann. Hartmann's unpublished letter (1544) to Duke Albert of Prussia on his discovery, is reprinted by Hellmann, *loc. cit.* By Norman the dip always is called "declination", whereas magnetic declination is called "variation." This terminology also was taken over by Gilbert. Gilbert's inclinometer is a mere copy of Norman's instrument, but Norman proves to be the more experienced instrument-maker. *E.g.* he makes the bearings of the needle's axle of glass. Gilbert neglects that excellent detail.

magnetism. "I will not offer", he says modestly, "to dispute with the Logitians in so many pointes as here they might seeme to overreach me in Naturall causes". So he restricts himself to a reference to "God in his omnipotent providence". He discusses (chap. 9) magnetic declination and its diversity at different places, stressing - again we remember Gilbert - that there is no "equal proportion" in it, as some navigators had believed who, "notwithstanding their travells mostley have *more followed their Bookes than experience* in this matter". He himself refers to the "18 or 20 years that I have travelled the Seas". He complains that most mariners have but confused ideas on declination because of lack of suitable instruments: "wherefore I have devised one very necessarie". The last chapter (10) discusses the different types of compasses in various countries and is the source of the corresponding chapter in *De Magnete* (IV, 8). It follows a second part containing astronomical tables for the use of navigators.

We have already become acquainted with the empirical temper of this simple instrument-maker who, no less than Gilbert, Francis Bacon, and Galileo, prefers observation to books. His intellectual attitude is expressed even more clearly in the remarkable preface to the book. It is addressed "to the Right Worshipfull, M. William Borough, Esquire, Comptroller of her Maiesties Navie". It starts with the anecdote of Archimedes who, while taking a bath, discovers the law of buoyancy, runs naked to the street, and shouts - Norman avoids Greek - "I have found it". Norman continues:

So I (although in other respects and points of learning and knowledge, I will not presume to compare with Archimedes ... nor with other learned Mathematicians, being myself an unlearned Mathematician) by occasion of my profession, making sundry experiments of the Magnet stone, found at length amongst many other effects this strange and Newe properties of Declining of the Needle: which forgetting or rather neglecting my own nakedness and want of furniture, to set forth the matter, I have heere in simple sorte proposed ... to the view of the world.

Again he cites an ancient anecdote, the story of Pythagoras and the hecatomb he offered after having discovered his theorem, and continues:

So that we see these men ... being carried and overcome *with the incredible delight* conceived of their own devices and inventions, though, they follow partly the peculiar contention of their privat fancies, yet they seme chiefly to respect either the glory of god or the furtherance of some publike commoditie .... And seing it hath pleased God to make mee the instrument to open this noble secret, that his name might be glorified, and the commoditie of my Country procured thereby, I thought it my dutie to adventure my credite and make my name the object of slaunderous and carping tongues rather then such a secrete should be concealed and the use of thereof unknown.

Continuing, Norman stresses the utility of navigation to his country and again explains his resolution to publish his discovery "to frame as it were a theorike" for the use of mariners, and to describe "whatever I could find by exact trial and perfect experiments".

Wherin, although I may seeme to have discouered my nakedness and want of eloquence and orderly Methode to utter my conceits withall, I trust the reader will either of his curtesie take all things for good, that is well ment, or of his grauitie, *not regarding the words but the matter*, dissemble my faults, and accept of my paines.

He mentions that he has communicated his findings before publication to a few learned friends and concludes with respectful words to William Borough as "your worships most humble Robert Norman". In his short preface to the reader he emphasizes also that he will "ground his arguments onlye upon experience, reason and demonstrations". "Many and divers ancient Authors, Philosophers and other" have written on the magnet, but he intends to write "contrary to the opinions of all them". This remarkable man who, twenty-five years before Galileo's first publication, speaks of the "incredible delight" of experimental discovery, was a craftsman. At the end of the first edition of his booklet a kind of advertisement was printed stating that the instruments described "are made by Robert Norman and may be had at his home in Ratcliff".[32] When the seamen of the sixteenth century went to sea, they laid the foundation-stone of the British Empire and when they retired and made compasses, of modern experimental science.

The note just quoted refers to Norman's own inclinometer and to two declinometers constructed by the mariner William Borough and described in Borough's *Discourse of the Variation of the Compass or Magneticall Needle*, that in all editions was annexed to Norman's booklet. Borough is mentioned in *De Magnete* (I, 1 p. 7) together with Norman as an inventor of magnetic instruments.[33]

Robert Norman is of great importance for our problem. Except for the Latin erudition, the quotations and polemics, and the metaphysical philosophy of nature, he has everything that is peculiar to Gilbert. Norman as well as Gilbert proceeds by experiment and, "not regarding the words but the matter", bases his statements on experience rather than on books. Moreover, the measuring-instruments and the details of the experimental technique, the most exact experiments, and many single empirical statements of *De Magnete* are already contained in his booklet. It is true that the compass-maker Norman is a craftsman and Gilbert a scholar; but Norman already feels "incredible delight" at his discoveries and is interested in knowledge for its own sake: neither his

---

32 Quoted from Hellman *loc. cit.* The note is omitted in the later editions, presumably because Robert Norman had died.

33 He was born in 1536, travelled to the White Sea, became Comptroller of the Queens Navy in 1583, and was commander of an English ship in the Armada battle of 1588. Socially he belongs to a higher rank of mariners than Norman and is superior to him in education. In the preface to his *Discourse* he urgently recommends mathematics to the seamen, emphasizing that there are sufficient books on that subject written in English. He mentions "Vitriuius" (*sic*), Albert Duerer, and the ship builder Mathew Baker, as outstanding representatives of the "mechanicall sciences" to which also navigation belongs. He praises the good maps of Abraham Ortelius and criticizes the bad ones of the Paris professor Postillus. Ortelius (1527-98), the most famous map-maker of the period, came from handicraft, but became geographer to Philip II of Spain. Guillaume Postel (1505-81) is a learned polyhistor. The navigator Borough with his relations to superior handicraft on the one hand, to practical astronomy, cartography, and a bit of mathematics on the other, illustrates rather well the soil out of which Gilbert's work has grown.

THE ORIGINS OF WILLIAM GILBERT'S SCIENTIFIC METHOD   89

experiment on the ponderability of magnetism nor his dilemma concerning "point respective" or "point attractive" has any practical bearing. In things that are farther away from his occupation he is a little less critical than his follower of higher birth: he modestly believes in the story of Paracelsus which is vehemently criticized by Gilbert. On the other hand he is more religious than Gilbert: where Gilbert takes to Neoplatonic theories of universal animation, he retreats to God's impenetrable providence and avoids further explanation. Socially this is the difference between the highly educated scholar of the late Renaissance and the retired mariner. As to scientific value, however, Norman's attitude does not compare at all unfavorably with Gilbert's. Far reaching theories are lacking in his book; but is it Gilbert's metaphysics of "distinguished spherical form" that brings about magnetism a useful scientific explanation? The modern scientist may miss it in Norman's paper as little as he does Gilbert's quarrelsome polemics and erudite quotations. By the absence of all these Renaissance paraphernalia the experimenting compass maker is even nearer than Gilbert to the sober objectivity of modern natural science. Or, if we may put it the other way round: modern science and the modern mind in general are nearer to the experimenting manual workers of early capitalism, in which they had their origin, then to Renaissance humanism, which still influences even Gilbert.

III

8. The last paragraphs have answered our main question. Gilbert's experimental method and his independent attitude towards authorities were derived, not from ancient and contemporary learned literature, but on the one hand from the miners and foundrymen, on the other from the navigators and instrument-makers of the period. Alchemistic experiments probably never were performed by Gilbert, for he always vehemently attacked the alchemists and derided their attempts to make gold.[34] A rather complete assortment of the sources of his scientific achievements has been given by himself in his discussion of the practical use of the magnetic needle. There (III, 17 p. 147) he explains that by means of the needle the content of iron can be diagnosed in ores. The needle is the main part in the compass, which is, as it were, "the finger of God", and has made possible the Spanish and English circumnavigations of the globe. By means of the magnetic needle veins of iron ore can be discovered, subterranean galleries can be driven in sieges, guns can be pointed at night, territories can be surveyed, and subterranean water-conduits can be constructed.[35]

---

34 He reproaches them (pref. fol. iij) with "veiling things in darkness and obscurity by means of silly words". They are called (I, 3 p. 19) "cruel masters of metals who torture and harass them by many inventions". They are "delirious" (p. 20) and their doctrine that metals can be changed into gold is "futile" (p. 24).
35 The considerable part played by military engineering in this enumeration might be striking. We have already met with gun making in Gilbert's discussion of metallurgy (I, 7 pp. 23f.), have been forced to mention naval warfare and privateering several times, and should meet with military engineering even

Altogether, the impression of Gilbert's originality is considerably impaired, when he is confronted with his sources and especially with Norman. In spite of that, Norman is virtually unknown today, whereas Gilbert is counted among the pioneers of natural science. But this proves to be less unjust when the rise of science is viewed as a sociological process. Unfortunately we can only give a sketchy and simplified exposition of that view here and, of necessity, must omit a part of the evidence bearing on the point.[36]

From antiquity until about 1600 a sharp dividing-line existed between liberal and mechanical arts, i.e., in the final analysis, between arts needing heads and tongues only and others needing the use of hands also. The former were considered as worthy of well-bred men, the latter were left to lower-class people. Thus the contempt for manual labor tended to exclude experiment (and dissection) from respectable science. The prejudice against manual labor, however, did not prevent the experiments of the alchemists. Alchemy is not an occupation as carpentering, or forging; it is made respectable by the charm of both magic and gold, and even well-bred people may practice it as a hobby. But no respectable scholar who was proud of his position as a representative of the liberal arts even thought of using the methods of the mechanical arts. The case of those craftsmen who aspired to a higher social level is different; they - e.g. the Italian artists of the fifteenth century - discussed the social qualifications of manual work again and again, and stressed that they were connected with mathematics, i.e. with science.

The social background and the professional conditions of the scholars of the fifteenth and sixteenth centuries can not be discussed here. Nearly all of them had academic degrees and were consequently more or less linked to the universities, or they were humanists. Though several humanists had obtained academic chairs, generally speaking the universities of the period were still dominated by the spirit of Scholasticism. Both the university-scholars and the humanistic literati were accustomed to deal with natural phenomena chiefly in so

---

more frequently if we discussed the investigations of Leonardo da Vinci, Tartaglia, Duerer, and Galileo. *[Military technology has contributed considerably to the rise of the experimental spirit and natural science.]* Its influence on Gilbert is comparatively rather slight.

36 On the following *cf.*: Leonardo Olschki, *Geschichte der neusprachlichen wissenschaftlichen Literatur* (vol. 1: *Die Literatur der Technik und der angewandten Wissenschaften vom Mittelalter bis zur Renaissance.* Heidelberg 1918; vol. 2: *Bildung und Wissenschaft im Zeitalter der Renaissance in Italien*, Leipzig-Roma-Firenze-Geneva 1922; vol. 3: *Galilei und seine Zeit*, Halle 1927). All these volumes abound in valuable information on the scholar-literature and the craftsman-literature of the period and contain many sociological aspects. The third volume contains statements, until now scarcely used, on the influence of contemporary technology on Galileo (on the relations of the artists to handicraft, mechanics, military engineering and mathematics *cf.* I, 30-447; on mathematics and mechanics III, 72-110; on Galileo III, 117-469). On a later period *cf.* Robert K. Merton: "Science and Technology in the 17th Century", *Osiris* vol. 4 (1938) pp. 360-632. On the prejudice against manual labor and its intellectual implications *cf.* Edgar Zilsel: *Die Entstehung des Geniebegriffes*, Tuebingen 1926 (pp. 112-130 the humanistic literati, 130-143 the inventors and discoverers, 143-154 the artists and artist-engineers, 310-315 two strata of intellectual activities). On the effects of the prejudice against manual labor on astronomy *cf.* 'Copernicus and Mechanics', this volume pp. 123-127. On the effects on anatomy *cf.* Benjamin Farrington: "Vesalio and the Ruin of Ancient Medicine", in *Modern Quarterly*, London, vol. 1 (1938) pp. 23 ff.

far as they had been treated before by the authorities of Scholasticism and humanism respectively. On the other hand, since the decay of the guilds and their traditionalism real observation of natural phenomena, and even some experimentation, were to be found among skilled manual workers. Very little, however, is known of their intellectual interests. Since they got no education but the practical one in the workshop of their masters, their observations and experiments must have proceeded rather unmethodically.

With the advancement of early capitalistic society two major intellectual developments occurred: on the one hand, by virtue of technological inventions, geographical discoveries, and economic changes, the contrast between present times and the past became so obvious, that in the second half of the sixteenth century rebellion against both Scholasticism and humanism began among the scholars themselves. Representatives of the learned upper ranks such as Telesio, Patrizzi, Bruno and Campanella, vehemently attacked Aristotle and the belief in "words", felt enthusiastic about nature and physical experience, but did not experiment. Merely speculative metaphysics was, as it were, the older brother rather than the father of modern experimental science (cf. above § 4).

On the other hand, among the ranks of manual laborers a few groups of superior craftsmen formed connections with respectable scholars. During the fifteenth century Italian painters, sculptors and architects had slowly separated from whitewashers, stone dressers and masons. As the division of labor was still only slightly developed, the same artist usually worked in several fields of art, and often in engineering too. The technical problems of their occupations led them more and more to experimentation. Many of them made contacts with humanistic literati, were told of Vitruvius, Euclid, and Archimedes, and a few of them, such as Brunelleschi (1377-1446), Ghiberti, Leone Battista Alberti, Leonardo da Vinci, Benvenuto Cellini (1500-1571), started writing diaries and papers on their achievements. Biringuccio's treatise on metallurgy, *Della Pirotechnia* (1540), Dürer's two treatises on descriptive geometry and fortification, of 1525 and 1527, in some respect even the papers of Stevin, belong to this literature of the artist-engineers. Another group of superior manual workers were the surgeons, who practiced dissection and made contacts on the one hand with artists interested in anatomy, and on the other with medical doctors. Others were the navigators, who formed connections with mathematicians, astronomers, and cosmographers and published treatises on navigation; and, finally, the makers of nautical and of musical instruments. These superior craftsmen were the predecessors of modern experimental science, though they were not regarded as respectable scientists by contemporary public opinion. So far as papers were composed by them, they were written in the vernacular, not in Latin, and were not read by most of the respectable scholars, even if they were printed. By their colleagues, however, the books, especially those navigation, were diligently read, as is proved by the five editions of Norman's and Borough's treatises between 1581 and 1614. One has only to recall the humble apologies in Norman's preface to realize the barrier between craftsmen-literature and scholar-literature at the end of the sixteenth century. Experimental science could not have come into existence before this barrier was demolished.

But a few learned authors, very few, comparatively, already showed an understanding of mechanical arts before 1600. The German physician George Agricola published Latin treatises on mining and metallurgy (1530 and 1556); the chaplain at the royal court of Madrid, Peter Martyr, wrote two Latin books on the great geographical discoveries of the period (1511 and 1530); the learned secretary of the Senate of Venice, Ramusio, did the same in Italian (1550); a few Portuguese and Spanish cosmographers, such as Nuñes and Pedro de Medina, wrote mostly vernacular books on navigation. But especially in England, and in the period of Gilbert, similar studies increased. The Oxford B.A. Richard Hakluyt (1552-1616) edited Peter Martyr and published his own widely read books on the great English voyages and discoveries; the prebendary of Winchester, William Barlow, wrote an English treatise on navigation (1597). The East India Company engaged Cambridge graduate William Wright as a lecturer on navigation to their master- mariners. Wright and two more mathematicians, the Oxford graduates Thomas Harriot and Robert Hues, published Latin and English books in the same field (1588, 1594, 1599).

All these half-technical, half-learned activities show that some branches of the mechanical arts had become so important economically that they began to engage and to interest a few scholars. But they dealt with metallurgy and mostly with navigation rather than with experiments. The first academically trained scholar who dared to adopt the experimental method from the superior craftsmen and to communicate the results in a book not to helmsmen and mechanics but to the learned public was William Gilbert, who was a personal friend of most of these English authors. This is Gilbert's achievement in history. It might have been as difficult for the physician in ordinary to Queen Elizabeth to overcome the prejudice against manual labor as it was for the craftsman Norman to raise and answer his theoretical problems - though the two achievements are of a rather different kind. By his understanding of the scientific importance of experiment Gilbert made it - or helped to make it - respectable among the ranks of the educated. A few years later two other scholars likewise followed the method of the superior craftsmen: Francis Bacon, who ranked the great inventors and navigators above the scholars of his period, and Galileo, who started from military engineering.[37]

---

37 Galileo had already experimented a few years before *De Magnete* appeared. He became acquainted with Gilbert's book rather soon. We have a letter from Gilbert to William Barlow, telling that Gilbert met with the Venetian ambassador who brought him a Latin letter of Johannes Franciscus Sagredus. Gilbert continues: "Sagredo is a great Magnetical man and writeth that he has conferred with ... the Readers of Padua and reported wonderful liking of my booke" (Barlow, *Magnetical Advertisement*, London 1616). The letter must have been written between 1600 and 1603. Sagredo was a friend of Galileo and later figures as one of the persons of the discourse in Galileo's great dialogues. No doubt, Galileo himself, who was then lecturer on mathematics at the University of Padua, is the "Reader of Padua." Thirty years later, Galileo praises Gilbert highly because of his new and true observations and his habit of examining all statements of authorities by his own experiments. The only thing he misses in him is a little more mathematical knowledge (*Discorsi, Opere*, Edizione nazionale, VII, 432).

But we must deal with an objection. Is it true that experimental science could not come into existence so long as liberal and mechanical arts were kept separate by the contempt for manual labor? The fact that Pierre de Maricourt had already performed experiments does not seem to fit in with our exposition. Yet it is significant that Pierre tries to come to terms with the prejudice against manual work. In chapter 2 of his treatise he emphasizes that the investigator of magnetic phenomena must not only know "the nature of things" and celestial motions, but that he must also be "industrious in manual work" (*industriosum in opere manuum*); only by "manual industry" will he be able to correct errors which by mere reason and mathematics cannot be avoided. Obviously Pierre can not stress the value of experimentation without immediately speaking of and defending manual labor.

Pierre, no doubt, was the best experimentalist of the Middle Ages.[38] He probably was not a monk but a nobleman and might have been in the orient as a pilgrim or crusader as his surname *Peregrinus* suggests. In 1269 he took part in the siege of Lucera in Apulia, probably as a kind of military engineer. Most probably he is identical with the *magister Petrus*, the *dominus experimentorum*, often mentioned in Roger Bacon. If this assumption is correct, we know a little of his scientific and social attitude. This Petrus was, as Roger Bacon puts it, keen for the experiences even of "laymen, old women, and country bumpkins"; he was interested in metal founding, the working of gold and silver, mining, arms and military engineering, the chase, surveying, earthworks, the devices of magicians, and the tricks of jugglers.[39] In short, he was interested in all branches of technology that his period had developed and was hampered in his interest by the social prejudices of neither clergy nor nobility. It is significant that in the report of Bacon himself some social scruples still are hinted at ("country bumpkins, old women, jugglers"). Unfortunately we do not know where Petrus' freedom from bias originates. Altogether Pierre's attitude rather confirms than disproves the importance of manual labor and the mechanical arts for the history of science. The extremely rare medieval experimentalists would need an extensive and careful sociological investigation. We have, however, to return to Gilbert.

The social rise of the experimental method from the class of manual laborers to the ranks of university-scholars in the early seventeenth century was a decisive event in the history of science. Natural science needs theory and mathematics as well as experiments and observations. Only theoretically educated men with rationally trained intellects were able to supply that other half of its method to science. With Gilbert, however, not much of the superiority of academic training as to the theoretical side of science can be noticed: his general speculations have not proved to be fruitful. It is different with Francis Bacon and Galileo. Bacon's far-reaching ideas on the advancement of learning and

---

38 On the following, *cf.* the papers on Petrus Peregrinus quoted in footnote 14.
39 *Roger Bacon, Opus tertium*, cap. 12 p. 46 (ed. Brewer).

scientific co-operation could scarcely have been formed by craftsmen, though they were nothing but generalizations of their own practice. Galileo, on the other hand, joined mathematics with experiment.

Why did Gilbert himself never reckon, why did he come to a standstill at the first beginnings of quantitative inquiry? Certainly that deficiency is connected with his subject matter. Magnetic and electric processes can be measured only by complicated methods and, in consequence, were first measured almost two hundred years after Gilbert by Coulomb. It is mechanics that was the birthplace of quantitative research, since mechanical processes can be measured comparatively easily. Therefore, authors dealing with mechanics, such as Stevin and Galileo - and centuries before them Archimedes - were the first mathematical physicists. Gilbert on the other hand, as we have seen, is remarkably little interested in mechanics. He almost appears to have been biased against it. In *De Mundo* (II, 10 p. 154) he criticizes mechanistic astronomers who think Ptolemy's spheres to be material. He objects to their hypothesis on the ground that by it the universe is made a great wheelwork and God a mechanic. In the eighteenth century a comparison like this scarcely could have served as an objection to a theory; on the contrary, similar comparisons were commonplaces in the period of mechanistic physics and deism.

Gilbert's pre-mechanical way of thinking and his predilection for a field where measurements are so difficult may be due to his individual characteristics. But they are connected also with the special conditions of his native country. Practically all quantitative investigations in *De Magnete* originate in nautical technique and the work of the compass-maker Norman; Gilbert's interest in iron-making and iron-foundries, on the other hand, does not result in any quantitative inquiry. It was English iron-making and English iron-manufacture, however, that were advancing fast in the late sixteenth century. Instructive inferences can be drawn from the rapid rise of iron-manufacture.[40] Blast furnaces were introduced in England in the middle of the sixteenth century; the first English wire-mill was built in 1568; iron cannon, which had previously been imported from abroad, began to be exported from England in the same period; in 1581 and 1585 two laws were passed forbidding the construction of more blast-furnaces, in order to prevent devastation of the forest, since blast-furnaces were heated with wood. Certainly these laws show that iron-manufacture was not yet the leading industry of England; wool-trade and cloth-making still were much more important. Altogether, in the sixteenth century iron had not yet reached its dominant part in technology. It still was used in making weapons and simple tools rather than in machinery. And just this point leads us back again to our problem.

The first machines were made of wood and the first mechanical insights, therefore, were acquired from wooden devices - levers, reels, windlasses, inclined planes. There the Italian artist-engineers and Stevin made their studies

---

40 Ludwig Beck: *Geschichte des Eisens*, Braunschweig 1893-95, vol. II, pp. 892 and 896.

and found quantitative relations and laws. Galileo, when experimenting on the law of falling bodies, made brass balls roll down an inclined wooden groove. Not before the eighteenth century did iron machines, and not before the nineteenth did metallurgy become subjects of calculation. In the preceding centuries, therefore, predilection for iron prevented rather than promoted application of mathematical methods. Thus England's natural, economic, and social conditions might form, not a sufficient, but a necessary condition for the characteristics of Gilbert's method. When reading *De Magnete* we must never forget that twelve years before its publication English ships and English iron guns annihilated the Spanish Armada, then the most powerful fleet in the world. England, the country of iron mines and advancing navigation, produced the first learned book on experimental physics. It dealt with the mariner's compass, magnets, and iron. And for that very reason it did not introduce mathematical methods into natural science.

6

THE GENESIS OF THE CONCEPT OF PHYSICAL LAW

Investigation of physical laws is among the most important tasks of modern natural science. The naturalist observes recurrent associations of certain events or qualities. He is convinced that these regularities, observed in the past, will hold in the future as well, and he calls them "laws of nature", especially if he has succeeded in expressing them by mathematical formulas. Knowledge of physical laws is of the greatest importance both to the theorist and to the engineer. Whoever knows a law of nature is able to predict and, consequently, to control events: without investigation of laws there is no modern technology. As Western civilization of the modern era is based materially on its technology, so it is distinguished spiritually from the cultures of all other periods and nations by making the investigation of natural laws the basic task of science. To primitive and oriental civilizations the concept of physical law is quite unknown. We shall see that it was virtually unknown to antiquity and the Middle Ages, and that it did not arise before the middle of the seventeenth century.

It is strange that, in spite of its importance, the genesis of the concept of natural law has not yet been thoroughly investigated. Yet this is but a symptom of the rather unsatisfactory state of research in the field of the history of ideas in general. We must not confuse, however, the juridical term "natural law" with the same term in the sense in which it is used by our naturalists. As is generally known, the juridical concept (*ius naturale, lex naturalis*) designates moral commands that are based not on statute law but on reason, divine commandment, and moral instinct, and are common to all nations. It asserts how reasonable beings *shall* behave, whereas natural laws, as they are studied by modern naturalists, state and describe as a mere matter of fact how physical processes *do* take place. Numerous historical inquiries on the former concept have clarified its development.[1] On the other hand, the historical remarks on the concept of physical law are as rare as they are poor. In this field the most valuable preliminary work has been done, up to now, by the authors of a few dictionaries.[2] We shall use the material collected by them and try to increase it in

---

[1] Cf. the article "Natural Law" in the *Encyclopaedia of the Social Sciences* (New York, 1933), xi, 284 ff. (Georges Gurvitch).

[2] The remarks in Ernst Cassirer, *Das Erkenntnisproblem* (2nd ed., Berlin, 1911), I, 367 ff. and especially in Franz Borkenau, *Der Übergang vom feudalen zum bürgerlichen Weltbild* (Paris 1934), 15-97 are not quite reliable. The article "law" in Murray's *New English Dictionary* gives most valuable material. Some material is to be found in Littré's French, in Liddell-Scott's Greek, and in Harper's Latin

essential points. Since the historical problem we are dealing with is intricate and the literature that must be taken into account is very extensive, we do not make a claim to completeness. The juridical concept of natural law will not be discussed in this article. Yet we cannot disregard it completely, since it cannot be neatly separated from the naturalist's concept in its embryonic stage.

The concept of physical law, as it is used in modern natural science, does not contain any ideas of command and obedience. Yet it obviously originates in a juridical metaphor. In a well governed state there will be laws which are for the most part observed by the citizens. Lawbreaking will occur comparatively seldom, and will be punished when detected. The more powerful the government and the cleverer the police is, the rarer it will be. Let us suppose now the government to be omnipotent and the police to be omniscient. In this ideal case the behavior of the citizens would completely conform to the demands of the lawgiver and laws would be always observed. With such an ideal state nature was compared in the seventeenth century. The observable recurrent associations of physical events, in which the philosophers and scientists of the period began to be interested, were interpreted as divine commands and were called natural *laws*. Thus the concept of natural law originated in theological ideas. Later these non-empirical components fell gradually into oblivion. Our historical investigation, therefore, will have to trace the idea of God as a lawgiver to nature and the influence of this idea on the rising natural sciences. Since one is, generally speaking, inclined to consider contemporary ideas as a matter of course and to ascribe them uncritically to thinkers of the past, we shall bring into prominence the differences from modern thinking before the seventeenth century. Finally we shall try to explain sociologically why the concept of physical law was lacking then and why it developed in the period of Descartes, Hooke, Boyle, and Newton.

2. The roots of our concept go back to antiquity. They consist in a few passages of the Bible and the *Corpus Juris*. A few other ancient ideas are of less importance.

The divine lawgiver is the central idea of Judaism. Since God in addition is the creator of the world, it is easy to understand that the idea arose of his not only having given the moral and ritual laws to his people, but also having prescribed certain prohibitions to the physical world. In a description of God's power and omniscience *Job* 28, 26 says that God made a *law* for the rain (and a way for lightning and thunder). The Hebrew text uses the word *chok*. This is derived from the verb *chokak*, meaning to engrave, and is the same term which

---

Dictionary. The *Thesaurus Linguae Latinae* has not yet proceeded to the article "lex". The *Vocabulario della Crusca* and Du Cange, *Glossarium Mediae et Infimae Latinitatis*, do not contain material referring to our problem. Hans Kelsen in his article "Die Entstehung des Kausalgesetzes aus dem Vergeltungsprinzip", *Journal of Unified Science (Erkenntnis)*, 1940, 69 ff., and his book *Vergeltung und Kausalität* (which will appear in Holland), derives the ideas of causality and physical law from the juridical idea of retribution. Kelsen's valuable paper could not be used in this article.

is used for moral and ritual laws in the Old Testament. The Septuagint translates very freely "he *numbered* the rain (ρίθμησ v)", the Vulgate literally gives *ponebat legem*. The same word *chok*, which however in this context means rather boundary, is used in *Job* 26, 10, which says that the Lord made a *boundary* (Septuagint: πρόσταγμα, Vulgate: *terminum*) to the water, until light and darkness come to an end. Likewise *Job* 38, 10 says the Lord set a *boundary* (*chok,* ρια, *terminos*), bars, and doors to the ocean. The following verse 11, without using the term "law", pronounces the wording of a divine command or, better, prohibition: the Lord says to the sea: "Hitherto shalt thou come but no further; and here shall thy proud waves be stayed". The Hebrew text uses the future to express the command, as is usual in Hebrew and is done also in the Ten Commandments. The Septuagint and the Vulgate too translate literally by the future.

There are a few more analogous passages in the Old Testament, *Psalm* 104, 9 says the Lord has set a *boundary* (*gevol, terminum*) to the waters that they may not pass over; that they turn not again (Hebrew: future) to cover the earth. *Proverbs* 8,9 even twice uses the word *law* (*chok*), for which in the translations, however, two different terms are used. It says the Lord gave his *decree* (*chok, terminum*) to the sea that the waters should not pass his *commandment* (*chok, legem*). And *Jeremiah* 5, 22 says the Lord has placed the sand for the bound of the sea by a perpetual *decree* (*chok,* πρόσταγμα, *praeceptum*), that it cannot pass it; and though the waves thereof toss themselves, yet can they not prevail; though they roar, yet can they not (Hebrew: future) pass over it.[3]

We have met with the most ancient stage of the concept of physical law in ten verses of the Old Testament. The influence of the Bible on occidental thinking is immense. These verses were quoted through centuries again and again, and have decidedly contributed to the formation of concepts in rising natural sciences. Viewing the text of the Vulgate (which before the rise of Protestantism was the really effective factor), we twice find the term *law* (*lex*) (*Job* 28, 26 and *Prov.* 8, 29) and once (*Job* 38, 11) the wording of a divine prohibition. In this passage and especially in *Jeremiah* 5, 22 the idea is distinctly implied that the sea, to which the divine command is addressed, wishes to offer resistance but, being too weak, is forced to bow before the supreme power of the Lord. This might be a survival of primeval animism and demonology. As subject to divine commands rain, lighting and thunder, winds, and earth and, especially, the sea are given: the laboratory, which is the very birthplace of the scientific concept of natural law still is far remote. The empirical background of the biblical idea, presumably, is the observation that there are certain permanent traits in nature: the waves of the sea advance and recede, when tossed up by a gale, but, eventually, the dividing line between sea and land remains unchanged; wind and water produce

---

3 The following *Job*-verses are somewhat less important: 28, 25 The Lord made the *weight* for the winds and the *measure* for the waters; 38, 5 He gave the *measures* to earth and stretched a *line* upon it; 38, 8 He shut up the sea with *doors*. -I am greatly indebted to Dr. Boas Kohn, librarian of the Jewish Theological Seminary of America, for his information on the Hebrew text.

considerable destruction, but, after all, life goes on as usual. These observations refer to certain empirical regularities and would not be so different from the statements made in modern physical laws, if only the regularities were specified. A statement of the circumstances with which, in spite of storms, the situation of the seashore is regularly associated, would make predictions possible and would be a geophysical law. Of course, the authors of the Old Testament were not interested in statements like that. They were inspired by the emotional idea that nature, being ruled by the Lord, *must* behave as it does, and they restricted themselves to the vaguest indications as to *how* nature behaves. The same idea of "must" participated in the formation of the modern concept of physical law, but it was supplemented by the exact description of the empirical facts. In the further development the idea of necessity gradually receded to the background and eventually vanished, the observable recurrent associations of events remaining as the only content of physical laws.[4]

3. To classical antiquity also the idea is not quite foreign that physical processes are superintended and enforced by God or gods as by judges. It is implied as early as in the oldest philosophical fragment in the Greek language that is literally left to us. In the first half of the sixth century B.C. Anaximander says[5] that all things arise from the indefinite, the primary substance, and return to it "according to necessity. For they pay fine and penalty to each other for their iniquity according to time's order". The inevitability of a certain physical process is expressed here in juridical terms. Yet gods as lawgivers or judges are not mentioned. They appear half a century later in Heraclitus. "The Sun", Heraclitus says, "will not transgress his measures; otherwise the Erynyes, the bailiffs of Dike (the goddess of justice), will find him".[6] In Anaximander the physical regularity which is interpreted by him half mythologically, half juridically, is still based on metaphysical construction; nobody had ever observed that all things spring from and return to the indefinite. In Heraclitus, on the other hand, the physical statement is based on actual observation, the regular course of the sun being an empirical fact. The regularity itself, however, is presumed as obvious and not described.

---

4 It is remarkable that in the great document of ancient Egyptian monotheism, in Akhenaton's major *Hymn to Aton*, the sun-god Aton is praised as the creator of the universe but not as lawgiver: neither moral nor physical laws are mentioned (cf. James H. Breasted: *The Dawn of Conscience*, New York 1933, 281 ff.). This is possibly connected with the fact that, apparently, some social conditions obstructed the development of legislation in Egypt, in contrast to Babylonia which produced the code of Hammurabi. Actually the Babylonian creation-story in the Gilgamesh epic conceives the sun-god Marduk as the lawgiver to the stars. Tablet 7 raises the rhetorical question: "who prescribes the laws for (the star-gods) Anu, Enlil, and Ea, who fixes their bounds?" and explains that Marduk "maintains the stars in their path" by giving "commands" and "decrees" (cf. Morris Jastrow: *The Civilization of Babylonia and Assyria* (Philadelphia 1915), 441 last line ff.).
5 Diels, *Fragmente der Vorsokratiker*, 5th ed., Berlin 1934, 12 B I.
6 *Ibid.*, 22 B 94. Fragment B 114 says: "all human laws are derived from the one divine law. This... is stronger than everything". In a somewhat dubious scholium to a medical poem of Nicander Heraclitus is said to have called fire and the sea slaves to the winds "according to divine law" (ατ θ ον νόμον *ibid.*, 22 A 14a).

With progressing rationalism the scanty indications of a juridical interpretation of the course of nature vanished again in the following period. In the period of the sophists the term "law" and "nature", ὑσις and νόμος, became even opposites, "law" designating everything that is, as a mere convention, artificially introduced by men. Democritus therefore did not know anything of "natural laws", though he attempted to explain all physical phenomena by causes. A century later Aristotle for the same reason never used the law-metaphor.[7] Plato uses the term "laws of nature" only once to characterize the behavior of the healthy in contrast to the sick human body.[8] As a characterization of the healthy and normal state the phrase occurs also in the second century A.D. in Lucianus.[9] The law-metaphor plays a certain part in the Stoics only. The Stoics were determinists and believed in fate and divine providence. Living in a period of rising monarchies they viewed the universe as a great empire, ruled by the divine Logos. Consequently the idea of a natural law was not unknown to them. For the most part it referred to moral prescriptions based on reason. This Stoic idea is the source of the juridical concept of natural law, which influenced jurisprudence and political philosophy through two thousand years. A few times, however, although the two meanings were never neatly separated, the idea was applied by the Stoics to physical processes too. Zeno, the founder of the school, speaks of natural laws in this ambiguous way. His disciple Cleanthes mentions the "law according to which the prince of nature steers the universe" three times in his well known hymn to Zeus. A few verses later the first passage speaks of the obedience of the firmament and the stars; the two other passages refer to moral law. His follower Chrysippus once compares the universe at length to a state and calls reason (λόγος) a *law* (νόμος) to nature.[10] In Cicero *On Laws*, however, the concept of natural law is not applied to physical objects.

The Stoics were not much interested in physical phenomena and never gave instances of natural law in its physical meaning. Such instances appear about the beginning of the christian era in Ovid. Ovid complains once of the betrayal of a friend; his faithlessness is so monstrous, that the rivers will flow uphill, the sun will go backwards, water will produce fire and fire water, in short, "all things will proceed reversing nature's laws (*naturae praepostera legibus ibunt*)".[11] Possibly the idea that the ordinary course of nature must be ascribed to laws is

---

7 Cf. Bonitz, *Index Aristotelicus*. In *Physics* II, 193a15, Aristotle points out that, if a wooden bed is dug in and sends up a shoot, the shoot is wood but not a bed. He contrasts the perishable and artificial shape of the bed with its permanent and natural material by calling the former a mere "arrangement according to law" (ατ νόμον διάθ σιν). This is the strict opposite to the terminology of modern science, in which laws always refer to the permanent traits of the physical processes.

8 *Tim.* 83e: when a man is sick, the blood picks up the components of the food "contrary to the laws of nature" (πα το ς τ ς ὑσ ως νόμους). Cf. Ast, *Lexicon Platonicum*.

9 *Amores* 22: "the legislature of nature" (τ ς ὑσ ως νομοθ σία) is observed among animals, as pederasty is unknown to them.

10 Zeno: Arnim, *Stoicorum Veterum Fragmenta*, I fg. 162; Cleanthes: *ibid*. I fg. 537 p. 121 l. 35, p. 122 l. 20, p. 123 l. 5; Chrysippus: *ibid*., II fg. 528, cf. fg. 919.

11 *Tristia* I, 8 verse 5. In *Met*. 15, 71 (Rim) Ovid says of Pythagoras that he knew all secrets of nature, the origins of snow, lightning, and earthquakes - and the "law according to which the stars move".

influenced by the Stoics. On the other hand the term "law" in Ovid designates hardly more than the opposite to disorder: in several passages unarranged hair is called "hair without law (*sine lege*)" by him.[12] A rather isolated passage in the Stoic Seneca, however, seems nearer to the modern usage of language. Seneca is not surprised that comets, being a very rare phenomenon, are "not yet subjected to certain laws (*nondum teneri legibus certis*)"; posterity will be surprised, he says, that we ignored such obvious things.[13] Possibly the Stoic idea of the divine law which is identical with the divine reason is here involved.

At any rate the law-metaphor was not quite unknown to the ancients. This is illustrated by the term *astronomy*. The Greek word *nomos* means law, and the science of the stars could not have been called astronomy if the idea had not existed that the order and regularity of the stellar movements were analogous to human law. The names astronomy and astrology originally were synonymous.[14] As early as the fifth century B.C. the term "astronomy" was familiar to Aristophanes.[15] Some authors, such as Aristotle, Archimedes, Polybius, and Hipparch, use the term "astrology" only. Others, such as Pappus and Seneca, prefer "astronomy". With the increasing influx of oriental superstition magical aspects eventually prevailed in the term "astrology". In the fifth century A.D. the Latin encyclopedias for monks explain "astronomy", literally translating the Greek terms, as the science dealing with the "law of the stars (*lex astrorum*)".[16] In the astrological literature of late antiquity sometimes laws of nature are mentioned in an entirely magical sense. Thus the astrologer Vettius Valens (about 150 A.D.), discussing astrological predetermination, speaks of the "legislation" of nature, fate, and the stars.[17]

4. On the whole one must take good care not to overestimate the similarity of the classical concept of nature and modern natural science. Deterministic ideas were known to the ancients. They were indicated as early as in Heraclitus' doctrine of the fiery Logos who rules the universe and expresses himself in the cyclic change of matter. They were explained in detail in the Stoic doctrine of fate. Nevertheless two points must not be overlooked. First, ancient determinists spoke much more frequently of the *logos* than of the *nomos*, more frequently of the *reason* than of the *law* of the universe. Secondly, the classical determinist

---

12 *Met.* I, 477; *Ars. amat.* III, 133. (On Ovid cf. Deferrari-Barry-McGuire, *A Concordance of Ovid*, Washington 1939, *s.v. lex.*).
13 *Natural. quaest.* VII, 25, 3-5.
14 On the names "astronomy" and "astrology" cf. Pauly-Wissowa, *Realencyclopädie d. class. Altertumswissensch.*, Stuttgart 1896, *s.v.* Astronomie (Hultsch). In addition cf. *Thesaurus Linguae s.v.* astrologia, astronomia, astronomicus, astronomus.
15 *Nubes* 194, 201.
16 Cassiodorus *inst.* 2,7; Isodorus *diff.* 2, 152.
17 *Anthologiae* (Kroll) 5, cap. 9 p. 219 1.26 ff. fate (μα μένη) *has given a law* (ν νομοτέθη ν to every being, surrounding it with an unbreakable wall; (The term μα μένη seems to indicate influence of the Stoics); 7 cap. 3 p. 272 l. 9 ff.: nature (ύσις) *gave a law* (νομοτέ ησ ν) and encompassed man with the wall of necessity; 9. cap. 7 p. 343 l. 33 ff.: the stars order the universe by their influence *without ever transgressing the boundaries of legislature* (νομοθ σίας).

doctrine had a tinge of myth and emotion rather than of science and experience. Heraclitus and the Stoics felt the development of the whole universe as necessary and enforced, but were not interested in single physical laws. How far remote from natural science the determinism of the Stoics was is revealed by their giving vaticination as a verification of their doctrine of fate.[18] The superstitious Stoics were determinists. On the other hand the ancient representatives of the scientific interpretation of nature, such as Democritus and Lucretius, who consistently advocated causal explanations of nature, did not use the law-metaphor. It is significant that, whereas modern translations of Lucretius speak of physical laws again and again, this term was unknown to Lucretius himself.[19] Lucretius, following Epicurus, stressed three basic principles of nature. In his poem they play a part analogous to physical laws in modern science: nothing can be produced form nothing, nothing can be annihilated, and the amount of motion in nature is constant.[20] But since all quantitative details are lacking, his "laws of conservation" (as one is tempted to call them) are extremely vague. Moreover, Lucretius does not call them *laws*, but speaks of *principles*. In Epicurean philosophy there are and can be no "laws" of nature, since the gods do not take care of the world.

There hardly is room for physical laws in ancient science either. Aristotle makes a few general statements approximately corresponding to laws of motion - sublunar bodies tend to their natural place, celestial bodies move in circles - but they are vague, incorrect, and formulated teleologically. And, of course, they were not called "laws" by Aristotle.[21] Peculiarly enough only three physical laws were correctly known to the ancients: the law of the lever, the optical law of reflection and the law of buoyancy. All three of them are discussed in Archimedes, who, however, never used the term "natural law". Although Archimedes, by far the most eminent physicist of antiquity, certainly verified all three laws by experiments (the law of buoyancy was even discovered by him experimentally), he does not explain them empirically. He rather follows the deductive method of Euclid, starts from postulates, and deduces and proves his physical statements, as if they were mathematical theorems.[22] Even from a mere external point of view his method is Euclidean, in so far as all theorems are

---

18 Cf. Arnim, *loc. cit.* vol. 2 cap 6 § 4 pp. 270-272.
19 The very complete *Index Lucretianus*, Gotoburgi 1911, by Johannes Paulson, gives only three passages in which the term "law" is used in a non-juridical sense. III, 692 (Munro) states that the human soul is not immortal but is subject to the "law of death (*leti lege*)"; V, 58 states that all things must perish and nothing can break" the laws of time (*aevi leges*)"; and V, 720 denies the existence of chimeras, stating that members of the body can combine only if they are adapted to each other; all animals are bound "by these laws" (*teneri legibus hisce*).
20 I, 149 ff.; I, 216 ff.; II, 71 (Munro).
21 The *Mechanica*, ascribed to *Aristotle*, is probably spurious. It knows the principle of the lever without using the term "law". Cf. above footnote 7. [This footnote is denoted with "20a" in the original text; Eds.]
22 Buoyancy: de *corp. fluit*. I theorems 3-7; lever: *de plan. aequ.* theor. 6 f. The law of reflection is used half a century before Archimedes in Euclid's *Optics*, theor. 19 (Euclid, *Opera*, ed. Heiberg-Menge vol. 7 p. 31); it was known to Archimedes (cf. scholium 7 to Euclid's *Catoptrics*, ibid. p. 348; Archimedes" *Catoptrics* is lost) and is given as theorem I in the (spurious) *Catoptrics* of Euclid.

numbered as the theorems in Euclid. In a mathematical treatise, however, there is obviously no room for the law-metaphor: Archimedes speaks as little of the "law" of buoyancy as Euclid speaks of the "law" of Pythagoras.[23] The deductive method in Archimedes probably originated in the same remarkable sociological phenomenon which also caused the poor state of physics in antiquity. Ancient civilization was based on slave labor and, in general, their patrons and representatives did not have occupations, but lived on their rents. In ancient opinion, therefore, logical deduction and mathematics were worthy of free-born men, whereas experimentation, as requiring manual work, was considered to be a slavish occupation. Archimedes himself gave expression to this contempt for manual labor and technology.[24] On the whole the development of physics was seriously impaired in antiquity by the contempt for manual work, technology, and experimentation. And where physics comes to a standstill in a rather embryonic stage the concept of physical law cannot develop.

Another literary document of antiquity has contributed a few ideas to the modern concept of physical law. This is the *Corpus Juris*. In the introductory sections, both of the Pandects and of the Institutes, the Stoic idea of natural law (*jus naturale*) is explained. In contrast to statute law natural law is based on mere reason, does not change, and is common to all nations. For the most part moral obligations - veneration of God, obedience to parents - are given as examples.[25] On the other hand it is stated that "nature has taught all animals the natural law". From it the intercourse of male and female, begetting and education of the offspring derive. Obviously in this explanation two different ideas are mixed. On the one hand a statement is made on matters of fact. The empirical fact that mammals propagate by sexual intercourse and take care of their offspring could be called a biological law in the modern meaning of the word. On the other hand these facts are interpreted as results of a sort of legal permission or command; nature or God being the lawgiver.[26] This confusion facilitated the application of the law- metaphor to physical facts even centuries later. In the history of ideas the *Corpus Juris* was almost as influential as the Bible.

5. The Christian Middle Ages did not make any contribution to the development of our concept. We need not enlarge, therefore, on medieval authors. Of course the Bible passages on God as the lawgiver of the universe were often quoted and paraphrased by the church fathers and Scholastics. The ideas also of the Stoics and the *Corpus Juris* on natural law (*jus naturale*) were exerting some influence. A passage in the Christian orator Arnobius (about 300 A.D.) is of some interest, since it gives a few instances of the physical regularities which were explained by the theologians through divine laws. In order to prove that the

---

23 Even today physicists still speak rather of the *principle* than of the *law* of Archimedes.
24 Cf. Plutarch, *Vitae*, Marcellus 14 and 17.
25 *Dig.* I, 1, 3; *Inst.* I, 2.
26 *Inst.* I, 2, 11: naturalia iura... divinā providentiā constituta.

Christian religion is not anything monstrous Arnobius[27] asks his audience whether "the laws initially established" have been overthrown since the time the new faith has spread. All instances are given in the form of rhetorical questions. He explains that the elements have not changed their qualities. The structure of the machine of the universe (presumably he has in mind the astronomical system) has not dissolved. The rotation of the firmament, the rising and setting of the stars, have not changed. The sun has not grown cold. The change of the moon, the turn of the seasons and of long and short days, have neither stopped nor been disturbed. It still rains, seeds still germinate, trees still produce and lose their leaves, etc. On the whole he takes his instances from the same field as the Bible and Ovid do; and, of course, there is no question of laboratory physics. A certain predilection for astronomical and cyclic processes is significant. About 400 A.D. St. Augustine[28] developed the concept of God's *eternal law* by which the universe is ruled. Augustine's eternal law stems from the biblical idea of the divine lawgiver, but is an entirely teleological concept. It is identical with the impenetrable providence of God and has nothing to do with the physical laws of the modern scientists.

An extensive discussion in Thomas Aquinas throws light on the Scholastic concept of law and may be analysed here, as far as our subject is concerned. Thomas combines seeming logical exactness with considerable empirical vagueness. A law, according to his definition,[29] is "a rule and measure of acts". Yet we shall see that in one case physical processes and phenomena are also covered by this concept. Thomas distinguished positive from natural law (*jus naturale*).[30] The former needs promulgation by the lawgiver (and consequently has no bearing on our subject); the latter does not, since "it is promulgated by the very fact that God instilled it into man's mind, so as to be known by him naturally".[31] Later, natural law is defined as "the participation of the eternal law in reasonable creatures".[32] With the "eternal law" (*lex aeterna*) we have transcended the province of human actions. It is "the type (*ratio*) of divine wisdom as directing all acts and motions".[33] Thomas explains that God governs the acts and motions of all creatures and is to the world as the artist is to his work. Since in every artist the type of the order (*ratio ordinis*) which is produced by him pre-exists, the type (*ratio*) of the divine wisdom bears the character (*rationem*) of law. So the idea is distinctly expressed that the whole of nature (not only human actions) is subject to law. Two points, however, must be brought into prominence. First, the whole idea that the type of the order of the world pre-exists in God is obviously Platonic. And, secondly, the order Thomas has in mind is a teleological, not a causal one: to the words "divine wisdom", quoted above, he makes

---

27 *Adv. Gentes* I, 2.
28 *E.g.*, de lib. arb. I, 6: de civ. dei XIX, 12.
29 *Summa Theol.* II, 1 qu. 90, art. I resp.
30 *Ibid.* qu. 71, art. 6, resp. 4.
31 *Ibid.* qu. 90, art. 4, obj. I resp. The agreement with the ancient concept of natural law is obvious.
32 Qu. 91, art. 2, obj. 2 resp. Here the term *lex naturalis*, not *jus naturale*, is used.
33 Qu. 93, art. I resp. (cf. qu. 91, art. 1).

the significant addition "moving everything to its due end (*ad debitum finem*)".³⁴ The distance from the modern concept of physical law is considerable.

Somewhat later³⁵ objections are discussed, stating that irrational beings cannot be subject to the eternal law, since it cannot be promulgated to them and since they do not participate in reason. The objections are refuted and it is expressly stated that "all movements and actions of the whole of nature are subject to the eternal law". What promulgation is to man "the impression of an inward active principle" (i.e. an Aristotelian entelechy) is to natural things. As the only instance, however, the sea is given, which, according to *Prov.* 8, 29, received a law from God. As to natural law Thomas shares the ambiguity with the *Corpus Juris*. On the one hand he gives moral precepts as instances (evil has to be avoided); on the other he quotes the *Pandects* and explains that sexual intercourse and education of the offspring among animals are based on natural law. Man has some natural laws in common even with inanimate things: not only man, but, as Thomas maintains, every substance strives to preserve its being.³⁶

On the whole Thomas combines the biblical idea of God as the lawgiver of the universe and the ancient concept of natural law with Platonic and Aristotelian ideas. His concept of eternal law, therefore, is entirely teleological and identical with the idea of divine providence. Moreover, our discussion is apt to give a distorted view of his interest in physical regularities. The passages of the *Summa Theologica* in which they are mentioned have been singled out from a very extensive exposition. In the edition of Pope Leo XIII (Rome, 1892) they fill just two pages, whereas his whole discussion of law extends over two hundred and seven pages. Actually the *Summa Theologica* is interested in all theological problems connected with the concept of law and deals with physical phenomena only in so far as they are mentioned in the Bible and the *Corpus Juris*.

6. In discussing authors of the modern era we have to show, first, that the concept of physical law was not known before the seventeenth century. Since numerous authors must be considered, we shall treat them in groups without strictly observing the temporal sequence. We may begin with a few theologians and jurists.

The widely read handbook of jurisprudence *Doctor and Student* by Christophe Saint Germain, published in Latin in 1532 and in English in 1530 and

---

34 The teleological character of the order of the world is stressed and identified with divine providence I qu. 22, art. I. *Ibid.* art. 2 obj. 1 and 3 it is explained that Democritus and the Epicureans denied providence (and with it, as we may add, order and eternal law) by ascribing the course of nature to "necessity of matter".
35 *Op. cit.* qu. 93, art. 5, obj. 1 and 2.
36 *Ibid.*, qu. 94, art. 2. It is not quite clear whether in Thomas physical regularities belong to natural or to eternal law. In the article just quoted a few of them are counted with natural law. In qu. 93, art. 5, on the other hand, they are counted with eternal law. Natural law is restricted to reasonable beings in qu. 91, art. 2, obj. 2.

1531, briefly repeats the opinions of Thomas and the *Corpus Juris* on eternal law, to which the universe is subject, and on natural law. Of the latter two meanings are distinguished, one referring to reasonable creatures only, the other to all creatures.[37] The first meaning is merely juridical, the second covers also biological and physical phenomena. Seventy years later, Richard Hooker advocates the same ideas more extensively in the first chapters of his well known treatise *The Laws of Ecclesiastical Polity*, published in 1592 or 1594.[38] Of course he knows and discusses God's eternal law which is identical with divine providence. As to natural law he separates the natural law of reasonable beings and the natural law which is kept "unwittingly by the heavens and elements".[39] He adds that the latter "hath in it more than men as yet attained to know or perhaps ever shall attain". The very laws of nature, which one century later became the most important subject of scientific investigation, are considered unrecognizable. Hooker quotes the Bible on God as the Lawgiver of rain and sea and gives an enumeration of natural laws reminiscent of Ovid and Arnobius.[40] The elements do not change their qualities; the celestial spheres, sun and moon, move regularly; the turn of the seasons, wind and rain, enable the earth to bear fruit. If this order were disturbed, "what would become of man whom these things do all serve?" Obviously Hooker's concept of natural law is still entirely anthropocentric and teleological.

Nothing essential about our problem is contained in Jean Bodin, the most eminent political philosopher of the period.[41] On the other hand an important advance to logical clarification of the law-concept was made in Suarez. In his *Tractatus de Legibus* (1612) the Spanish Neo-Scholastic consistently clings to the distribution between "morals" and "nature"[42] and restricts the term "law", in its proper meaning, to the former. Suarez opposes the definition of law in Thomas Aquinas because it disregards this distinction [43]. "Things lacking reason", he says,[44] "properly, are capable neither of law nor of obedience. In this the efficacy of divine power and natural necessity ... are called law by a metaphor". The wording of the Scripture (he quotes the well known passages) is said to be in accordance with this explanation. In the section on eternal laws"[45] the Bible passages are interpreted in the same way and the statement that

---

37 15th edition, London 1571, chap. I p. 3 and chap. 5 p. 5 f. On the first editions cf. S.E. Thorne, "St. Germain's Doctor and Student", *The Library*, IVth series, vol. X (1930) pp. 421-426.
38 Book I, chap. 3. *Works* (ed. Keble, 7th ed., Oxford 1888) I 200 ff.
39 Chap. 3 p. 206.
40 *Ibid*. p. 207.
41 *De la République* (1577) does not give a general analysis of the law-concept. *Universae Naturae Theatrum* (Francofurti 1597) restricts itself in the preface (fol. 3) to a few generalities on the unchangeable course of the celestrial spheres. *Methodus ad facilem historiarum cognitionem* (1566) mentions "some eternal law of nature" according to which everything undergoes a cyclic change: vices follow virtue, ignorance science, darkness light (VII, 36). The astronomical pattern is manifest.
42 *Moralia et naturalia* II, 2 § 12.
43 I, 1 § 1.
44 I, 1 § 2.
45 II, 2 §§ 4, 10, 12, 13.

irrational and even inanimate beings can be subject to the eternal law is called "a mere mode of expression".[46] The *Corpus Juris* too, when speaking of natural laws among animals (which, actually, are led "by natural instinct"), makes use of a metaphor".[47] "The real natural law inheres in the human minds only".[48] Certainly, Suarez knows as little of natural laws in the modern meaning of the word as Thomas did, but his concept of law is considerably more modern. The intellectual change from Thomas to Suarez will be explained in our last paragraph.

7. Beyond the ranks of theological and juridical writers about 1600 natural law is scarcely mentioned. In 1570 John Dee, the alchemist to Queen Elizabeth, mentions that nature "abhorreth empty space so much, that, contrary to ordinary law, the Elements will move or stand".[49] Montaigne in his *Essais* (1582) uses the term natural law only once and in its juridical sense.[50] Shakespeare once makes Falstaff speak jokingly of natural law in its juridical meaning. In *Cymbeline* (1609), on the other hand, he calls it "nature's law" that the human embryo remains nine months in its mother's womb.[51] Shakespeare thus adds one more instance of a prescientific natural law to the instances in the Bible, Ovid, Arnobius, and Hooker.

The term is used more frequently in Francis Bacon. In his *Advancement of Learning* (1605) Bacon discusses the pyramid of the sciences and gives knowledge of "the Summary law of nature" as its "vertical point".[52] He expresses, however, his doubt whether this knowledge can be attained by man. The theological origin of the idea is revealed by a Latin quotation, speaking of "the work operated by God from the beginning to the end". In the *Novum Organum* (1620) the term "law" is very often used synonymously with "form". "When we speak of forms", Bacon says,[53] "we mean nothing else but those laws and determinations of the pure act which set in order and constitute a simple nature. ... The form of heat and the law of heat are the same thing". These "laws" or "forms" were treated as rather mysterious entities by Bacon himself. They are nearer to alchemy then to modern science, are considered by Bacon as the very essences of things and qualities, and are, obviously, survivals of the Aristotelian and Scholastic *formae substantiales*. The only question is how Bacon came to introduce the term "law" for this medieval concept. As the passage in the *Advancement of Learning* indicates, the Bible suggested this expression. Thus

---

46 II, 2 § 12: *quaestionem esse de modo loquendi*.
47 I, 3 § 8.
48 I, I § 9
49 Preface to Billingsley's Euclid translation, Sig. dj ro.
50 *Essais* 2, 12; there is no natural law, since the customs of the various nations are different (ed. Villey, Paris 1930, vol. II p. 494 f.).
51 *Henri IV*, III, 2 last lines; *Cymbeline* V, 4 verse 37.
52 Works (Spedding and Ellis) VI 222.
53 *Nov. Org.* II, 17 (Fowler, 2nd ed., Oxford 1889), p. 389. 6 f. *cf. ibid*. I, 51 p. 228 f., I, 75 p. 268; II. 2 p. 346; II, 4 and 5 p. 348 ff.; II, 17 p. 399; II, 52 p. 597.

Bacon's terminology again reveals the theological roots of the concept of physical law. How far remote, however, Bacon still is from this concept is illustrated by the remarkable fact that he was ignorant even of the law of the lever.[54]

8. Now we have approached the period of rising natural science and it is time to look for the concept of natural law among its pioneers. The result, however, differs considerably from expectation.

Copernicus (1543) speaks of the "machine of the world founded by the best and most regular artificer"[55], but never of laws of this machine or of the solar system. The same holds of William Gilbert, who was among the earliest adherents of Copernicus in England. In his *De Magnete* (1600), when discussing the precession of the vernal point, he once speaks of a "rule and norm of equality" that may be ascribed to complicated astronomical movements by some hypothesis.[56] This corresponds almost exactly to the modern concept of physical law, though the term is not used. The isolated passage must not be overestimated. *De Magnete* is the first printed book on experimental physics by a scholar. Gilbert makes careful and numerous empirical observations, but still restricts himself in his theoretical explanations to metaphysical generalities on the animation of the globe and the magnet. In his extensive discussions of magnetic phenomena he once makes three rather vague statements which may be called magnetic laws.[57] They are, however, given in a short chapter, entitled "some problems", are not referred to further, and are called neither laws nor rules.

An abundance of physical laws is to be found in Galileo. In his manuscript *Le Mecaniche*, composed in about 1598 when he was a young professor in Padua, he discusses the lever, the windlass, and the pulley, and gives the conditions of equilibrium in quantitative terms.[58] Yet the term "law" is never used. His *Discourses and Mathematical Demonstrations on two new Sciences* (1638) laid the foundation stone of modern mechanics and mathematical physics in general. In this work[59] he discusses the dependence of the period of a pendulum on its length, the dependence of the number of vibrations of a string on its length, tension, cross section, and specific gravity, in quantitative terms. He does not

---

54 *De Augmentis* V, 3, 10.
55 *De Revolutionibus*, preface to Paul III. Thorn edition (Curtze) p. 5 l. 32.
56 *De Magnete* VI, 9 p. 237.
57 *Ibid.* II, 32 p. 99. Since they are for the most part overlooked, they may be discussed here. They state that equal loadstones approach each other with equal "incitation". The same holds for both magnetized and non-magnetized iron bodies. -In Gilbert's time theoretical mechanics in general and an exact concept of force in particular were not yet developed. His three laws compare magnetic forces dynamically by the „incitation" of the movement they cause, "incitation" probably being something vague between velocity, impulse, and kinetic energy. He restricts himself to the case of equality of forces and does not give a real measurement. His laws correspond approximately to Newton's third law (action equals reaction).
58 *Opere*, edizione nazionale, II, 147 ff.
59 *Ibid.* VIII, 139 ff. and 143 ff.

express these relations by mathematical formulas, but paraphrases them in words. And he never calls them "laws" or "rules", but occasionally refers to the law of the pendulum as a "proportion".[60] The same work contains his greatest achievement, the statement of the law of falling bodies, and his discussion of projection.[61] Again the terms "law" and "rule" do not occur. The results are given in the form of numbered theorems, propositions, lemmata, and corollaries, connected by mathematical demonstrations. Though the investigation is a model of experimental research, its literary exposition clings to the traditional deductive form of Archimedes and Euclid.[62]

Apparently Galileo did not know the term "natural law". When he occasionally mentions the law of the lever in the *Discorsi*, he paraphrases it by a long sentence and refers to it, a few lines later, as the "ratios" (*ragioni*) of the lever and as "that principle" (*questo principio*).[63] It is significant that the modern English and German translations often speak of physical laws, when Galileo expresses himself differently. When the translator speaks of the most perfect laws of nature, Galileo only speaks of the "most orderly world" (*mondo ordinatissimo*); when the translator denies that anything can happen against the laws of nature, Galileo only says "against nature" (*contro a natura*).[64] Galileo came nearest to the law-metaphor when he discussed theological objections. A few years before the condemnation of the Copernican doctrine by the church he defended the rights of free investigation in a letter to Castelli (1613). The Holy Writ, he writes,[65] and Nature both originate in the Divine Word, the former as a dictation of the Holy Ghost, the latter as "an executor of God's orders" (*ordini di Dio*) - orders, not laws.

The law-metaphor originates in the Bible, but what was new in Galileo's investigation was not influenced by the Bible. It cannot be verified here that Galileo's concept of science sprung from the method of contemporary technology, since the verification would imply an analysis of the origin of modern natural science in general. The superior craftsmen of the sixteenth century, the artists and military engineers, were accustomed not only to experimentation, but also to expressing their results in empirical rules and quantitative terms. The substantial forms and occult qualities of the scholars were of little use to them. They looked for serviceable, and, if possible, quantitative rules of operation when they had to construct their lifting engines,

---

60 *Ibid.* 139, l. 29 ff.
61 *Ibid.* 197 ff. and 288 ff.
62 *Ibid.* 266 Archimedes, Euclid and Apollonius are referred to. The deductive expositions are given in the Latin, the experiments in the Italian sections of the *Discorsi*. This is a survival of the social prejudice against manual labor. Respectable science deduces and uses the Latin language; experimentation is the business of vernacular craftsmen. *Ibid.* pp. 156-165 ff. he discusses quantitatively how the strength of a pillar depends on its breadth, weight, and length. The results are not called "laws", but are numbered as propositions and corollaries in the margin.
63 *Ibid.* VIII, 152 l. 3 ff., 13 and 16 (*Dialogo*).
64 *Ibid.* VII, 43 l. 6 (*Dialogo*) and VIII, 60 l. 14 (*Discorsi*).
65 *Ibid.* V, 282 l. 30. Literally conforming his letter to the Grand Duchess of Tuscany (1615) *ibid.* 316 l. 25.

machines, and guns. In the manuscripts of Leonardo da Vinci (about 1500) over and over again such quantitative rules of operation are given. They are usually formulated in the manner of cooking recipes: "If you want to know", Leonardo says explaining the drawing of a bent lever, "how much more than MB AM weighs, look how many times CB is contained in AD" etc.[66] The mathematical function-concept, applied to physical phenomena, appeared for the first time in the literature of mankind in a prescription for gunners. In 1546, eighteen years before the birth of Galileo, Tartaglia, in a booklet on gunnery, fortification, and applied mathematics, pointed out that an elevation of 25° gives a gun a certain range; if the elevation is 30° the range is "much greater", if 35° "greater", if 40° "somewhat greater", if 45° "a bit greater", if 50° "a bit smaller", if 55° "somewhat smaller" and so forth. One can make a table of the ranges, Tartaglia continues, and give it to the officer; the officer can tell the gunner how to level the gun, but the table itself can be kept secret, just as "the apprentices can carry out the prescriptions" according to the directions of the apothecary. Tartaglia was a quite poor, self-educated mathematics teacher and adviser to gunners, architects, and merchants, ten pennies a question. He was not a university-scholar but belonged with the superior artisans.[67]

These quantitative rules of the early capitalistic artisans are, though they are never called so, the forerunners of the modern physical laws. Galileo set the investigation of functional relations between physical quantities as the main task for science.[68] The concept of physical law and its paramount scientific importance was perfectly familiar to him. But the term "law" was never used by him, since he cared more for his experiments than for the writings of the theologians and the *Corpus Juris*.

Stevin and Pascal proceeded in a way similar to Galileo's: both were entirely familiar with the concept without ever using the term "natural law". Stevin (1585 and 1608) views the mechanical problems with the eyes of an engineer. Still he explains important laws of statics (the mechanical advantage of the inclined plane, the principle of Archimedes) in the deductive way of Euclid, giving numbered definitions, postulates, and propositions.[69] Pascal (1663) expressly rejects the doctrines of abhorrence of vacuum and of occult qualities.[70] He knows that all machines satisfy the principle of work, that the heights of two liquids in a communicating tube are inversely proportional to their densities, and

---

66 Ravaisson-Mollien, *Les Manuscripts de Léonard de Vinci*, Paris 1891, vol. 6, Ms. 2038 fol 3r. On the other hand Leonardo, discussing the propagation of light, says in a more poetical and theological vein: "O marvellous Necessity ... by a supreme and irrevocable law every natural action obeys thee by the directest possible process ... thou by thy law containest all effects to issue from their causes in the briefest possible way" (*Codice Atlantico*, ed. Pinnati, Milan 1901, III, 1161).
67 *Quesiti et inventioni* I, 1; "ten pennies" (scudi) *ibid.* III, 10.
68 Galileo himself declared the problem of the curve of projection (which puzzled the gunners of the period) as the starting point to his study of the law of falling bodies. *Opere*, letter 2300 to Marsili (1632), XIV, 386.
69 *Hypomnemata Mathematica*, 1608, vol. 4, *Statics*.
70 *Traité de l'équilibre etc. Œuvres* (ed. Brunschvieg-Boutroux III, 224 and 254).

THE GENESIS OF THE CONCEPT OF PHYSICAL LAW 111

discusses the principle of Archimedes - all this without ever speaking of laws.[71] Occasionally he mentions meteorological "rules" for the variations in the height of the barometer.[72] His ignorance of the law-metaphor is remarkable, since he was intently dealing with theological problems; his physical and his religious interests seem to have been separated by an impenetrable wall. The doctrine of natural law in the juridical meaning, however, was known and agreed to by him.[73]

9. Kepler seems to be the first naturalist who, occasionally, used the law metaphor. His well known three laws of planetary movement, however, never are called laws by him. The first and second, given in his *Astronomia Nova* (1609), are paraphrased in long expositions[74]; the third, published in *Harmonices Mundi* (1619), once is called a "theorem".[75]

On the other hand Kepler frequently compares the inverse proportion of velocity and solar distance of a planet (which is the basis of his second law) to the analogous proportion between the force and arm of a balanced lever. And in this context he sometimes speaks of the "law", more frequently however of the "ratios", of the lever.[76] Sometimes he uses the term law as almost synonymous with measure or proportion. Once he draws a diagram in order to clarify the question "which laws are required" in representing a planetary orbit. Or he remarks that the earth receives "the laws of its celerity and slowness" in proportion to its approach to and its movement away from the sun.[77] Other passages are nearer to modern terminology. He discusses the spread of forces from the sun and points out that the force diminishes either with the second or with the third power of the distance; this follows "from the very law of emanation", for the force, although being immaterial, is "not free from geometric laws".[78] The background of these expositions is formed by theological ideas. Kepler ends a long astronomical discussion with the remark that some "geometrical incertitude" is implied in the problem. "And I do not know", he adds,[79] "whether this is not repudiated by God himself, who, up to now, is always found (*deprehenditur*) to be proceeding in a mathematical way". From this the Pythagorean doctrine in Kepler's *Harmonices mundi* follows quite consistently, stating that God ordered the universe according to the principle of "geometrical beauty". In a letter to Fabricius (May 1605) Kepler reports that he

---

71 *Ibid.* 163, 171, 178.
72 *Fragments, (Œuvres,* II, 520).
73 *Pensées, (Œuvres* XIII, 216 no. 294).
74 III cap. 59 f.
75 V cap. 3 (*Opera*, ed. Frisch V, 280).
76 *Astr. Nov., Opera* III, 391: *lege staterae; Epitome Ast. Cop. Opera* vol VI, 373: quae sunt huius celeritatis et tarditatis leges et exempla? Exemplum genuinum est in statera. Rationes staterae: *Astr. nov. Opera* III, 300, 390, 391 and *Ep. Astr. Cop., Opera* VI 405.
77 *Astr. Nov. Opera* III, 315: quibus legibus opus sit ad ... orbitam repraesentandam; *ibid.* 149: leges celeritatis et tarditatis suae accipere ex modulo accessus ... et recessus.
78 *Ibid.* 303.
79 *Ibid.* 397.

has most laboriously treated the irregularities of the planetary movement "until they were at last accommodated to the laws of nature".[80] It can hardly be doubted that these laws of nature are nothing else than the divine principles of mathematical beauty.[81]

Kepler was at the same time a Pythagorean and a devout Protestant. His first work, the *Mysterium Cosmographicum* (1596), explained the solar system by means of the five Platonic bodies. His *Harmonices Mundi* (1619), which gave the third law among numerous mathematical relations without any physical importance, advocated the harmony of the spheres. He considered it his scientific task to reveal the mathematical order of the universe, to describe its beauty, and to praise God as its founder. Thus he changed the divine laws of the Bible into geometric prescriptions and used the term "law" almost as synonymous with ratio or proportion. He is distinguished from the numerous Neo-Pythagoreans of the late Renaissance by his care for empirical observation and his mathematical genius which succeeded in discovering the regularities in apparently most irregular phenomena. His interpretation of these laws, however, still is animistic. After having stated "the laws and quantity of the variation" of the planetary velocity, he raises the question "whether the laws are such, that they probably can be known to the planet". He explains extensively that the movement of the planet probably results from the "wrestling" of its animal and its magnetic faculty, the "mind" of the planet perceiving a certain angle and reckoning its sine. This sense-perception without eyes does not seem impossible at all to him. For an analogy he refers, on the one hand, to the sublunar bodies adjusting their behavior to the stars and, on the other, to his own mother who has born all of her children under the same constellation without making use of eyes.[82] Obviously Kepler's concept of law is quite near to astrology.[83]

The concept of natural law occurs fully developed in Descartes. In his *Discours de la Méthode* (1637) Descartes starts the short exposition of his new philosophy of nature with the declaration that he has found "laws which God has put into nature". God has impressed the ideas of them on the human mind in

---

80 *Opera* III, 37.
81 In his *Ad Vitellionem Paralipomena* (1604, *Opera* II) he gives many optical laws (not yet the law of refraction). The term "law" is never used; they are numbered as propositions and corollaries in the manner of Euclid.
82 *Astr. nov.* cap. 57 (*Opera* III, 392-397); Cf. *ibid.* cap 39 pp. 317-320. The astrological passage p. 319.
83 The question how the planets manage to move regularly results from the elimination of the solid spheres by Tycho Brahe, as Kepler himself states (*ibid.* p. 319). The same problem had been discussed a few years before (1591) by Patrizzi (*Nova de Universis Philosophia*, Pancosmia 12; 2nd. ed. Venice 1593, fol. 91 col. 3). Though Patrizzi does not speak of laws and contrasts only the "order of the world" to chaos, he is quite near to an embryonic concept of natural law. Like Kepler he refers to the *animae rationales* of the stars obeying "God's providence" and compares them to manouvring soldiers, obeying the order of the officer. God corresponds in this metaphor to the officer and the natural laws to his orders - Patrizzi's military metaphor reveals the importance of social changes for the history of ideas. He himself gives ancient Spartans and Macedonians as examples. A medieval author, however, could not have thought of the military metaphor, since battles in the feudal period consisted of a multitude of duels with very little discipline. Obviously Patrizzi is inspired by the new infantry tactics which, in early capitalism, had developed from the armies of Swiss mercenaries.

such a way, that their universal validity cannot be doubted.[84] In the following[85] it is explained that God, after the creation of matter, let nature develop from chaos in accordance with these laws. Even if God had created several worlds the "laws of nature" (*loix de la nature*) would be valid in all of them. The laws themselves, however, are not given in the *Discours*. When discussing the circulation of the blood Descartes only mentions that "the rules (règles) of nature are identical with the rules of mechanics".[86] As an appendix to the *Discours* Descartes published his *Dioptrique*. There the laws of reflection and (for the first time) of refraction are discussed. In connection with the latter the term "law" is used too.[87]

The *Principia Philosophiae* (1644) is the new work announced in the *Discours*. There Descartes explains in the second book[88] that the product of mass and velocity remains constant in nature, since God and his operations are perfect and immutable. "And from this immutability of God", Descartes continues, "some rules or laws of nature which are the causes ... of the various motions, can be understood".[89] He gives three laws, the first and second expressing the law of inertia, the third stating that in every impact "one body gives as much of its movement to the other as it loses". They are alternately and repeatedly called laws and rules. The immutability of God and his operations and the creation of the world are mentioned several times.[90] In order to make quantitative calculation of movement possible Descartes adds seven "rules" of impact which, however, are partly incorrect.[91] He closes the section with the remark that in his opinion no other "principles of physics" are necessary in the explanation of all phenomena in nature.[92] Being a mechanist, he believes he has exhausted not only all mechanical but all physical laws by his enumeration.

Descartes discusses natural laws in a few more places. In the third book of the *Principia* he states it is a "law of nature" that all bodies moving in circles try to recede from the centre.[93] He immediately explains that he does not intend to ascribe minds to moving bodies by this statement and thus shows how carefully he avoids the vitalistic concepts of the Middle Ages.[94] A quantitative determination of the centrifugal force, however, is not yet given. Near the end of the work he states, summarizing, that it has discussed "what must follow from the mutual impact of the bodies according to mechanical laws, confirmed by certain and

---

84 Disc. 5 (*Œuvres*, ed. Adam-Tannery, VI 41).
85 *Ibid.* 42 f, cf. 45 l. 11 ff.
86 *Ibid.* 54 l. 26 f.
87 (*Œuvres* VI, 100 l. 27 f.) The law of refraction was discovered by Snell in 1621 and first published by Descartes.
88 Princ. II § 36. (*Œuvres* VIII 61).
89 *Ibid.* II § 37 (p. 62). In the *Discours* also Descartes had stated that his laws of nature are derived from no other principle but God's perfection (VI 43 l. 5).
90 *Ibid.* II §§ 37-42.
91 II §§ 45-52.
92 II § 64.
93 III § 55.
94 III § 56.

everyday experiments".[95] His laws or rules of impact are discussed frequently in his correspondence, among others in a letter to Christian Huyghens.[96]

Descartes was a consistent mechanist. Convinced that in the last analysis all physical phenomena consist in movement and impact, he strictly denied any teleological, anthropocentric, or animistic explanation of nature. The soullike substantial forms of the Scholastics were discarded by him. On the other hand he was a devout Catholic. Adapting the traditional ideas of God and soul to the new mechanistic science, and stressing the idea of indestructibility, he created a new concept of substance, able to cover both matter and mind. To substantial souls he clung as firmly as he eliminated all soullike components from the physical world. Thus he introduced into human thinking a dualism of matter and mind, of outer and inner world, which in similar rigor had nowhere and never before existed. There is hardly any other philosopher as characteristic of the modern era and Western culture as is Descartes.[97] When we compare with other cultures and omit details, all modern Western philosophers appear more or less as Cartesians, since all of them deal with the mind-body-problem and the problem of the external world. At the end of the nineteenth century only, since the breakdown of mechanistic physics (and with the fading of religious orthodoxy), the influence of Cartesian metaphysics is beginning to decline.

The Cartesian concept of the world combined the basic ideas of the Bible and the new physics. By the same combination of ideas he became the most important pioneer of the concept of natural law which influenced the thinking of the modern era as strongly as his dualism. Like Galileo, he took over the basic idea of physical regularities and quantitative rules of operation from the superior artisans of his period. And from the Bible he took the idea of God's legislation. By combining both he created the modern concept of natural law.[98] Galileo understood the scientific importance of the interdependence of physical quantities at least as clearly as Descartes and made use of more and more complicated mathematical functions. But the law-metaphor was unknown to him. Kepler did speak of laws, but his law-concept was too animistic to influence rising physics greatly. Descartes only combined the law-metaphor with a mechanistic concept of natural law. His influence both on philosophy and on physics can be found in Spinoza, in the physicists of the Royal Society, and in Newton.

10. Spinoza gives a chapter on divine law in his *Tractatus Theologico-Politicus* (1670). In this he distinguishes the laws "depending on necessity of

---

95 IV § 200.
96 "Laws" and "rules" alternately; No. 114 to Huyghens (II, 50); No. 371 to Clerselier (IV, 183 ff.); No. 179 to Mersenne (IV, 396); No. 514 Burman (VI, 168); No. 566 to Morus (VI, 405).
97 Cf. Edgar Zilsel, "Problems of Empiricism", § 3.
98 The relations of Descartes to contemporary technology recede to the background in his writings. Yet in his *Discours* he stresses the utility of his principles and refers to the "various crafts of our artisans" (*Œuvres* VI 62). The Cartesian dualism presupposes the New Testament (concept of soul), his law-concept the old (God, the lawgiver).

nature" from the laws resulting from human decrees. As an instance of the former he gives the principle of the conservation of the quantity of motion following Descartes. He amplifies, however, the extent of the Cartesian law-concept from physics to psychology by adding the law of association by contiguity (as it is called today). Moreover he stresses universal determination: "everything is determined", he says, "by the universal laws of nature".[99] This idea is emphatically repeated in the chapter on miracles. The immutable and universal laws of nature are mentioned again and again and are expressly identified with the "decrees of God". Miracles which apparently contradict them are denied and explained by human ignorance.[100] The opposition of "natural law" and miracle, repeated in the following period over and over again, apparently occurs here for the first time. In the *Theologico-Political Treatise* Spinoza is still trying to hide his pantheism. He speaks, therefore, of the decrees of God in a rather ambiguous way. Since he did not believe in God's personality, however, he noticed, in contrast to Descartes, that application of the term "law" to physical things is based "on a metaphor (*per translationem*)".[101] Possibly this insight was also influenced by Suarez, who was known to Spinoza.

The determinist explanation of mental phenomena is carried out in Spinoza's *Ethics*. The famous preface to the third part begins with the statement that human affects too "follow the common laws of nature". Alluding to the origin of the term "law", Spinoza explains that man in nature does not form a state within the state and expressly denies Descartes' doctrine of free will. Actually "the laws and rules of nature, according to which everything happens and is transformed, are the same everywhere and always". Therefore he will deal with human behavior in a strictly causal way without any valuation - a modern author would have said, in the manner of the natural sciences. Since Spinoza, however, considered the deductive method of Euclid the most scientific, he says he will treat human actions as if the question were of lines, planes, and bodies. In the following this program is carried out *more geometrico*. Yet Spinoza is, of course, not able to give quantitative laws of psychology. The laws of nature are occasionally mentioned in later sections also of the *Ethics* and in his correspondence also.[102] One passage is interesting, since it occurs in a letter to Oldenburg, the secretary of the Royal Society, and since it expressly states that all physical processes follow "the laws of mechanics". Like Descartes and virtually all physicists of the period he is a consistent mechanist.

On the whole Spinoza has taken over the theistic concept of natural law from Descartes and has reinterpreted it in a pantheistic way. At the same time he has extended it to the province of mental phenomena. His ethical ideals being

---

99 *Tract. Theol.-Pol.* cap. 4. *Opera* (ed. Vloten-Land in 4 vol., The Hague 1914) II 134.
100 Cap. 6 (*ibid.* 156-170 passim).
101 Cap. 4 p. 135. -In his first work (1663), the *Renati Des Cartes Principia Philosophiae*, he gives the Cartesian seven "rules" of motion (II, prop. 24-31). In the annexed *Cogitata Metaphysica* he often speaks of the "decrees" of God (I, cap. 3).
102 *Ethics* 4, app., cap. 6f. -Letter no. 13 (to Oldenburg, *leges mechanicae*). Cf. letters no. 31, 33, 42.

entirely Stoic, he is a determinist. Yet his determinism is neither magical nor theological but mechanistic, as it is in Hobbes and as it became in the natural sciences of the following period. Spinoza is the first author combining general metaphysical determinism with the modern concept of natural law.

In 1638, six years before publication of his *Principia*, Descartes had written to the young Christian Huyghens on his laws of impact.[103] With these "laws" or "rules" Huyghens occupied himself in various manuscripts for many years, since he noticed the incorrectness of Descartes' statements. He discovered the conservation of kinetic energy in elastic impacts and published his results in a letter to the *Journal des Scavans, Sur les régles du mouvement dans la rencontre des corps* (1669). After his death a treatise of his on the same subject *De Motu corporum ex percussione* appeared in his *Opuscula Postuma* (1703). In all these papers the laws of impact are alternately called "rules" and "laws".[104] A Latin translation of his letter to the *Journal des Scavans* was published also in the *Philosophical Transactions* (1669), the new journal of the newly founded Royal Society. In the brief English introduction Oldenburg, the secretary of the Society, speaks alternately of the "laws" and of the "rules" of motion.[105]

On the same problem two Latin papers of Wallis and Christopher Wren had appeared one year before (1668) in the *Philosophical Transactions*. The whole discussion was started by the Royal Society. Wallis' paper has the title *A summary account given by Dr. John Wallis of the General Laws of Motion*. Wren's article is called *Lex naturae de Collisione Corporum*.[106] As these papers show, the term "law" first became customary among physicists with the laws of impact. In this the influence of Descartes is obvious, though the papers of Huyghens, Wallis, and Wren, no longer contain theological remarks. Possibly the terminology of the *Transactions* was also influenced by Spinoza, the friend of Oldenburg. The first volumes of the *Transactions* occasionally speak of natural laws in other contexts too. They report on a paper of a French Gentleman, Mr. Auzout, who believed he had found laws of cometary movement.[107] And they mention "odd laws" of variation of the barometer and the "laws of refraction" in optics.[108]

The last two passages refer to two new physical instruments of the eminent microscopist and experimentalist Robert Hooke and are almost literally taken from the preface to his *Micrographia* (1665). Hooke was Curator of the Society

---

103 *Cf.* above footnote 95. Huyghens was then 19 years old.
104 Huyghens, (*Œuvres Complètes* (ed. Soc. Holl. d. Sc.) XVI 95 (ms. of 1652), 104 (ms. of 1654), 139 (ms. of 1656), 181 (letter to the Journ. d. Sc.), pp. 33, 91 (de motu corp.). -In his *Horologium Oscillatorium* (Ibid. XVIII 69 ff.) and his *De vi centrifuga* (ibid. 366 ff.) the term "law" is not used, though important physical laws are stated for the first time.
105 *Phil. Trans.* IV (1669), 925. -The Royal Society was founded in 1663; the *Trans.* appeared first in 1665.
106 *Ibid.* III (1668) 864 ff.
107 *Phil. Trans.* I 4. Auzout speaks of "laws" also in his French pamphlet on the comet of 1664/65, printed in 1665 in Paris (pp. 1 and 7). -He believes that the comet in question moves in a circle. He is proud of having found this hypothesis upon three observations only (*Phil. Trans.* I 19).
108 *Phil. Trans.* I 31 f.

and had to prepare the experiments for their meetings. He discovered the so-called law of Hooke, stating that the stress of an elastic body is proportional to its strain. In his *Lectures de Potentia Restitutiva* (1678) he calls it a "Rule or Law of Nature".[109] This is the first time that a physical law, referred to in modern textbooks under the name of its discoverer, is called a law by the discoverer himself. As early as in 1662 Hooke used the term "law" occasionally also in his notes on his experiments on Boyle's law.[110]

Sir Robert Boyle was among the eminent members of the Royal Society. He published his law (the volume of a gas is inversely proportional to its pressure) in his *Defence of the Doctrine touching the Spring and Weight of the Air* (1660) without using the term "law". It is always called a hypothesis by him.[111] On the other hand he frequently speaks of natural laws in his theological writings. In his *Free inquiry into the vulgarly received Notion of Nature* (composed in 1666) he declares the term "law", when applied to inanimate things, "an improper and figurative expression", explaining this at great length.[112] In the explanation he strangely assumes that the law-metaphor ascribes teleological tendencies to physical objects. When an arrow, shot by a man, he says, moves towards the mark, "none will say that it moves by a law but by an external ... impulse".[113] Nevertheless he himself speaks in what follows very often of the "laws of motion prescribed by the author of things".[114] He confesses to belong to the "modern naturalists and divines", explaining the phenomena through "physico-mechanical principles and laws". Animistic interpretations of nature, therefore, are combatted by him, but he immediately adds that "sometimes" there are miracles.[115] The paper contains numerous biblical quotations and a long polemic against Descartes.[116] It is significant that twenty-one years before the publication of Newton's *Principia* the memory of the metaphysical character and theological origin of the concept of natural law is completely alive in a treatise of a physicist of the Royal Society.

Newton's *Philosophiae Naturalis Principia Mathematica* (1687) has definitely made the term "law" a familiar component of the scientific vocabulary. Particularly his famous three "laws of motion", given at the beginning of the work,[117] were taken over by all physicists of the following period. The reference to Wren, Wallis, and Huyghens, and their laws of impact, in this section[118] reveals the origin of the terminology. The term "law" is applied by

---

109 Reprinted in R.T. Gunther, *Early Science in Oxford*, Oxford 1938, VIII 334 and 336.
110 *Ibid*. VI 83.
111 *Works* (ed. Birch) I 158, 162.
112 *Ibid.* V 170 ff.
113 P. 171.
114 P. 177, *cf.* pp. 194, 225, 151, 252.
115 P. 215.
116 P. 242.
117 *Opera* (ed. Harsley) II. 13 ff. They state, as is generally known, the principle of inertia, the proportionality of force and change of momentum, and the equality of action and reaction. The section is named "*Axioms or Laws of Motion*".
118 P. 23. As in Descartes, Wren, Wallis, and Huyghens, the laws of impact are alternately called

Newton also to his gravitation-formula. The "laws and measures of gravitation" appear as early as in the preface. And at the end of the *Principia*, just after the famous refusal to invent hypotheses about the cause of gravitation, Newton states: "it is sufficient that actually gravitation exists and acts according to the laws given by us."[119] In several problems the mathematical and formal side is conspicuous in his concept of law. He looks for "the law of the centripetal force", given the orbit of a moving body. In one special case the solution is: the force is proportional to the distance from the center. On the other hand a different "law" (the law of gravitation) results, if other orbits are given.[120] Here "law" is obviously almost synonymous with "proportionality" without any tinge of metaphysics.

Still theological components have not vanished in Newton's physics. Though he never mentions the divine origin of the natural laws, he declares that only creation of the world by an intelligent Entity can explain the remarkable coincidence of the directions and planes of the planetary movements. The planets stay in their orbits "by the laws of gravitation, but they could not, initially, receive the regular position of the orbits by these laws".[121] There follows a long exposition, explaining with classical and biblical citations and a footnote on the etymology of the word "God", that everything is in God. Altogether in the *Principia* theology has retreated from the laws to (as the modern physicist would put it) the initial conditions. However, Newton would have certainly admitted, if he had been asked, that the laws too were established by God. Of course Newton knew also Descartes. It cannot be explained, as he states, by Cartesian vortices "that the movement of the comets follows the same laws as the planets".[122]

Newton realized the novelty and scientific importance of the concept of physical law. He starts his work in the preface with the statement: "the modern scientists, omitting the substantial forms and the occult qualities, have undertaken to explain the phenomena of nature by mathematical laws". The essential point of the modern scientific method is explained here with surpassing clarity. And he concludes his work regretting the imperfect state of knowledge in the field of cohesion, nerve activity, and electricity. It is not yet known, he writes, "by which laws" these phenomena must be explained. Thus the first and the last sentence of the *Principia* deal with natural laws. Still the term "law" occurs more rarely than in a modern textbook on physics. It does not occur at all in his *Lectiones Opticae* and his *Opticks*, not even in the sections on reflection and refraction.[123]

We need not trace the origins of our concept further. Whoever knows the immense influence exerted by Newton's *Principia* on the science and even the

"laws" and "rules".
119 *Opera* III, 174. His third letter to Bentley says: "Gravity must be caused by an agent acting constantly according to certain laws" (IV 438).
120 I, sect. 2 prop. 10, probl. 5. *Cf.* I sect. 3, prop. II, probl. 6 and prop. 12 f.
121 *Lib.* 3, scholium generale (III 171).
122 *Ibid.*, 170.
123 *Opera* III, IV.

whole literature of the following period, will not be surprised at the rapid spread of the idea of natural law in eighteenth-century physics, philosophy, and deistic theology. The concept was simply taken over in the shape in which it appears in Newton. As a consequence, the physical meaning of the term "natural law" gradually displaced the juridical. Voltaire, who contributed most to the popularization of Newton's ideas on the continent, is already entirely familiar with the idea that nature is governed by laws, successfully investigated by scientists.[124] No doubt Newton's *Principia* (1687) is the turning point in the rise of this idea. Whereas Locke's *Essay concerning Human Understanding* (1690) does not yet know the concept of natural law in its physical meaning, it already is a matter of course in Berkeley's *Treatise concerning the Principles of Human Knowledge* (1710). In 1655 Hobbes' *De Corpore* does not speak of laws. In 1642 his *De Cive* and in 1645 his *Leviathan* discuss natural law in its juridical meaning only.[125] In 1748, on the other hand, Montesquieu's *L'esprit des lois* dedicates the first chapter to physical laws.

11. Finally, we must try to give an explanation of the development described. Why has it taken place in the way and at the time given? We cannot explain here why at the time of Galileo the idea of mechanical regularities arose. This explanation exceeds our present task, since it is linked with the much more general problem of the origin of experimental science and the quantitative spirit, and will be attempted at another place. Here it may be indicated only that in all civilizations experimentation originates in handicraft. In the period of nascent capitalism experimenting artisans began to look for quantitative rules of operation. The roots of these mechanical rules, therefore, must be searched for in the sociological and technological conditions of handicraft in the early modern era. They rose to science in Galileo.[126]

However, why were these mechanical regularities eventually interpreted as divine laws of nature? This is not a mere question of terminology. Without the metaphysical components, contained in the law-metaphor, they could hardly have obtained their scientific and philosophical impetus. Reference to the strength of religious tradition does not give a sufficient answer to our question. The idea of laws given by God to nature, does not occur in all religions. It is lacking in ancient Egypt,[127] and special investigations would be required to decide whether similar ideas were known in Persia and India. In Babylonia with its Hammurabi code and in the Old Testament with its abundance of ritual and

---

124 *Elements de la philosophie de Newton* (1738) 3, I: (*Œuvres*, Firmin Didot, V, 721), les lois de l'attraction; 3, 3 (p. 726); lois de la chute de corps trouvées par Galilée; 3, 5 (p. 730): lois de la gravitation, règles de Kepler. *Essai sur la nature du feu* 2, 3 (ibid. p. 776): eight "laws". His *Dictionaire Philosophique* (1764), however, gives the juridical meaning only s.v. "loi naturelle" (VIII 21 ff.).
125 Locke, *Essay* I, 3 § 13, mentions the juridical meaning only. Berkeley, *Principles* § 30. Hobbes, *De Cive* cap. 2-4, *Leviathan* I, chap. 14 f., II chap. 26.
126 *Cf.* Edgar Zilsel, "The Sociological Roots of Science", *The American Journal of Sociology*, XLVII (1942) 545ff. [This volume pp. 7-21, Eds.]
127 *Cf.* § 2 footnote 4.

other law it has developed, but everyone who knows how in Christianity different ideas were emphasized or receded to the background in different periods, will not doubt that the divine commands to nature in the *Book of Job* could have easily stayed uninfluential in the history of ideas. In fact God's "eternal law", as we have met with it in Thomas Aquinas, was not a leading idea of medieval Catholicism. The idea of divine providence certainly was important, as far as human fates were concerned, since it gives consolation and hope. As far, however, as it implies eternal laws of nature, it was confined to being mentioned by learned theologians. The Middle Ages perceived the reign of God much more in miracles than in the ordinary course of nature. Comets and monsters were of greater moment to medieval piety than the daily sunrise and normal offspring. How was it that in the modern period the idea of God's reign over the world shifted from the exceptions in nature to the rules?

The expressions "reign over the world" and "law of nature" spring from a comparison of nature and state. Is it not almost a matter of course that the concept of the divine reign changed with changes in the structure of the state? In the feudal state of the Middle Ages government and law differed entirely from the corresponding institutions of the modern era. Thomas Aquinas lived in a period when Italian feudalism was already disintegrating under the influence of the rising money economy. Yet he mentions traits of human law, in his discussion of eternal and natural law, which would hardly fit the physical laws of modern science. There are, he says, "special laws" for the various estates of society; to priests it is "law" to pray, to princes to govern, to soldiers to fight.[128] This would still agree with physical laws: the laws of mechanics differ from the laws of electricity. But Thomas thinks the individual can occasionally change his estate and "law" by order of his lord; *e.g.*, a "soldier" (a nobleman) can be turned out of the "army" (of nobility) and can become subject to rural or mercantile law.[129] To this there is no analogy in modern physics. The "laws" Thomas is here speaking of are, obviously, the bonds of feudalism, varying according to the estate of the individual. They are not based on statute law but on sacred tradition and do not derive from rational regulations of a legislator. Which paragraph of which code orders the prince to rule and the priests to pray?

The feudal state was an extremely loose organization. The bonds by which it was tied together were irrational and considered a matter of course. If the prince issued regulations they were most frequently privileges given to single noblemen, monasteries, and towns, corresponding rather to exceptions than to rules. The medieval interpretation of nature seems to correspond to this organization of the state: the Lord does miracles, they are noteworthy; the regular course of nature, on the other hand, is sacred but a matter of course. At any rate, the idea of a comprehensive multitude of rational physical "laws" could

---

128 *Summa Theologica* II, I qu. 95, art. 4, resp. 2.
129 *Ibid.* qu. 91, art. 6 resp.

not have arisen in feudalism, even if the corresponding physical facts had been known.

It is generally admitted that the Stoic doctrine of the one *Logos* ruling the universe is correlated with the rise of monarchies after Alexander the Great. The analogy might hold for the modern concept of natural law. It will be remembered that Patrizzi had explained the orderly course of the stars by comparing them to soldiers obeying the command of their officers:[130] the loose knightly armies of feudalism had been displaced by the mercenary armies of early capitalism with their rational discipline. The application of the law-metaphor to physical phenomena has probably been produced by the analogous change of the entire state. Money economy disrupted the bonds of feudalism and traditionalism, made national regulations and statute law necessary, and increased immensely the power of the prince. Even in England, where Roman law was not introduced, this process took place under the Tudors; it reached its peak, however, in seventeenth-century absolutism on the continent. It is not a mere chance that the Cartesian idea of God, the legislator of the universe, developed forty years after Bodin's theory of sovereignty. Perhaps it is not even a coincidence that both thinkers were French: France was the native country of centralized absolutism. At any rate the doctrine of universal natural laws of divine origin is possible only in a state with rational statute law and fully developed central sovereignty. Possibly the change in the structure of the state also gives the explanation why the metaphorical character of the term "law", when applied to unreasonable beings, was not noticed before Suarez.[131] Under feudalism even animals and things could be summoned and punished. Thomas hardly thought of legal actions against animals, when discussing God's eternal law; but only rational statute law is, with necessity, restricted to rational beings. Man is a social being. He seems to be inclined to interpret nature not only according to the needs but also after the pattern of society. Yet one difficulty in our sociological explanation must be mentioned. How could medieval theologians speak of the legislature of God, when the power of the prince was very limited? The idea, however, had not originated in feudalism. It had been conceived under entirely different sociological conditions. Its authors were Jews who had outgrown their past of Bedouin clan-organization centuries ago, and its sociological pattern was the despotism of ancient oriental states. The idea could be preserved in a rudimentary form through two thousand years, even through a period in which it did not fit the sociological conditions, till it awoke to new life in early capitalistic absolutism. This fact, and there are numerous analogies, is very important for the theory of history. Ideologies are extremely conservative. They never can be explained by present conditions alone, but mirror the whole past too. At any rate historical problems are very complex. Even if this

---

130 *Cf.* § 9 above, footnote 79.
131 After Suarez (1612) this insight was to be found in Spinoza (1670) and Boyle (1686). In Thomas (about 1270) it is still lacking.

sociological explanation should be falsified by future investigations, the material here collected on the genesis of the concept of physical law remains.

7

COPERNICUS AND MECHANICS

Copernicus overthrew the medieval conception of the solar system by starting from the scanty reports on heliocentric theories in antiquity, by specifying the implications of these geometrically in every detail, and by thus furnishing the exact foundations for ephemerides that far surpassed the exactness of the older tables of planetary movements based on the theory of Ptolemy.[1] His outstanding contribution to astronomy was a mathematico-geometrical one. It is, however, sometimes not sufficiently noticed how far removed Copernicus still is from modern physical and especially mechanical thinking. A few remarks on this point, therefore, may be useful. They refer to the first book of *De Revolutionibus Orbium Coelestium* (1543), in which Copernicus explains the basic ideas of his theory and where, consequently, Pythagorean and Scholastic ideas predominate. Ancient and medieval philosophic ideas recede into the background in the following five books (II-VI) in which the mathematical details are explained.[2]

1. Copernicus uses again and again concepts of value in his general arguments. The third sentence of Book I asks the rhetorical question: "What is more beautiful than the sky? ... Because of its preëminent excellence most of the philosophers have called it the Visible God". In chapter 8 he supports the statement of the immobility of the sun in the following way (p. 24, ll. 2 *ff.*): "Furthermore the condition of immobility is considered more noble and divine than the condition of change and instability which, therefore,[3] is more fitting to the earth than to the universe. I add that it would seem rather absurd to ascribe movement to the containing and locating and not to the contained and located, which is the earth". In chapter 10 he explains (p. 30, ll, 1 *ff.*): "The sun is stationed (*residet*) in the middle of the universe. In this most beautiful temple who could put this lamp in another or better than the one from which it can illuminate the whole universe at once?" The sun, he continues, therefore is called by some "mind" and "ruler", and he ends by quoting the chief authority of occult science, alchemy, and Neo-Platonism: "Trismegistus calls the sun the Visible God, Sophocles' Electra, Him who sees everything. The sun, indeed, sitting on a royal throne rules (*gubernat*) the family of stars moving around it".

---

[1] Cf. Angus Armitage: *Copernicus. The Founder of Modern Astronomy.* London 1938, pp. 90 and 161*f.*
[2] All quotations from *The Revolutionibus* refer to the Thorn edition, 1873 (ed. M. Curtze).
[3] The idea that immobility is nobler than movement is Platonic and Pythagorean (cf. the well-known Pythagorean table of values, Diels, *Fragmente der Vorsokratiker* 45 B 5). Ultimately it goes back to the Eleatic school (Xenophanes, Diels *FVS* 11 B 26).

2. Copernicus is inclined to apprehend inanimate objects as living beings striving to reach aims. Sometimes he expresses himself in an almost animistic way, more often he gives teleological explanations in the more rational way of Aristotle and the Scholastics. A few sentences after the passage just quoted he remarks (p., 30, line 9): "The Earth conceives from the Sun and is impregnated with annual birth". In chapter 1 he explains the spherical form of the universe as follows (p. 11): "This form is the most perfect one, does not need any joint (*nulla indigua compagine*), ... and is the most capacious figure. ... All objects strive (*appetant*) to be bounded in this way. This is seen in drops of water and other liquids when they wish (*cupiunt*) to be bounded by themselves". Gravity he explains in the following way (chap. 9, p. 24, line 25): "I think gravity is nothing else but a natural appetency (*appetentia*) given to the parts by the divine providence of the maker of the universe in order that they may establish their unity and wholeness (*ut in unitatem integritatemque suam se conferant*), by combining in the form of a sphere. It is probable that this affection (*affectionem*) also belongs to the sun, the moon, and the planets in order that they may, by its efficacy, remain in their roundness (*ut eius efficacia in... rotonditate permaneant*)". On the phenomenon of terrestrial gravitation he says (Chap. 7, p. 19, line 28): "The element of the earth is the heaviest, and all heavy things are driven towards it, striving (*contendentia*) to its innermost center".

3. Closely related to this teleological conception of nature is the opinion of Copernicus that objects of the same kind exert "sympathetic" influences on each other. In chapter 8 he discusses the fact that the surrounding air rotates with the earth and gives two explanations which he considers to be equally admissible, a medieval-sympathetic one and a modern mechanical one. The air rotates, he explains (p. 22, ll, 18 *ff*.), "either because it is mixed with earthen and watery matter and, therefore, follows the same nature as the earth (*eandem sequatur naturam quam terra*), or because the motion of the air is acquired and the air participates in it without resistance, since the air is contiguous to the constantly rotating earth". A few lines later, discussing loose heavy objects (which rotate with the earth as well), he repeats the "sympathetic" explanation alone (p. 22, line 31): "Since the objects which are depressed by their weight are mainly earthen, there is no doubt that the parts retain the same nature as their whole (*eandem servent partes naturam quam suum totum*)". It becomes perfectly clear that in the opinion of Copernicus, "equality of nature" is the point that matters in the whole argument when he discusses flames: they participate in the rotation (p. 22, l. 33) "because this fire is earthly and is nourished mainly by earthen matter".[4]

---

[4] It may be mentioned that the medical prescriptions of Copernicus also - he was for a time physician in ordinary to his uncle, the bishop of Ermland - show an entirely medieval spirit. For examples cf. M. Curtze, *Inedita Coppernicana*, Leipzig 1878 (*Mittheilungen des Copernicus-Vereins*, Heft 1) p. 56 ff. E.g., Copernicus thinks that the seeds of water-cress cause "unhealthy humidity" because water-cress grows in humid places. *Loc. cit.*, p. 64, 15.

4. The teleological, half-animistic conception of nature appears also in his theory of motion, which is based on the Aristotelian distinction of "natural" and "artificial" movements. Copernicus explains the falling of bodies by the Aristotelian theory of "natural place" (*locus naturalis,* chap. 8, p. 23, 1. 10). He continues (p. 23, 1. 13 *ff.*): "Rectilinear movement belongs with objects which wander or are expelled from their natural places. ... Nothing is so contrary to the order of the universe and the form of the world as for a thing to be out of its place (*extra locum suum ... esse*). Rectilinear motion, therefore, occurs only if things are not rightly ordered (*rebus non recte se habentibus*)". Obviously Copernicus fully accepts the theory of Aristotle and classical astronomy[5] that celestial bodies move in circles and that this movement is something "natural", whereas rectilinear motion belongs only to terrestrial bodies and is "artificial", as it were.

The medieval idea that everything natural is endowed with an, as it were, spiritual power which is lacking in artificial and imperfect objects and processes leads Copernicus to a discussion of centrifugal force which contradicts modern mechanics in a remarkable degree. Already Ptolemy had objected to the rotation of the earth that by it all objects would have to be thrown off the earth.[6] Copernicus has to defend his theory against this objection. He does it as follows (chap. 8, p. 21, 1. 5): "Things governed by nature produce effects contrary to those governed by violence. Things upon which force and impetus are conferred must dissolve and they cannot subsist for a long time; but what is done by nature is rightly ordered (*recte se habent*) and is preserved in its best composition. Ptolemy, therefore, is wrong in fearing lest the earth and all terrestrial things might be dispersed in a rotation brought about by the efficacy of nature. This is something quite different from art or what human ingenuity can carry on." Obviously Copernicus thinks centrifugal force appears only in "artificial" not in "natural" rotation.

The modern answer to Ptolemy's objection, the argument that the effects of centrifugal force may be neglected compared with gravity, would not have been entirely out of the way. Copernicus himself uses the analogous argument against the objection that the revolution of the earth around the sun must bring about parallactic shiftings of the fixed stars. There he argues quite correctly that these cannot be observed (with the insufficient instruments of his period, as we have to add) because of the great distance of the fixed stars (I, chap. 10, p. 30, 1. 24).[7] Certainly positions of stars could already be measured in antiquity, whereas in the time of Copernicus no way was available of measuring centrifugal forces and comparing them quantitatively with gravitation. The lack of methods of measurement rather often has resulted in metaphysical explanations of physical phenomena. At any rate the quoted passages may have shown sufficiently how

---

5 Aristotle, *De caelo* I, 23; Ptolemy, *Almagest* III, 3.
6 *Almagest* I, 7.
7 Copernicus gave the same argument previously in his *Commentariolus*: M. Curtze, *op. cit.,* p. 6, Quarta petitio. Translated by Edward Rosen, *Three Copernican Treatises,* 1939, p. 58, assumption 4.

much Copernicus is imbued with Pythagorean, Aristotelian, and Scholastic metaphysics.

A correct quantitative theory of centrifugal force was developed for the first time by Huyghens, one hundred and twenty years after Copernicus. Galileo, however, already ninety years after Copernicus, discussed the centrifugal force connected with the rotation of the earth in an entirely unmetaphysical way. Certainly his explanation[8] is not yet correct - he thinks the centrifugal force must at any rate be smaller than gravitation, however fast the earth would rotate, and produces a would-be geometrical proof of this assertion - but he knows that the centrifugal effects in question cannot be observed for the reason that they are too small. The idea that "artificial" rotations behave differently from "natural" ones is not even mentioned by him. This is highly important, for in the last consequence the entirely non-mechanical distinction between "natural" and "artificial" movements excludes experimental research on natural objects. Also with Galileo some teleological ideas still persist, but they form nothing but the general background of his explanations. He almost always uses purely mechanical arguments when he proves his single statements and is strongly opposed to explanations of natural phenomena by means of sympathy and antipathy.[9]

Copernicus is interested in the exact formulation of the mathematical regularities of celestial movements; he is a Pythagorean, and advances not one real mechanical idea. Galileo, on the other hand, is a mechanist: in his dialogue on the theory of Copernicus he is so little interested in the exact details of the planetary movements that he does not even mention the laws of Kepler.[10] He considers it much more important to support the basic ideas of Copernicus by new observations, to show that there is no fundamental difference between celestial phenomena and terrestrial mechanics and physics, and to refute the pre-mechanical ideas and objections of the Aristotelians of his period.

The difference between Copernicus and Galileo is not a difference of individual psychology only, and even less can it be explained by the mere difference of time.

Kepler, who was a contemporary of Galileo, was, as is generally known, at least as Pythagorean and thought at least as teleologically as Copernicus. There rather seems to be a difference between astronomy and mechanics as to their historical evolution and sociological origins. The very first astronomers were Babylonian priests and this connection with priesthood was never quite interrupted; and from antiquity through the Middle Ages up to the end of the sixteenth century, astronomy belonged to the "liberal" arts, as contrasted with the "mechanical" ones. This might explain why metaphysical, Pythagorean and teleological ideas could persist in astronomy until Copernicus and Kepler. It is scarcely mere chance that Copernicus starts his work (I, p. 9) with a eulogy of astronomy "which is the

---

8 *Dialogo sopra i due massimi sistemi del mondo*, 1632. *Opere*, Edizione nazionale VII, 221, 7 *ff.*
9 Cf. *Dialogo*, Ed. naz., VII, 436, 17 *ff.*; *Discorsi*, Ed. naz., VIII, 116.
10 The dialogue appeared in 1632, Kepler published his laws in 1609 and 1619.

chief of the liberal arts, is most worthy of free men, and rests upon almost all kinds of mathematics". And it is not mere chance that, by enumerating these, Copernicus gives mechanics as the last one. For mechanics belonged to the "mechanical arts", to those which required the use not only of head and tongue, but also of hands, and therefore were left to lower-class people. It may be that in the modern era the experimental method and the elimination of teleological and animistic by causal thinking originated in those ranks of mechanicians and craftsmen. Certainly scientific mechanics and physics did not appear in modern times before the way of thinking of the craftsmen was adopted by academically trained scholars of the upper class, as happened in the period of Galileo. A more extensive inquiry, however, than could be given in this short note on Copernicus, would be necessary to verify this sociological explanation.

8

THE GENESIS OF THE CONCEPT OF
SCIENTIFIC PROGRESS AND SCIENTIFIC COOPERATION*

INTRODUCTION: THE PROBLEM

Science appears, fully developed, only in Western civilization and the modern era. It is best understood if it is compared with other creations of the human mind playing analogous roles in other cultures: with the magic of the primitives, the theology of certain oriental cultures and that of the Arabic and Christian Middle Ages, the combination of metaphysics and rhetoric characteristic of education in classical antiquity, and the humanism of the Renaissance and classical China. Science shows certain differences from and certain conformities with each of them. It deals with worldly subject matters; so do magic, humanism, and classical philosophy. In contrast to magic, theology, and humanism it is not bound to authority; only classical philosophy was to a similar extent founded on individual thinking. In contrast to magic, science proceeds rationally - like the rest of its kin. But the rationality of science essentially differs from the rational methods of the classical philosophers and, particularly, the scholastics and humanists. Only science rationally investigates recurrent associations of phenomena, called "laws", and has developed quantitative methods. And only science uses experimentation and systematically checks its findings with experience. All of these characteristics of science have been discussed in other places. Here the genesis of another trait, not less characteristic of science, is to be analyzed.

The modern scientist sees science as a great building that grew up, stone by stone, through the work of his predecessors and the contemporary fellow-scientists, a structure that will be continued, but never completed, by his successors. In this he wants to co-operate. Either the mere theoretical aim, construction of the building, is the object of his activity; or he follows a utilitarian view, links the progress of science with progress of civilization, and

---

* [A highly edited version of this essay was published as 'The Genesis of the Concept of Scientific Progress', in *JHI*, 6, pp.325-49. The editors have been able to procure the considerably more extensive original. Zilsel was apparently required to radically shorten this longer version. In so doing he skipped much of the historical detail and tried to make his point as succinct as possible. As we find that both essays have their own merits, we have composed this combined version to avoid printing both essays separately. We have added all those passages of the shortened essay not contained in the original more extensive version. These passages are denoted with an asterisk(*) at the beginning and end of the respective passage. Eds.]

has in mind the benefit to mankind produced by the practical application of the scientific theory. No modern scientist, however, would dare to give personal advantage or fame as his incentive. This means that science, both in the theoretical and the utilitarian interpretation, is regarded as the product of a co-operation for non-personal ends, a co-operation in which all scientists of the past, the present and the future have a part. In the present this idea, or ideal, is considered almost self-evident. Yet no Brahmanic, Buddhistic, Moslem, or Catholic scholastic, no Confucian scholar or Renaissance humanist, no philosopher or rhetor of classical antiquity knew of this ideal. It is a specific characteristic of the scientific spirit and of modern Western civilization. [It appeared for the first time fully developed in the works of Francis Bacon.] We shall trace its genesis and shall try sociologically to explain its rise. The title of our article, chosen for brevity's sake, does not cover the entire problem since it does not mention the impersonality of the scientific aims. Certain parts of our problem, the idea of progress and the rise of scientific societies and periodicals in the seventeenth century, have already been thoroughly investigated.[1] These studies have enabled us to restrict the scope of this article. We shall use them and hope that the new context will shed new light on some problems. Since contemporary ideas, for the most part, are considered a matter of course we shall bring the lack of the scientific ideas in the past into prominence. In intellectual history the absence of ideas is as important as their presence. And we shall, at the end, advance a few general remarks on the use of sociological methods in the history of thought.

It was shown by J.B. Bury that before Bacon only scanty rudiments of the concept of progress are found in Western scholarly literature. Since Bury, however, is interested in the idea of cultural rather than of scientific progress, he deals chiefly with statements about the general course of history: his discussion of the pre-Baconian period gives an illuminating analysis of the classical conception of cultural history but does not investigate the classical conception of science. Moreover, the sociological origin of Bacon's ideas is not traced in his excellent book. We shall therefore try to show that the modern idea of progress through co-operation stems, like many other elements of modern scientific procedure from the superior artisans of the fifteenth and sixteenth centuries. To elucidate the social causes of this specifically modern intellectual development we must, for purposes of comparison, also give a brief analysis of the classical concept of science. This introductory account, however, must be restricted to an outline and will be documented elsewhere.

---

1 Jules Delvaille: *Essai sur l'Histoire de l'idée de progrès*, Paris 1910 and especially, J.B. Bury: *The Idea of Progress*, London 1924. Martha Ornstein: *The Role of Scientific Societies in the Seventeenth Century*, 3rd. ed., Chicago 1938; Harcourt Brown: *Scientific Organizations in Seventeenth Century France*, Baltimore 1934; Michele Maylender: *Storia delle Accademie d'Italia*, Bologna 1926-30.

I. CLASSICAL ANTIQUITY

The role of theoretical knowledge

"I would rather have discovered one causal explanation than have received the throne of the Persian empire".[2] This statement of Democritus shines like a torch across a distance of almost two thousand and four hundred years. It is one of the rare personal confessions of classical authors and is among all the sayings of the antiquity nearest to the spirit of the modern natural scientist. We must not overlook, however, that Democritus neither mentions followers who will build on his findings nor predecessors, nor co-operation with contemporaries. Democritus's enthusiasm for causal research is, moreover, a rather isolated phenomenon in classical antiquity. Half a century later Plato expressly blamed Anaxagoras for his indulgence in non-teleological explanation and it would seem that his accusation was taken to heart by the Aristotelians and Stoics of the subsequent period.[3] On the other hand Democritus clearly expresses enthusiasm for secular knowledge and this, at least, is not an exceptional phenomenon in classical antiquity. Archimedes who, in his joy at his discovery, ran naked through the streets, who, absorbed in his calculations, was slain by the Roman soldier, must have known this enthusiasm. The anecdote making Pythagoras, out of gratitude for the discovery of his theorem, sacrifice a hecatomb mirrors the same zeal for scientific knowledge. If one is aware of the difference between the causal, experimental, and quantitative method of modern science and the methods of classical philosophy one may even say that secular knowledge is not less an ideal for all classical philosophers than for modern scientists. Not until late antiquity did philosophy through the influx of oriental ideas approach theosophy and theology.

Yet, especially after Socrates, the search for knowledge had a somewhat different tinge than in modern science. Among the post-Socratic philosophers the Epicureans were nearest to the aims of natural science. They stressed the necessity of causal explanation as much as Democritus. This, certainly, sounds most scientific. But apart from the methodological difference (the theories of Epicurus and Lucretius were metaphysical constructs without any experimental foundation) the Epicureans did not philosophize out of scientific interest in nature but in order to free man from superstition and fear of the other world: the imperturbability of the sage was their real aim. Similar ideals stood behind the search for truth in the other post-Socratic schools which, moreover, were not interested in causal theories. To the Cynics, Cyrenaics, and Stoics - and to Socrates himself - knowledge was, basically, a mere prerequisite of self restraint. All of them strove not for science but for wisdom. Only the pre-Socratics, the Platonists, and Aristotelians searched for knowledge for its own sake. Thus the impersonal ideal of theoretical knowledge was known only to a minority of the

---

2 Fragment B 118 (Diels).
3 *Phaedo* 98 B. A similar objection in Aristotle, *Metaph*. I, 4; 985 a 18.

philosophical schools and to the mathematicians, astronomers, and geographers (of whom, however, we know much less than of the philosophers). An individualistic ethical ideal, the ideal of the perfect sage, was, apparently, of greater importance to classical thinking than mere theory. If, on the other hand, classical theorists wanted to apply theory to practical life apart from morals, they restricted themselves to politics (like the Pythagoreans, Sophists, Socrates, Plato, and the Stoics), war machines (like Archimedes) and medicine. The Baconian idea that science is of value because, by its application to technology, it furthers the material welfare of mankind, was unknown to classical authors. In his Republic Plato demands astronomical instruction for the rulers of the ideal state.[4] He bases his demand on metaphysical reasons and rejects with contempt the plea of the usefulness of astronomy for agriculture, navigation, and strategy. The aristocratic philosopher regards the utilitarian argument as a concession to "the multitude".

The different roles assigned to theory in classical antiquity and the modern era are accounted for by basic differences in the social structure. In classical society a considerable fraction of the rough manual work was done by slaves. Slaves cannot be entrusted with handling complex devices and there is no inducement to introduce labor saving devices where there is an abundance of cheap labor. Hence in cultures based on slave labor there is no machinery. Also in classical antiquity: slavery obstructed the use of machines in the production of goods and in traffic. Only warfare machines played a major part. Since, consequently, an industry based on discoveries and computations of theorists did not exist and since the achievements of the classical theorists dealt with metaphysics and history, astronomy and mathematics, geography and natural history, but hardly influenced every day life, a utilitarian interpretation of the ideal of truth in the manner of Bacon could not develop. The classical theories were actually of no practical use. These facts, however, had even further-reaching consequences. As a comparative analysis reveals they obstructed also the development of the ideas of scientific co-operation and continuity and of the general idea of progress.

Professional men and institutions for research
Classical civilization was carried by a rather thin upper class, living without occupation on their incomes. The few philosophers and authors who stemmed from or were in sympathy with the lower middle classes should not disturb this picture. Consequently, in those circles which determined ancient public opinion, earning money by professional work was regarded as a not quite respectable activity. The writers, philosophers, and theorists were either aristocrats living on the yields of their estates, or beggars (as the Cynics), or they lived in the household of noble patrons as secretaries and librarians or tutors to their sons. Their pay, for decency's sake, bore rather the character of gifts than of salaries.

---

4 *Rep.* 7, 527 d.

There were, however, also independent teachers of philosophy who drew their pupils from the noble youth and sometimes received high honoraria. Some wandered from city to city, as the Sophists in the fifth century B.C., others had permanent residences, as the Stoic Posidonius in Rhodes in the first century B.C. Of course there were also professional physicians and attorneys. The attorneys were trained oratorically rather than juridically as, on the whole, eloquence played a greater part than in modern society. Politics belonged to the chief occupations of the members of the upper class and since the political career required the ability to make speeches there were numerous teachers of eloquence whose pupils belonged to the same circles as those of the philosophers. Philosophical and rhetorical instruction were, for the most part, hardly separated. Most of the philosophers were more or less near to rhetoric; only Plato was a consistent opponent of oratory. Apart from the physical and military training a combination of metaphysics and rhetoric was the backbone of the education of a well bred young man and took the place of modern scientific instruction.

Naturally classical civilization had produced technological skills. The technicians were called "architects" and had to direct the construction of public buildings, temples, and fortifications, port facilities, bridges, and military roads. Also the construction of war machines and clocks belonged to their tasks. Some of those classical engineers had some mathematical and philosophical knowledge and composed books on the technological and esthetic principles of their profession. Most of them, however, were probably (rather little is known of them) hardly more than foremen and supervisors of slaves. Neither in number nor influence nor education can the classical "architects" compare with modern engineers.[5] On the whole in classical society physicians, attorneys, and teachers of rhetoric and philosophy were the only professional men of importance. Particularly the only rudimentary development of engineering and the non-existence of university professors is conspicuous.[6]

Three famous schools are, comparatively, nearest to modern universities: the Academy, founded about 387 B.C. by Plato, the Lyceum, founded about 340 B.C. by Aristotle, and the Museum in Alexandria, founded half a century later by the first of the Ptolemies. Juridically the Academy, the Lyceum, and the museum were *thiasoi*, i.e. religious associations, devoted to the worship of the Muses. According to classical law this was the only way to secure the character of a juridical person to an association of private people. In each of these institutes there was an altar of the Muses where the schoolhead officiated as priest. The three schools were devoted rather to theorists intending to remain

---

5 Except for three works, composed by Philo of Byzantium, Hero of Alexandria, and Vitruvius respectively, all classical treatises on engineering are lost. Ctesibius, the son of a barber, was the most eminent classical engineer, but it is not even certain whether there were one or two technicians of this name (cf. Hermann Diels: *Antike Technik*, 3rd. ed. Leipzig 1924, p. 198, footnote 2). On the classical engineers in general cf. Pauly-Wissowa: *Realencyclopedie der klassischen Altertumswissenschaft*, 2nd.ed. s.v. "architectura".
6 To the following paragraphs cf. Pauly-Wissowa, *loc. cit.* s.v. "Museion" and U.v. Wilamowitz - Moellendorf: "Antigonos von Karystos" (*Philol. Untersuchungen*, Heft 4), Berlin 1881, 263-291.

there permanently than to the education of young aristocrats. In all three of them the members lived "on the campus" but there was not much opportunity for co-operative research.

In the Academy mathematics and philosophy, of course only the official philosophy of the Academy, were the only subjects studied and taught.[7] The head of the school and the more advanced of the members took part in the teaching. Also the Academy edited a few books: the works of Plato and the lost dialogues of Aristotle who in his youth had been a member. Compared to the Academy, Aristotle's Lyceum was an institute for empirical studies. In the Lyceum not only metaphysics and logic but also several more empirical branches of knowledge, from biology to constitutional history and theory of poetry, were investigated and taught. A rich library was at the disposal of the members, but there were no laboratories and even the library was the personal property of Aristotle.

A few decades after the death of its founder the Lyceum declined and the Museum in Alexandria became the center of empirical research. Among the fellows of the Museum there were mathematicians, astronomers, geographers, philologists, and physicians. Philosophy played a minor part and was not, as in the Academy and the Lyceum, restricted to one official line of thought. The Alexandrian military engineers, being artisans, were, of course, not members. The fellows received salaries from the Ptolemaic kings, later from the Roman emperors but, probably, were not obliged to lecture though there were students at the Museum. Some members, like the philosopher Strato of Lampsacus and the geographer Eratosthenes, were also tutors to Ptolemaic princes. The famous library, a common dining room, and a hall for disputations were parts of the Museum. The Museum had a few astronomical instruments but it is improbable that they belonged to a permanent observatory.[8] Experiments and dissections were performed but disputations and symposiums of the scholars, partly in the Museum, partly at royal dinners, probably played a greater part. The Museum declined one hundred years after its foundation but rose again under the Roman emperors, primarily as an academy of philologists and orators. It existed through nearly seven hundred years. Similar institutions on a smaller scale existed at the courts of Asiatic princes, the largest in Pergamum. How far the professional men and the institutes for advanced studies developed the spirit of scientific co-operation and the idea of scientific progress must now be discussed.

---

7 In the presence of a visiting physician Plato once also discussed a botanical problem, namely the question of the place of the pumpkin in the system of plants (cf. Wilamowitz, *loc. cit.* p. 283 f.). If we may, however, draw an inference from Plato's exposition of his natural philosophy in the *Timaeus* and from his explicit rejection of careful observation and measurement in astronomy and acoustics (*Rep.* VII, 530b and 530e), we may assume that the natural sciences, as far as they were discussed at all, were treated in a way rather different from the modern one.

8 Cf. Ernst Zinner: *Geschichte der Sternkunde*, Berlin 1931, p. 149. Probably there were no permanent observatories at all in classical antiquity.

Philosophers and scientists on their aims

In societies with an upper class without economic occupations social prestige is of great importance. Prestige takes the form of fame if the privileged group is secular. This applies also to classical antiquity. It is sufficient to mention that fame, which in the Homeric period was attached primarily to warlike deeds, with the rise of an urban culture also expanded also to intellectual, literary, and artistic achievements. In all civilizations in which authors neither belong to the economically independent upper ranks, nor are salaried officials, nor live on the sale of their writings to a broad public, they depend on wealthy patrons and discharge the sociological function of increasing the prestige of their protectors. Also in classical antiquity the dispensing of prestige and fame became the very basis of existence for those authors who did not live on their incomes. Particularly in the post-Alexandrian period the men of rank liked, for the sake of their social prestige, to attach literati to their households. Obviously the Ptolemaic kings and the Roman emperors also subsidized the Alexandrian Museum to increase their fame. The engineer Philo of Byzantium (third century B.C.) even explains the eminent achievements of the Alexandrian constructors of war machines by the rich subsidies from kings "loving fame and the art".[9] One should assume that kings subsidize artillerists rather for the sake of power than fame. We may be happy that the post-Alexandrian princes strove for glory not only by warfare and politics but also by aiding scholars. Also in our time analogous motives might sometimes play a certain part in the foundation of institutions for research. In classical antiquity, however, such motives were admitted quite openly and authors themselves frequently gave desire for fame as a respectable if not the only incentive for every intellectual activity. To this there is no analogy in our period.

A few examples may illustrate this frankness. In the first century B.C. Cicero explains at length that in human life there is no worthy aim but fame and honor. "It must be openly declared", he says, "all of us are impelled by the desire for fame and he is the best whom it inspires most". At another place Cicero states that "it is honor that nourishes the arts and men are compelled to studies by fame".[10] Also in the first century B.C. the historian Sallust points out that most men live only for their bellies, "spending their lives unlearned and uneducated"; those only really live "who strive for the glory of an eminent deed or a good art".[11] One century later Pliny the Younger states that there is nothing greater for men than glory, praise, and immortality.[12] All these remarks relate primarily to literature, poetry, or historiography; science and philosophy are not mentioned in this context. However, the authors make quite general statements on the "studies and arts", on "learning" and "education". The "studies" could be

---

9 *Belopoika* 3, ed. Diels-Schramm, *Abhandlungen d. kgl. preuss. Akademie d. Wissenschaften, phil.-hist.-I-Klasse* 1918, no.16, p. 9.
10 *Pro Archia* 6 cf. 11 and Tusc. I, 2, 4.
11 *Bellum Cat.* II, 8f; cf. *Bell. Jug.* I, 3 ff.
12 *Ep.* III, 21.

regarded in classical antiquity as a means of obtaining literary fame because science and philosophy were hardly separated from literature. Also in classical historiography the fame of the persons described, sometimes also of the historiographer himself, played a considerable part.[13] We must, however, study the opinion of the philosophers and scientists on fame before we can draw conclusions.

Heraclitus advocates the ideal of fame without restriction. "The best", he says, "prefer one to all mortal things - eternal fame. But the multitude eat their fill like cattle".[14] Obviously this is an aristocratic outburst of anger. It would be anachronistic, however, to ascribe desire for literary fame to a thinker of the sixth century. Heraclitus, the prophet of the divine Logos, might rather have felt like the founder of a religion. Probably he was proud of having recognized the hidden ruler of the universe and thought that he had deserved as much fame for this feat as the aristocratic warriors and politicians. With literature - and science - this attitude has little in common. Also a century later, in a period in which literary men were no longer a rarity, the same aristocratic appreciation is found in Plato. Men, Plato states, are willing to endure all dangers rather out of desire for fame than for their children's sake; neither Alceste nor Achilles, nor Codrus would have sacrificed their lifes "had they not believed that they would thereby win a deathless memory for valor".[15] A modern author would never have given fame but only the life of the husband, the memory of the friend, the benefit to his country as the objectives of Alceste, Achilles, and Codrus respectively. Plato is nearer to the modern valuation of fame in his Seventh Letter. There he blames his former pupil Dionysius, the tyrant of Syracuse, for having published a philosophical work out of "ignominious ambition". In contrast to this unworthy behavior, in the soul of a true philosopher a flame kindles from the continuous application to his problems like a spark from a fire. "Could I have done anything more beautiful in my life", Plato asks, "than writing what is of great benefit to mankind and bringing forth to light for all the true reality?"[16] The passage well conforms with modern opinion on the tasks of a decent scientist. Aristotle probably held similar opinions. He was the perfect type of a scholarly theorist and regarded knowledge as the summum bonum and the only activity worthy of the deity. In his extant works he never speaks of fame as of an aim of the philosopher. Possibly, he did not think such an idea worth discussion. And contempt of fame was even a part of the school doctrine of the Cynics, Sceptics, and Stoics. All of them counted fame with the external goods to which the true

---

13 Herodotus I,1; Sallust, *bell Jug*. II, 4; Dionysius of Halic., *antiqu. Rom*. I,1, 3 & I, 6, 3. Thucydides, on the other hand, is not much interested in fame. Against fame as an object of the historiographer Josephus Flavius, Jew. War I, 1.
14 12 B 29 (Diels).
15 Sympos. 208 C ff. Plato's statement is intended to support his famous doctrine of the love of the eternal. In Plato's opinion this love connects man with the imperishable Ideas and is the motive of the true philosopher: the immortality of fame is only a symbol for the eternity of the Ideas. But it is significant that Plato could use this symbol to express a metaphysical doctrine.
16 341 C f. and 344 E.

sage is indifferent.[17] Cicero who, as we have seen, highly praises fame when he speaks as an orator and literary man, despises it when he remembers the stoic school doctrine. As far, however, as the numerous philosophical literati are concerned, Cicero hit the point with his remark that also those who write books against fame do so for their fame's sake as proved by their giving their names in the titles of their books.[18] Actually, the frequent philosophical declamations against fame must not be taken too seriously and only prove the strength of the desire for fame in classical society. Before a public of business people modern preachers preach against the treasures which are eaten by rust and moths; before a public of literati and aristocrats without economic occupations classical philosophers preached against fame.

It is characteristic of the role of science in classical culture that most of the scientific writings are lost and, with rare exceptions, virtually nothing is known about the lives of the authors. We know also very little about the personal attitudes of the scientists to their research. The mathematical works of Euclid, Archimedes, Apollonius, Diophantus, as far as they survive, deal impersonally with the scientific problems. Psychological motives, however, are occasionally discussed in the medical literature. The Hippocratic school, obviously in order to distinguish themselves from the quacks, were anxious to maintain a high moral level. They demanded unselfishness from their members and set up "love of men" and "love of the art" as professional ideals. According to Hippocrates these two motives are essentially connected with each other.[19] Manifestly the goals of the Hippocratics were no less impersonal than those of modern medicine and modern science. Also six centuries later Galen stated that neither fame nor profit are the goals of the true physician: he adds that the physician Diocles (fourth century B.C.) had taught this doctrine and Hippocrates, Empedocles, and "many ancients" had practised "out of love of men".[20] Galen's exposition is a polemic against a contemporary physician, Menodotus. Menodotus had openly advocated fame and profit. Galen rejects this attitude, yet without special indignation.

Desire for fame was also criticized in the middle of the first century A.D. by the Roman admiral and naturalist Pliny the Elder. In a lost section of his *Roman History* Livy had pointed out that, having obtained sufficient fame, he could put an end to his writing, were it not for his restless mind "feeding on work". At this remark, characteristic of the attitude of the classical literati, Pliny takes offense. His objection is that Livy should have written "rather for the glory of Rome than for his own"[21]; it would have been a greater merit had he persevered in his work for the work's sake and worried less about his own mind.

---

17 Against fame: Antisthenes Diog. Laert. 6, II; Crates ibid. 6,9 and Stobeus 10,59; Diogenes: Epictetus I,24,6; Timon: Euseb. prep. ev. XIV, 18,19; Elder Stoa: Arnim *stoic. vet. fragm.* I 537 v.28,I 559 f. and III,42.
18 *Pro Archia* 11.
19 IX 258, Littré.
20 V. 751 Kuehn.
21 *Hist. nat.* I, 16 Mayhoff.

The astronomical literature is nearest to the modern attitude. The only extant work of the greatest astronomer of antiquity, Hipparchus (about 125 B.C.), is a polemic against an astronomical poem of Aratus. The brief treatise is dedicated to one Aischrion, obviously a pupil or friend of the author, who had asked him questions about Aratus' poem. As Hipparchus states in the initial sentences,[22] he wishes to rectify the incorrect descriptions of the constellations in Aratus "because of Aischrion's desire for knowledge and the general benefit of the readers". He does not write "to create sensation by refuting others (for this is vain and reveals a petty soul; on the contrary I think we must be grateful to everyone who took pains for the general benefit's sake)"; he rather publishes his book "lest Aischrion and the other lovers of knowledge go astray in the theory of cosmic phenomena". There follows a philological comparison of Aratus' poem with two older astronomical works on which it was based, an extensive criticism of Aratus, and finally, as an original contribution of the author, a list of fixed stars with the times of their rising and setting. The book differs from a modern astronomical treatise only in the lengthy philological section and the fact that a scholar of Hipparchus' standing criticises an astronomical poem at all. Manifestly science in classical antiquity was less separated from literature than in our time. Yet Hipparchus' concept of the impersonal objectives of scientific publications conforms with modern ideas.

About 150 A.D. Ptolemy expresses a similar attitude in the *Almagest*. His book was written, as Ptolemy briefly remarks in the last sentence, "only for the benefit of the theory, not for ostentation". The *Almagest* is based to a great extent on the works of Hipparchus, older by almost three hundred years. Whether the final remark is also influenced by Hipparchus, or whether similar statements were customary in astronomical literature cannot be decided since most of the original works are lost.

We have collected enough material to draw a picture. The idea that theoretical work must not be done out of personal ambition, that truth benefits man or, at least, the "lovers of knowledge" was known to classical scientists and philosophers. The attitude of Aristotle, the statements of Plato, Hipparchus, and Ptolemy leave no doubt of that. Also the scientifically most eminent among the physicians advocated impersonal professional ideals. On the other hand literati, historians, and even Plato state so openly that in all human activities personal fame is the final goal that a substantial divergence from the modern attitude must be ascribed to classical public opinion. Manifestly the impersonal ideals inspiring modern scientists were considerably less acknowledged. The statement that classical authors were greedier for fame or readier to admit selfish motives than modern ones would be a poor explanation of this fact. Psychological differences, if they appear as mass phenomena, have sociological causes. Originally the classical ideal of fame was an inheritance from the military nobility of the Homeric period. Up to the end of antiquity public opinion was

---

22 *In Arati et Eudoxi Phaenomena Commentaria*, ed. Manitius, I,1,5.

dominated by this ideal because most of the intellectual leaders were aristocrats without economic occupations and the people with economic occupations had hardly any part in the intellectual life. Authors, as far as they were not aristocrats themselves, were supposed to increase the fame of their aristocratic patrons. The Alexandrian scholars were, basically, court scholars. There were no public professors and the "architects" could not compare in social influence and number to the modern engineers. The impersonal ideals of such professions would have counterbalanced the fame ideals of the aristocrats, literati, and rhetors in public opinion and could also have strengthened the ideals of "love of men" and "love of art" developed in the medical profession. In antiquity, however, the institutions and motives that among modern scientists have checked the unrestricted growth of the desire for prestige were lacking: there were no laboratories and institutions for research where scientists could co-operate and no scientific technology which engineers could help to promote. It is rather noteworthy and testifies to the strong appeal of theoretical research - and the honesty of human nature - that even under such sociological conditions men like Hipparchus could set up the ideal of impersonal interest in objective truth as a measure of the behavior of a decent scholar. The high moral standard of the Hippocratic school is equally noteworthy.

Progress and co-operation in philosophy, medicine, astronomy, and engineering
It is especially the lack of co-operation that distinguishes classical from modern theoretical activities. Of course, even classical theorists were no solitaries, but their mutual relations were rather agonistic or emotional than co-operative. A comparison of the philosophical banquets described in Plato and Xenophon and the co-operation for objective ends in a modern laboratory illustrates this difference. Meetings of classical philosophers or scholars were social gatherings in which contests were combined with less spiritual entertainments. Even the more sober philosophical debates (there was an abundance of dialectic and "eristic" in classical philosophy) are to modern scientific research as athletic contests are to the division of labor in a modern factory. On the other hand the Greek term to "co-philosophize" (συμ ιλοσο v) and to "co-philologize" (συμ ιλολογ v) designated hardly more than joint studies of friends, as Cicero and Atticus, based on emotional harmony. Emotional bonds between philosophers were perhaps even more developed than in our time; between teacher and pupil they sometimes assumed erotic forms, but these have little to do with our problems. It was regarded as a matter of course that philosophers formed "schools". These "schools", however, were held together rather by common devotion to a master or an opinion than by common research and were therefore called *diadochai*, i.e. successions (namely of master and followers) or *haireseis*, i.e. selections (namely of opinions). Thus it could happen that in the Christian era "heresies", i.e. religious sects (which do certainly not join for research work), and philosophical "schools" were designated by the same term. Even the more firmly organized schools, as the Pythagorean brotherhood in Italy which was intermediate between a religious sect and a political party, or the Academy,

the Epicurean and Peripatetic schools in Athens, which were *thiasoi*, never carried out investigations or edited publications conjointly.

The same is true for classical science. No publications, no astronomical or geographical investigation which are the work of several collaborating scientists are known.[23] Even the learned compendia of the Roman period (Varro, Pliny, Celsus) and the encyclopedias of late antiquity (Boëthius) were composed by single polyhistors. There is no evidence that the Alexandrian Museum conjointly carried out investigations. Laboratories, the birth places of scientific co-operation in the modern era, existed neither in the Alexandrian Museum, nor in the Academy, nor in the Lyceum. As far as the fellow scholars of the museum did not work each for himself they might have contented themselves with dinners and debates. And of course, there were in antiquity no scientific periodicals in which new findings could have been discussed.

References to other authors are frequent in the philosophical literature. Many of them are polemic and the polemic often reached a degree of mordancy that would have been impossible if the insight had existed that there are various approaches to truth and that they are made only step by step. The personal invectives against all famous philosophers of the past in the satirical poem of the sceptic Timon may be an exception even in classical antiquity. Yet they have no counterpart in modern literature.[24] On the whole, classical scepticism, with its sterile and merely formal arguments against the possibility of truth, could develop only in a period ignorant of the idea of scientific or philosophical progress.

Among the philosophers the sense of indebtedness to predecessors is, comparatively, best developed in Aristotle. Aristotle's extant works contain his academic lectures. In them the philosopher, like a modern professor, often inserts critical reports on the treatment of his problems in the previous literature. For the modern historian these reports are very valuable, enabling him to reconstruct the philosophical development. Aristotle himself, however, hardly realizes that the concepts have really developed, i.e. were gradually refined and enriched. And he hardly manifests the insight that his own theories would be

---

23 Probably, the oldest part of the Septuagint is the only product of literary collaboration in classical antiquity. The composition of the Septuagint is told in the *Letter of Aristeas*. According to this report the high priest of Jerusalem sent, upon request of king Ptolemy Philadelphus (288-247 B.C.), seventy-two translators to Alexandria, six from each of the twelve Jewish tribes. "They arrived at an agreement on each point by comparing each other's work" (*loc.cit.* tr. H.St.J. Thackeray, London 1917, § 302). The story of the isolation of the translators in separate rooms and their miraculous literal agreement is a later addition of the church fathers. But also the *Letter of Aristeas* is legendary and spurious. The seventy, or seventy-two, scribes are, probably, a reminiscence of the seventy elders accompanying Moses to the slopes of Mount Sinai. Yet it is possible that under the early Ptolemies the Jewish community in Alexandria had a version of the Pentateuch officially composed or approved by a commission of scribes. The Talmud gives the number of the translators as five (*loc.cit.* p.90). The translations of the other parts of the *O.T.* were gradually added in the course of the subsequent centuries. They certainly have different authors.

24 Vehement polemic: Heraclitus, (Diels, *Vorsokratiker*) 12B 40 and 42; Xenophanes *ibid.* 11 B 11; Epicurus: Usener, *Epicurea* 227 ff. and 238. Timon (Diels, *poet.philos.*) 9 B 25, 27, 28, 36, 38, 41, 51, 54.

impossible without the work of his predecessors.²⁵ After the Alexandrian age collections of philosophical doctrines and of biographies of philosophers became frequent. The work of Diogenes Laertius, however, the only one which is extant, nowhere shows understanding of the gradual development of philosophical thinking.

In science use of the works of predecessors was customary. In many cases, however, we know of it only by inference. A systematic work like Euclid's *Elements* must have been based on the results of many older mathematicians though these are not mentioned in the text. Also the classical grammarians and philologists, whose treatises for the most part are lost, certainly used the writings of their predecessors. In the post-Alexandrian period many compendia were composed in which extracts from the learned literature of the past were compiled. They are not interested in the gradual progress of knowledge. The *Natural History* of the Roman admiral Pliny the Elder (d. 79 A.D.) is perhaps the most learned compendium that has come down to us. Pliny, though not a historian of science, feels at least indebted to his sources. He gives long lists of the authors used and explains that he considers it a matter of "gratitude and honesty" to give the names of people by whom "one has profited". He adds, however, that most authors copy literally without quoting the source.²⁶ A few decades earlier, Celsus begins his compendium of medicine with a brief report on a few great physicians of the past. The survey concludes with the following terse statement: "Primarily through these men medicine has increased".²⁷ Celsus' historical introduction manifests a certain interest in the history of science. Yet there is still a long way from the "increase of medicine" in Celsus to the modern concept of a steady progress of science, almost as long a way as from the "profit" for which Pliny is grateful to his predecessors.

Greek medicine was not ignorant of the continuity of science. At least the Hippocratic physicians in the fourth century B.C. must have been conscious of the fact that much can be learned from the experience of predecessors. Otherwise they would not have recorded and published case histories of their patients. They were members of a corporation with a highly developed sense of professional dignity and a strict code and, obviously, sought to establish a tradition of knowledge in their corporation. In the oath which every candidate had to take before his admission he had also to swear that he would give medical instruction only to his sons, the sons of his master, and to pupils who had sworn

---

25 In *Met. I,3* Aristotle, after having explained his distinction of the four kinds of causes, starts the historical discussion with the words (983 b 1) "Setting to the discussion of substance we want to review also our predecessors" since those too discussed causes. The review is called (983 b 4) "a kind of preparatory work" (τ προ ρμου) which will serve either to confirm the completeness of Aristotle's classification or to discover further kinds of causes. The historical paragraph concludes with the terse statement (987 a 27): "this much can be taken from the predecessors and others". On the other hand Aristotle derives Plato's metaphysics from a combination of Heraclitean and Socratic ideas (*Met*.I,6 and 9).
26 I, 21 p. 7 Mayhoff.
27 I p.3 Daremberg. Celsus also begins his expositions of pharmacy and surgery with brief historical surveys (V,1 and VII,1).

to the regulations of the corporation.[28] Possibly the famous Hippocratic aphorism "art is long, life is short"[29] implies even the quite modern idea that medical knowledge grows only with the passage of the generations.

We know much less of the later physicians and their opinions on the gradual increase of knowledge. Once Sextus Empiricus reports that the empirical physician Menodotus (second century) advocated the use of the writings of the medical predecessors, "because the art is so great that one Man's life does not suffice to find everything".[30] This sounds modern and is, probably, an allusion to the Hippocratic aphorism just discussed. It is rather a matter of course that physicians who are not mere medicine men utilize the experience of their predecessors. Yet systematic endeavors to increase the fund of knowledge hardly can be ascribed to the classical doctors. Also the progress actually achieved must not be overestimated, in spite of some important discoveries of post-Hippocratic physicians. In the post-Alexandrian period there were two medical schools vehemently combatting each other: the "rational" physicians indulging in philosophical speculations, and the "empirics", rejecting all theories. In Greek these two schools were designated by the same term *haireseis* as the philosophical schools - and later the Christian sects.[31] Scientific progress is neither as fact nor as ideal compatible with scientific "sects" co-existing through centuries.

Among the philosophers one will look in vain for an analogy to the Hippocratic insight to the continuity of knowledge. This difference is accounted for both by methodological and sociological reasons. Medicine, in contrast to philosophy, has practical aims and proceeds empirically. On the other hand the Hippocratic doctors, also in contrast to the philosophers, were professional men. They belonged to a corporation with a professional tradition and were apprenticed in their youth by their masters who in most cases, probably, were the fathers of the medical apprentices. In both respects the classical physicians conform to the practices of artisans although they were by no means regarded as manual workers.[32] Craftsmen, at least if they require superior skill, necessarily experience that working methods can be gradually improved and, if they are sufficiently educated, they express this experience. Actually, perhaps the clearest statement on continuity and progress of knowledge in extant classical literature is found in a treatise of Philo of Byzantium who, as a military engineer, was a superior artisan. In his treatise on war machines, Philo reports that the early war machines were rather poor; only the later engineers in Alexandria put artillery on

---

28 IV, 628 Littré.
29 IV, 458 Littré. The context, however, only stresses difficulties facing the doctor who wants to make a diagnosis. Cf. also the quotation from Menodotus below.
30 Sext.Emp. 64,17 ff. In Sextus strangely enough a lengthy discussion of the reliability of historical reports follows.
31 Cf. Galen, Kuehn I,64 ff. and 106 ff.
32 In the Hippocratic Collection the physicians are counted among the *demiourgoi* (I,571 f., Littré) and it is mentioned that they also work with their hands (ibid.); but their difference from the *banausoi* is stressed (IX,234, Littré). *Demiourgoi* are people who are paid for their work, *banausoi* are common manual workers. Galen points out at length that medicine belongs to the liberal arts (I,38 f. Kuehn). Among the fellows of the Alexandrian Museum there were several physicians but no engineers.

a sound basis, as he expressly states, "partly by learning from the errors of the earlier constructors, partly by observation of later trials".[33] The remark is brief but is exceptional in classical literature and exactly describes an essential point of the procedure of modern experimental science. On account of the structure of classical society, however, this method did not ascend to the ranks of the scholars and classical technology made little actual progress.

Among the scholars only the astronomers used the scientific work of their predecessors systematically. There is, however, no evidence that they followed the example of the engineers or other artisans. Since Babylonian astronomy had also been based on centuries old records of celestial phenomena they were possibly rather influenced by the model of the Babylonian priests. However this may be, Hipparchus (about 125 B.C.) gave a more accurate determination of the length of the year and the month and discovered the precession of the equinoxes by comparing his own observations with Babylonian records and observations of earlier Greek astronomers. He also composed a list of fixed stars with their positions which was intended as a basis for the scientific work of the astronomers to come. His star catalog was not the first of its kind. Similar lists had also been composed by earlier Greek astronomers, the last one by Eratosthenes a hundred and twenty years before Hipparchus, and Hipparchus' catalog served as basis to Ptolemy's more complete one two hundred and fifty years later. In classical literature these star catalogs have a parallel only in the case histories of Hippocrates. Manifestly, in classical astronomy science was seen as a task for many generations. This conception is rather clearly expressed in the introduction to Ptolemy's *Almagest*. We want to increase the love of astronomy, Ptolemy says, "acquiring knowledge reached by sincere investigators in the past and contributing an addition ourselves as large as the time elapsed since them up to us permits".[34] Though in a modern publication the passage would not be conspicuous it is, next to the quoted remark of Philo of Byzantium (who, however, was not a theorist), the clearest expression of the idea of the continuity of science in antiquity. It also seems to be the only passage in extant classical literature in which an author speaks of his work as of a mere "contribution" to the body of science.[35] Classical authors were, as is usually said, more immodest

---

33 *Belopoika* 3, cf. above footnote 9.
34 I, 1, p. 7 f., Heiberg.
35 Since the terms used by Ptolemy (συν ἥκην ση ἐρειν) twice contain the preposition σύν (together with) the idea of co-operation is distinctly implied. It is not impossible that Christian ideas play a part in the modern phrase "to contribute something to science". Petrus Lombardus begins the prologue to his *Sentences* (about 1150) with the statement that he has undertaken a work beyond his strength because, in his poverty and weakness, he wants "to throw something in the poor box" like the poor widow in *Luke* 21,2; the wish to defend the Catholic truth against errors made him overcome his scruples. The comparison of his book with the contribution of the poor widow, naturally, does not imply any connotation of scientific co-operation or progress but only expresses Christian humility. Yet, Peter's allusion to the gospel might be important. Dante too introduces Peter as the man "who with the poor widow offered his treasure to the Holy Church" (*Par.*X, 107 f.). Since in the late Middle Ages the *Sentences* of Petrus Lombardus were the most widely read book it is quite possible that in the genesis of the modern phrase "scientific contribution" not only the development of scientific co-operation but also the idea of Christian humility participated. A history of this phrase would be illuminating. On this

than modern ones. The statement that, for sociological reasons, they were ignorant of the idea of scientific co-operation and continuity (and of cultural progress in general) describes the difference more precisely. Among the theorists only the astronomers overcame to some extent the individualistic attitude of the classical authors. Also the emphasis upon their impersonal objectives in Hipparchus and Ptolemy will be remembered.

This exceptional position is, probably, explained partly by the nature of astronomy, partly by historico-sociological reasons. Many astronomical problems are concerned with time intervals that greatly exceed the life span of an individual. It is both easier and more necessary for astronomers than for other theorists to overcome individualism. On the other hand, Greek astronomy was dependent on the astronomy of the Babylonian priests. In all civilizations which, for religious and agricultural reasons, are interested in the calendar, astronomical observations are recorded and kept by priests. Everywhere priest-scholars are guardians of tradition. It is quite possible therefore, however paradoxical it may sound, that the astronomers in the individualistic Greek city society could anticipate the modern concept of scientific continuity because they were influenced by pre-individualistic and traditionalistic methods developed by priest scholars in an oriental agricultural civilization. At any rate, astronomy was the least individualistic among the theoretical sciences of antiquity.

In an astronomical context also may be found what appears to be the only classical passage in which the future advance of knowledge is predicted. It is the Stoic Seneca (about 50 A.D.) who, discussing the eclipses of the sun and the orbits of the comets, remarks that generations to come will clear up the problems involved and will even marvel at our ignorance of the explanation.[36]

We must, however, not overestimate the progress actually achieved in classical astronomy. About 120 B.C. Hipparchus proposed astronomical methods to determine geographical longitudes and latitudes. Two hundred and fifty years later the same proposals were repeated by Ptolemy without any attempt having been made to carry them out in the mean time.[37] An accurate determination of geographical positions would have required organized co-operation of astronomers and surveyors, backed by public authorities. Since in antiquity neither warfare nor commercial traffic required particular accuracy, such a joint effort never came into existence and maps were always based on rather inaccurate dead reckoning. In theoretical astronomy progress was considerably greater. Ptolemy's mathematical astronomy exceeded by far the naive speculations of Thales. Yet, the heliocentric ideas of Aristarchus could remain an isolated paradox without influence on the theoretical development. They were sometimes referred to but an attempt never was made to develop them to a degree of accuracy that would have allowed a quantitative comparison with the geocentric theory. Progress was even less, fluctuations, setbacks, and

---

phrase in Descartes, see *Oeuvres* (Adam-Tannery) VI, 63.
36 *Nat quaest.* VII,25 and 31. Cf. I.B. Bury: *The Idea of Progress*, London 1924 p. 13.
37 Cf. Pauly-Wissowa, *Realencyclopaedie* VII, 1679 (Rehm).

sects combatting each other were even more frequent in the other natural sciences.

In Athens and other Greek cities at certain religious festivals running contests were held in which the racers, in the manner of a modern relay, would pass on a burning torch. This classical torch race is often used in modern literature as a simile for the continuity of culture: the cultural achievements of the past are passed on from generation to generation like a torch. However, this beautiful simile is not classical. Plato once discusses marriage and its biological ends. There he uses the torch race as a metaphor, but the torch symbolizes life passed in succession from parents to children. Nowhere in the context are culture, or philosophy, or science mentioned. With the same strict biological meaning the metaphor also appears in two other classical authors. No classical writer, however, ever came to conceive culture as a shining torch carried forward in a holy running contest. Antiquity was familiar with the idea of the continuity of life; the concept of scientific and cultural progress arose only in the modern era.[38]

The "philosophical" character of classical science and classical individualism
Progress of knowledge is not the aim of primitive theorists. If one starts from traditional mythology and endeavors to replace it with one's own speculation, one might well believe that one has obtained, in a few basic insights, the final truth about the universe. The idea that knowledge progresses only gradually implies an element of self criticism unknown to naive thinkers. If, on the other hand, people with practical aims - artisans detaching themselves from the authority of the workshop tradition - try to improve working methods by their own experience, they must realize that improvements are achieved only step by step. It takes time - and presupposes certain favorable sociological conditions - until theorists take over the method of trial and error and are ready to admit that their conclusions are susceptible to further improvement. Actually the classical constructors of war machines knew that "art is long, life is short". The classical philosophers, on the other hand, from the pre-Socratic metaphysicians up to the neo-Platonists and neo-Pythagoreans, were much too convinced of the truth of their doctrines to believe in the progress of knowledge. The pupils would simply adopt the systems of their masters. If, however, in the course of time, the school doctrines would change, this was interpreted rather as a return to the original intentions of the masters than as philosophical progress. Naturally the theological metaphysicians of India also believed in the final truth of their systems.

---

38 Plato, *Laws* VII, 776 B; G. Kaibel: *Epigrammata Graeca*, Berlin 1878, 231; Lucretius II, 74 ff. Munro. In the modern era a torch and two hands with the motto is used as signet by the publishing house Harper Bros., New York. The verse of Lucretius "vitai lampada tradunt" serves as an inscription on the wall of the office of the Rector of the University of Vienna. It would be interesting to establish when and where the metaphor assumed its modern meaning.

Compared with the agricultural and priestly civilizations of the Orient on the one hand and modern capitalism on the other, classical culture may be called a half-scientific one. Science in classical antiquity was much less separated from philosophy than in the modern era. Even medicine which, in contrast to most of the other sciences, had not branched off from metaphysics but stemmed from surgeons and priests, became early involved in philosophical speculation. Only cosmography and political historiography always stayed apart from it. The separation from metaphysics was, comparatively, most complete in mathematics, astronomy, and philology. Also at the peak of the evolution, however, there were fewer special sciences and many fewer scientific specialists than in our time. In higher education philosophical and rhetorical instruction were always dominant. Also the rise of science did not last very long. More than half of the eleven centuries from Thales to the end of the Roman Empire are but a period of scientific standstill and decline. Metaphysics not only preceded but also survived science by several centuries.

Systematic increase of knowledge, even in the spring of classical culture was not among the aims of the philosophers and scientists. Yet it would be an overstatement to characterize scientific activity as a merely individualistic undertaking. On the whole classical science had reached, approximately, that degree of steadiness and co-operation which in our era is characteristic of philosophy. Findings of predecessors were used most systematically by the astronomers, but schools combatting or ignoring each other occurred frequently, the fund of undisputed knowledge was small and the theoretical achievements were due to a few eminent individuals - as in modern philosophy. Modern philosophy has in the very last decades even developed certain institutions - periodicals and congresses, reference works composed by experts in collaboration - to which classical science did not proceed. These philosophical innovations, however, are quite recent and were introduced after the model of the corresponding scientific institutions. Whether they will give scientific exactness to philosophy, whether they will abolish the strange phenomenon that there are "schools" in philosophy as in oriental and western theology or in painting and music, remains to be seen. In science, "schools", if they develop, do not long survive their founders. Modern science seeks to increase the generally recognized fund of knowledge. These efforts materialize in institutions of research, periodicals, and handbooks, and require great numbers of trained research workers, not all of whom need be eminent. All this is true primarily for modern natural science, to a less extent too for modern historical and philological research, but is found to be absent in antiquity.

Methodically the "philosophical" character of classical science is explained by the predominance of metaphysical speculation. Science exists only where there are differences between experts and other people: the one can verify their statements, the other cannot. It is not a mere coincidence that the term "expert" is derived from "experience". Speculation is subjective, experience inter-subjective. Since the classical physicists, biologists, and many physicians preferred speculation to experience they required less co-operation and there was less scientific progress - although this explanation does not apply to mathematics. Sociologically, the more individualistic character of classical theory and the predominance of speculation are due to the aristocratic structure of classical society. On the one hand classical aristocrats cherished individual fame. On the other, people living on their incomes speculate and make speeches; in their eyes only *banausoi*, i.e. artisans, and slaves work with their hands, make experiments, and are interested in the experiments of other people. Actually co-operation for objective ends and the ideal of progress has developed in the ranks of modern artisans. Even if this sociological explanation is not accepted, the individualistic character of classical thinking cannot be overlooked. Classical ethics dealt extensively with individualistic virtues, with self control, wisdom, valor, and moderation. The social virtues, like compassion and justice, primarily treated in modern ethics, were artificially derived from the individual ones or not discussed at all. A society whose moral philosophers, without exception, do not start from the society but from the individual can, in scientific research, not produce co-operative methods and the ideal of cultural progress for the benefit of mankind.

The ideal dominant in the cultural activities of classical antiquity is not free of impersonal elements.[39] Fame relates both to the author and his work. Only the modern era separated personal abilities completely from objective accomplishments. The seventeenth century substituted progress for fame. Progress is an impersonal ideal since, as far as it considers individuals at all, it regards only their accomplishments. But on the other hand the modern era, and only the modern era, created worship of genius. In genius the accomplishments of the eminent individual have become almost irrelevant. Thinkers whose ideas contradict each other can be equally worshipped as geniuses and an artist's genius manifests itself even in an unfinished sketch. Comparing it with the classical ideal one may say that the concept of genius is obtained through abstraction from all impersonal elements in fame and idealization of the personal gifts of the famous personality. Thus cultural progress and genius are complementary ideas. In classical antiquity both were unknown and the ideal of fame was in their place.

II. FROM THE MIDDLE AGES TO FRANCIS BACON

Scholasticism, the Alphonsine Commission, Humanism
With omission of all details, we want to give a sketch of the further development, sufficient to show the real origin of the concepts of scientific progress and scientific co-operation. To the church fathers and scholastics these

---

39 [We have reconstructed this sentence to the best of our ability. Zilsel's original sentence reads: "The ideal dominant cultural activities in classical antiquity is not free from impersonal elements". Eds]

concepts were, of course, unknown. Revealed truth is not susceptible of improvement and even the idea that its correct interpretation gradually progresses is foreign to theological thinking. Remarkably enough, in all civilizations theologians place not only revelations but also their authoritative interpreters in the past. At the universities, arising in the late Middle Ages, there was no scientific co-operation. The medieval universities were institutions for the professional training of clerics, notaries, and medical doctors rather than for research. They neither had libraries for common use nor, of course, laboratories and the dominant method of disputation bred quarrelsomeness rather than co-operative spirit.

However, the Middle Ages produced one co-operative scholarly organization. Soon after his accession to the throne in 1252, King Alphonso X of Castile and Leon appointed a commission of Arabic, Jewish, and Christian scholars to translate astronomical and astrological works from Arabic into Spanish. Among numerous other treatises they published a Spanish version of Ptolemy from the Arabic, the *Libros del Saber de Astronomia (Books on Astronomical Learning,* 1276), and the famous Alphonsine Tables.[40] The *Libros del Saber,* as the entire work a collection of translations, was meant as an account of all astronomical instruments and their use and also contains five treatises on sundials and water clocks going back to classical tradition. The Alphonsine Tables, which were generally used by the astronomers up to the time of Kepler, are based on Ptolemy. In their preface (the tables themselves are lost) the two authors state that, since many celestial motions are exceedingly slow, their determination cannot be carried out by one man but only by many "working through long periods one after the other". "We shall", the preface concludes, "correct everything dubious or requiring correction and accept everything certain or almost certain".[41] Manifestly the scholars of king Alphonso were as conscious of the continuity of science as Ptolemy and the other classical astronomers. The idea, however, that science progresses was foreign to them. Although many versions are freely revised the entire work of the Alphonsine commission is a traditionalistic cyclopedia without new scientific conclusions. Also the Alphonsine Tables do not seem to have been much superior to Ptolemy's.

On the other hand the cyclopedia is, apparently, the first scholarly publication in history composed by several authors.[42] Significantly enough it deals with astronomical questions. That the ideas of scientific continuity and scientific co-operation first appeared in astronomy is due partly to the nature of problems, partly to classical and Babylonian tradition. As to intellectual descent, the work belongs more to the Arabic than to the medieval Christian civilization. A

---

40 Cf. George Sarton: *Introduction to the History of Science II* (Baltimore, 1931) 835-837. The following quotations from the modern print of the *Libros del Saber,* Madrid 1863-67.
41 *Loc.cit.* IV, p. 111.
42 In the preface to the *Libros del Saber* written by the king himself he states that his Jewish physician in ordinary and his cleric were employed with the translation and that "they are to have as help" three more scholars whose names are given (*loc.cit.* I,7).

complete sociological analysis would, therefore, require a historical investigation of these ideas in the Arabic culture. The Alphonsine commission seems to have had no connection with any university. Since they wrote Spanish, their work is not addressed to university-scholars or clerics. The practical interests of navigation may have played a part in its composition. A treatise on armils, contained in the collection, stresses the fact that astronomers must be familiar with the methods of the artisans making the astronomical instruments.[43] Yet only an extensive investigation of the relations of Arabic astronomers to the instrument makers could decide the sociological import of this remark. Personally Alphonso "the Learned" had a predilection for establishing scholarly foundations; being more interested in them than in politics he lost his throne through a revolt of the nobility after a few years.[44]

The Renaissance humanists, the first secular scholars of the modern era, likewise did not stand for scientific progress, since with few exceptions (Benedetto, Accolti, Pico della Mirandola), they considered the classical authors unsurpassable. Viewed sociologically the humanists descended from the political secretaries of the late medieval Italian princes, municipalities, and the curia. These secretaries were supposed to increase the prestige of their employers by the polished Latin of their diplomatic correspondence and their speeches. Also when in the fifteenth century the humanists had turned into court literati, free writers, and lecturers, they felt themselves as "dispensers of glory" who by their writings made their protectors, and themselves, immortal. The open admission that fame is the goal of all literary activity appears even more frequently in the humanist than in classical literature, either because the purely individualistic conception of the literary profession was actually more developed in the Renaissance, or only because more literary testaments are extant. Since the desire of the humanists for fame has often been described we give only two less well known examples, relating to the fields of mathematics and physics. Analogous passages occur very frequently in the more literary writings.

The first Greek print of the *Arithmetic of Diophantus* appeared in 1621, the editor, the French humanist Claude Gaspard Bachet, dedicated his work to an influential lawyer. In the preface he explains that "in honorable emulation for fame" he is thinking day and night how he might become as famous as his patron. After having examined all disciplines he decided to choose mathematics "since it wonderfully delights the mind and since in mathematics the subtlety of the intellect especially comes to light. This Diophantus will give evidence whether I have deserved fame beyond the ordinary mathematician".[45] The court mathematician of the Duke of Savoy, Benedetti, who often but erroneously is called a forerunner of Galileo, was not strictly a humanist but even stressed his

---

[43] *Ibid.* II, p. 4.
[44] He founded the University of Sevilla (1254), completed the incorporation of the University of Salamanca (1254), and had the first national history of Spain compiled in Spanish.
[45] Reprinted in Behria Botfield: *Prefaces to the First Editions of the Greek and Roman Classics*, London 1861, p. 656.

lack of regular education. Yet he imitated the humanists as literary models. In 1585 he published his Latin correspondence on mathematical and physical problems, as he states in the preface, to make the addressees immortal "as far as it is in my power".[46] The idea that publication of a scientific correspondence could further science is foreign to the literati of the period. *Analogous expressions of glory worship occur very frequently in the more literary writings of the humanists.[47] On the other hand, in the humanist literature of the Renaissance there seems to be no case in which an author states that he is publishing his treatise in order to make further investigations possible. We shall see how differently the contemporary craftsmen express themselves.*

It is evident that the individualistic professional ideals of the humanists were incompatible with scientific co-operation. At the end of the sixteenth century Henri Estienne, the Younger, out of scholarly rivalry did not allow his son-in-law, Casaubonus, to use his library.[48] Both were outstanding classical scholars and highly moral Huguenots, far remote from the lax morals of the Italian literati of the Filelfo and Panormita period a hundred and fifty years earlier. However, rivalries, quarrels, and personal intrigues accompanied the literary and scholarly careers of all humanists. In the literature of the Renaissance there exists no scholarly encyclopedia or dictionary composed by several authors in collaboration. Only the great humanist printers and publishers in Venice, Basel, and Paris, as the first Aldo Manuzio (d. 1514) in Venice, who employed numerous classical scholars as editors and proofreaders, must have achieved a certain amount of co-operation among their learned assistants.[49] Printing, however, was a mechanical art and the publishers, though themselves classical scholars, were not literary dispensers of glory but businessmen. Their printing houses were among the biggest and best organized plants of the sixteenth century.

Superior artisans 1400-1600

In the workshops of the late medieval artisans co-operation resulted quite naturally from the working conditions. In contrast to a monk's cell or a humanist's writing chamber, a workshop or dockyard is a place where several people work together. On the other hand the guilds stressed continuity rather

---

46 Joh. Bapt. Benedictus: *Diversarum Speculationum Mathematicarum et Physicarum Liber*, Turin 1585 p. 204. Lack of regular education: *Resolutio omnium Euclidis Problematum*, Venice 1555, preface, seventh page.

47 Examples: Filelfo: *epistolarum familiarum* (Venice 1502), 12 r, 54 v; Emile Legrande: *Cent-dix lettres grecques de Filelfe* (Paris, 1892), p. 63 Georg Voigt: *Die Wiederbelebung des classischen Altertums*, 3rd ed. (Berlin 1893), p. 334, p. 446, p. 527; E. Zilsel: *Die Entstehung des Geniebegriffes* (Tübingen, 1926), pp. 111-123.

48 Cf. J.H. Sandys, History of Classical Scholarship, Cambridge 1908, II, 205.

49 The first of them was Aldo Manuzio. He founded the *Neo-Academia* in Venice about 1500; most of the members were his assistants. The humanistic "academies" of Ficinus in Florence, of Pomponius Letus in Rome, and of Fontanus in Naples, all of them in the later half of the fifteenth century, held debates and banquets but did not do real research. Cf. Michele Maylender: *Storia delle Accademie d'Italia* (Bologne, 1926), I, 125 ff; IV, 249 ff, 320 ff, and 327 ff.

than progress of craftsmanship. The apprentice learned the workshop tradition from his master and was taught to honor it as the master had honored the tradition of his master's workshop. Rising capitalism and economic competition, however, broke the power of guild tradition. Only the artisan who had either invented some commercial or technological innovation or who understood the value of an invention of another fellow became a capitalistic manufacturer. Thus the inventive genius of the individual gradually came to the fore. The professional ideals of the early capitalistic artisans can be inferred with some probability. As petty manual laborers they could not well strive for literary immortality like the humanists. Social conditions directed them to more impersonal goals. If they wanted to justify their work and their inventions ideologically they had to refer to the glory of God and the Saints, of the craft and the workshop - and to the usefulness of their craft and the public benefit. Actually these ideals are in fact expressed in several treatises composed by superior artisans such as artists, instrument makers, and gun makers. Sometimes the authors even uttered the intention to further, through their treatises, the craftsmanship of their colleagues. Such statements reveal the social root of the modern ideal of progress. To modern ears they may sound rather trivial. We must not forget, however, that in classical, scholastic, and humanist literature, apart from a few astronomical writings, statements on the necessity of the gradual improvement of knowledge do not exist. Naturally, only members of the most highly skilled crafts wrote treatises and only few of these craftsmen-authors conceived the idea of progress with any clarity. Even in the sixteenth century a considerable number of the manual workers, particularly outside Italy, were illiterate.[50]

A first handbook of a secular craftsman was composed about 1400 by the Florentine painter Cennini. It, as all similar treatises, written in the vernacular, circulated as a manuscript among the painters of the Quattrocento and is still imbued with the medieval guild spirit. *It deals with the making of pigments and the various techniques of painting; but being the treatise of an artisan it is also concerned with the paintings of chests and the make-up of women.* Cennini wrote his booklet as the title *or better, incipit,* explains, in the reverence of God, the Virgin, and the Saints, "and in the reverence of Giotto, Taddeo, and Agnolo and for the use and profit of any who wants to enter the craft". Giotto was Taddeo's, Taddeo Agnolo's, and Agnolo Cennini's master: the author was well conscious of the continuity of craftsmanship. On the other hand Cennini states that he "will also make note of what he has tried out with his own hand".

---

50 Johann Neudörfer (*Nachrichten von Künstlern und Werkleuten*, Nuremberg, 1547; reprinted by G.W.K. Lochner, *Quellenschriften zur Kunstgeschichte*, X Vienna, 1875) discusses two illiterate masters even among the contemporary Nueremberg craftsmen: the locksmith Hans Bulmann (d. 1549), the constructor of an astronomical clock, and the carpenter Georg Weber (d. 1567), the maker of complex wooden clockwork (pp. 65, 79). They were, however, manifestly exceptions (2 among 111 masters) and their illiteracy is mentioned as a curiosity.

The addition hardly implies the idea of progress. Yet it shows that Cennini no longer considers the mere workshop tradition as sufficient.[51]

Almost a century later (1486) the printed treatise of a late Gothic master-builder clearly advocates the advancement of craftsmanship. Mathias Roriczer, *an architect of the cathedral of Regensburg,* is the author and his booklet is dedicated to his previous employer, the bishop of Regensburg.[52] Although Roriczer treats the tradition of the craft with great reverence, his geometrical constructions are, he states, of his own invention. At the end of the dedication he emphasizes that he has not written the book for fame's sake but for the public benefit "and to better wherever something is to be bettered, and to amend and explain the arts". The public benefit (*der gemeine Nutzen*) is three times mentioned as the aim of the author. The dedication, probably, only expresses what all of the more inventive craftsmen of the period might have felt, but neither Plato nor Aristotle ever stated that their work was intended to mend philosophical shortcomings and to improve the state of philosophical knowledge. A hundred and twenty years later Francis Bacon proclaimed the advancement of knowledge for the benefit of mankind as goal of the scientist. It is this ideal of progress, naturally in an embryonic form and restricted to the craft of the masons which appears in Roriczer's treatise. "Progress" however, has its limitations in the booklet of the Gothic master-builder. According to the regulations of the guild of the masons the secrets of the art had to be kept from laymen. But since Roriczer himself printed his treatise on his own printing press he could, and probably did, confine its sale to the members of the craft.

*A few years before or after Roriczer's booklet, between 1484 and 1489, a similar but briefer treatise appeared in Nuremberg. It has no title and its author is an otherwise unknown Hans Schmuttermayer of Nuremberg.[53] The author treats not only the same topic but also has almost the same aims. In the initial lines he states that he is writing his book "for the betterment and adornment (*zu Besserung und Zierungen*) of the holy Christian church buildings, to... the instruction of all masters and journeymen who use this high and liberal art of geometry... so that they may better *(bass)* apply their imagination to the true reason of tracery. And not for my own honor's sake but rather to the praise of our ancient predecessors and the inventors of this high art". Although Schmuttermayer must have belonged to the craft it is well established that he was not a member of the masonic guild in Nuremberg where he was born and

---

51 Cennini's *Libro dell'arte*, tr. D.V. Thompson, Jr. (New Haven, 1933); chests p. 170, make up 125, "with my own hand" 1. *The Schedula Theophili* (11th century) is composed by a monk, the sketch-book of Wilars de Honecourt (about 1255) was written only for Wilars' own workshop; in both works progress is not mentioned. For various reasons progress is not mentioned either in a few early modern treatises on the crafts, Ghiberti's *Commentarii* (1477), Leonardo's *Book on Painting* (c. 1496), Biringucci's *Pirotechnia* (1540), and Palissy's *Recepte véritable* (1563) and *Discours admirables* (1580).
52 *Von der Fialen Gerechtigkeit* (How to build turrets correctly), ed. A. Reichensperger (Trier, 1845). The quoted passages in the dedication, p. 13.
53 Reprinted with an introduction in *Anzeiger f. Kunde d. deutschen Vorzeit*, Neue Folge XXVIII (1881), 66-78. The quotation, p. 73.

where his book was also printed. This is possibly an explanation of the fact that he published guild secrets.

While the two architects discussed still belonged to the Gothic style, their younger contemporary, Albrecht Dürer, is a representative of the Renaissance. Yet Dürer also was, of course, like all the artists of the period, a superior artisan. Dürer wrote three treatises. The first, *Unterweisung der Messung mit dem Zirkel und Richtscheit* (Instruction in Measurement with Compass and Rule), printed in 1525, deals with problems of practical and theoretical geometry and perspective. In the dedication to his learned humanist protector, the councillor of Nuremberg Pirckheimer, Dürer states that the German painters lack geometrical instruction. He composed the treatise "to benefit not only painters but also goldsmiths, sculptors, stonedressers, cabinetmakers, and all those requiring measurement". Nobody is forced, he adds as an apology, to use his doctrine. "I know, however, that he who accepts it will not only get a good start but will reach better understanding by daily practice; he will seek farther (*weitersuchen*) and find much more (*gar viel mehr*) than I now indicate".[54] Even more clearly is the idea of progress expressed in his book *On Human Proportion*, printed in 1528. The treatise gives extensive quantitative data on the proportions of the human body. In the dedication to Pirckheimer Dürer remarks that some people might blame him, because he, a non-scholar, teaches a subject in which he received no instruction. Yet, "risking slander", he published the book "to the public benefit of all artists and to induce also other experts to do the same so that our descendants may have something which they may augment and improve, so that the art of painting, in the course of time, may again attain and reach its perfection". Nobody is forced, he adds, to follow his doctrine everywhere, "for human nature has not yet so weakened that another could not invent something better". He goes on to point out the importance of original invention and expresses the conviction that the art will again become perfect "as in olden times"; then the German painters will not be inferior to any other nation.[55] Dürer manifestly put weight on co-operation and progress. In a letter to Pirchheimer he expressly requested the humanist to compose his preface to *On Proportion* so that it would not contain any "talk of glory" *(Ruhmredigkeit)*, and to state that Dürer "begs those having something instructive to say on art, to publish it".[56] Progress, however, is not mentioned in his third treatise, *Etliche Unterricht zur Befestigung der Städt, Schloss, und Flecken* (Some instruction in the fortification of cities, castles, and towns), printed in 1527. In the dedication to the King of Hungary and Bohemia Dürer only states that the treatise was written "to the benefit of your Majesty and other princes".[57] Whether the Habsburg was delighted at the idea that other princes also might learn how to fortify their cities

---

54 Ed. Moritz Thausing, *Quellenschriften z. Kunstgeschichte* III (Vienna, 1872).
55 *Ibid.*, 63 f. Dürer's opinion that classical painting was "perfect" is borrowed from humanism. Of course no classical painting was known to him.
56 *Ibid.*, 61, items 1 and 7.
57 *Ibid.*, 54.

is dubious. Two remarks may be added to the three prefaces. Since, in contrast to Cennini, Roriczer, and Schmuttermayer, Dürer had to reckon also with non-artisans as readers, he apologizes for his writing books as a non-scholar. We shall frequently meet with analogous apologies. And it is, secondly, rather improbable that Dürer had ever read the booklets of the two Gothic architects. The fact that he has the same aim, the progress of craftsmanship, is due to the same sociological conditions, and possibly to oral tradition among the craftsmen rather than to literary influence. This applies also to the authors below. Dürer died in 1528.*

In the sixteenth century kindred ideas were expressed by several craftsmen, the more clearly expressed the better instructed the authors were. In 1547 one Kaspar Brunner, a Nuremberg master of ordnance, previously a locksmith and clockmaker, stated that he wrote his treatise on gunmaking and gunnery "for his generation and others to come".[58] Naturally, his remark is concerned not with fame but the advancement of gunfounding. *Since in the middle of the sixteenth century the technique of gunmaking was still a more or less strictly kept guild secret, Brunner did not publish his manuscript but only presented it in four copies to the council of Nuremberg. About 1400 the poem by an anonymous German master-gunner had stressed secrecy; about 1530 the master gunner of the Duke of Bavaria, Franz Helm, refused, for secrecy's sake, to publish his handbook on gunnery. On the other hand the *Feuerwerksbuch*, composed by an anonymous master gunner before 1425, was printed in 1529 and, in a French version, in 1561.[59] This fading of guild secrecy shows how the idea that technology must be furthered through publication made headway in the sixteenth century.*

In 1578 one William Bourne of Dover in England pleaded the gradual progress of "the arts and sciences" as an excuse. Bourne was an expert on measuring instruments but no longer a member of any guild. Originally an innkeeper, he served as a gunner, did some surveying, and wrote several treatises on gunnery, surveying, and navigation. His spelling is bad but he knew the mathematical and nautical literature as far as it existed in English and understood his subjects rather well. In 1578 he published a booklet, *Inventions or Devices*. It is addressed to "all Generalles and Captains" and for the most part deals with military engineering. As in all his writings Bourne in the preface humbly apologizes for his lack of learning and the imperfection of his inventions. And there the idea of progress appears. "In any arte or science", he states, the first inventions are imperfect; "yet they that came after them brought it into perfection". He expresses his hope that "there may be some further matter gathered of his inventions" and, of course, stresses his "good will to profit the commonwealth".[60]

---

58 *Gründlicher Bericht des Büchsengiessens* (Extensive account of gunfounding), printed *Archiv f. d. Geschichte d. Naturwissenschaften und Technik*, VII (1916). The quoted passage, p. 171 fol. 174 b.
59 Cf. Max Jähns: *Geschichte der Kriegswissenschaften* (München, 1889), I, 382, 384, 408, 591, 608.
60 *Loc. cit.*, Preface to the Reader, about the end. On Bourne cf. B.G.R. Taylor: *Tudor Geography*

In 1581 the London instrument maker Robert Norman, a retired mariner, published a treatise on the dip of the magnetic needle discovered by him. The book describes nautical instruments on sale in Norman's workshop, contains astronomical tables and is intended to further navigation.[61] In a few remarkable sentences Norman discusses "the incredible delight" of the discoverer of new facts at his discoveries. Nevertheless he rejects these personal motives as "private fancies". He is rather convinced, as twenty-five years later Francis Bacon was, that new discoveries profit society. In Norman's opinion scientific discoverers "chiefly respect either the glory of God or the furtherance of some publike commoditie". He considers it his duty to publish the discovery of the dip and to make his name "the object of carping tongues rather than such a secrete should be concealed and the use thereof unknown". "Men that will search out the secrets of their arts and professions and publish the same to the behoofe and use of others", he states, "must not be condemned". Norman hardly felt restrained by medieval guild secrets. He deems it necessary to justify his publication because he was no scholar. *Yet his apologies are characteristic of the early modern era. As we shall see Tartaglia, Peter Apian, and even Descartes deem it necessary expressly to vindicate the publication of their new ideas.*

The two most important pioneers of scientific mechanics before Galileo, the Italian Tartaglia and the Dutchman Stevin, were not artisans. Both were familiar with classical mathematics. Tartaglia published, from Latin translations, Italian versions of Archimedes and Euclid, Stevin a French version of Diophantus. However, they had no academic training but came to science from military engineering and commercial problems. Their writings contain important remarks on scientific progress and scientific co-operation.

Tartaglia (1499-1557) was a self-educated man, the son of a mail coach groom brought up in direst poverty *with little instruction*. He made his livelihood as mathematical adviser to gunners and merchants, at ten pennies a question, and as a mathematics teacher. When his customers would give him a worn-out cloak for his lectures on Euclid instead of the payment agreed upon he had to litigate with them. This remarkable man published many books and treatises in the vernacular on gunnery, military engineering and mathematics, *which greatly influenced nascent modern science and in which, for the first time in history, an embryonic form of the function concept is used to state a physical law.[62]* In the dedication of his *Quesiti et Inventioni* (1546) Tartaglia states the motives of his publications. The man, he points out, who has found something new "and wants to own it for himself alone deserves no little blame. For if all our ancestors had behaved this way we should today little differ from

---

(London, 1930), 153 ff. *Ibid.*, 155, an example of a triangulation carried out by Bourne. The method of triangulation had been invented in 1533 by Gemma Frisius of Louvain.

61 *The Newe Attractive* (London, 1581). Quotations from the second edition (London, 1592), Preface to the Reader about the end, and Dedication to W. Borough, about the beginning.

62 On his youth, *Quesiti et Inventioni* VI, no. 8; "ten pennies (scudi)", *ibid.*, III, no. 10; "function concept", *ibid.*, I, no. 1 (cf. E. Zilsel: 'The Genesis of the Concept of Physical Law', this volume, pp. 96-122; the worn-out cloak: *Travagliata Inventione* (Venice, 1551), sig Fij v.

irrational animals. To avoid this blame I have decided to bring to light my *Questions and Inventions* for everybody".[63] Scientific research is here clearly recognized as a service to the public and the idea of progress is also implied. Tartaglia could not have taken these ideas from the scholars of his period, since they were unknown to them. Though he was not regarded as a real member of the craft by the military engineers and gunfounders with whom he must have been in constant contact, his concept of science reveals their frame of mind.[64]

Sixty years later Simon Stevin (1548-1620) had a similar, though more successful, career. Originally a bookkeeper of the municipalities of Bruges and Antwerp, Stevin turned to military engineering, became technical adviser to Maurice of Nassau, and died as quartermaster-general of Holland. His scientific education was higher than Tartaglia's but he entered the University of Leyden as a man of thirty-five. Like all engineers of his period he was much nearer to manual labor than his modern colleagues. Once he reports how he learned the technical terms of his profession from dikers, carpenters, masons, and metal workers.[65] *He introduced decimal fractions and the parallelogram of forces and first stated the condition of equilibrium on the inclined plane.*

Stevin had the same progressive and utilitarian concept of science as Tartaglia. This becomes evident in the preface to his collected papers on applied mathematics[66]. There Stevin states that he has published the papers in order to make possible "the correction of his errors and the addition of other new inventions profitable to the public". *He adds two other motives. Through publication he also wants to forestall plagiarists and to further the use of the mother tongue in the scientific literature of the great nations.*[67]

Stevin also advocates scientific co-operation. He points out that in the present the experiments are lacking "which are the solid bases on which the arts must be built. For this experience, however, the joint effort and the work of many people

---

63 2nd ed. (Venice, 1554), fol. 4 r. In the dedication of his *Travagliata Inventione* (Venice, 1565), he claims the benefit of Venice as the purpose of his publication. His printer, Troiano, states in the dedication of Tartaglia's version of Euclid (Venice, 1565), that he published the posthumous manuscript "to benefit the world and to make illustrious this author's name" (Tartaglia's? Euclid's?). This is a nice compromise between the modes of speech of the humanists and the craftsmen.
64 Advising gunfounders, *Ques. et Inv.*, I, no. 22 f.; military engineers (architects), *ibid.*, II, no. 9.
65 *Hypomnemata Mathematica*, I, 41 (*Oeuvres Mathém.*, ed. Girard/ Leyden, 1634, II,126). On Stevinus cf. George Sarton, *Isis*, 21 (1934), 241-262.
66 They were published at the same time in the Dutch original (*Wiscontighe Ghedaechtnissen*) and in Latin and French versions (*Hypomnemata mathematica*, *Mémoires Mathématiques*).
67 Without this "the arts and sciences can not reach the perfection of the learned century". - Stevin believes in the existence of a pre-Greek "learned century" in which science florished because it was based on experience rather than on belief in authority (*Hyp. Math.* I, 11 ff.; ed. Girard II, 106 ff.). Only scanty remnants of this golden age are, according to Stevin, extant in the Hermetic literature, but it can be awakened to new life "since human ingenuity has by no means diminished". Stevin is convinced that science can and must be steadily improved. He is still ignorant, however, of the progressive interpretation of history and places the golden age of science in the past. On the existence of the "learned century" he refers to his scholarly friends, the jurist Hugo Grotius and the humanist Joseph Scaliger. From them he also must have picked up the strange esteem for Hermes Trismegistus. His own scientific analyses are strictly mechanistic and entirely free from any influence of occult science - in contrast to all natural philosophers of the period.

are required". As an example he cites astronomy. One man cannot carry out the necessary observations day and night for years, whereas if several observers collaborate, "the error or negligence of the one is compensated by the accuracy of the other". Observations of one observer, however good they may be, are always open to doubt and cannot be accepted as the basis for a theory by other astronomers. Only observations of various observers, if they agree well, can be relied upon. This applies to the observations of the Landgrave William of Hessen and Tycho Brahe. Many observers at many places are required also because the sky is sometimes overcast at one place. There may arise rivalry among the observers, each striving to make his observations best; "but only great advancement for the arts proceeds from this, though ambition also has its pitfalls".

King Alphonso, however, who spent thousands of ducats on his Tables did not achieve much. Yet his effort was praiseworthy. He failed only because his scholars, instead of making new observations, simply followed the hypotheses of Ptolemy. And what is true for astronomy is true for all sciences: science, concludes Stevin, "requires the joint efforts of many people". In the literature of mankind this is the first detailed exposition of the necessity of scientific co-operation. Tycho Brahe (d. 1591), and William IV of Hessen (d. 1592), mentioned in this exposition, were the first in Europe to employ a staff of assistants at their observatories in Denmark and Prague and at Cassel, respectively. It is significant that Stevin can give only astronomical examples although his exposition is quite general; it will be remembered that a certain amount of co-operation had been traditional in astronomy since antiquity. Yet Stevin's appeal for advancement of knowledge and co-operation is, undoubtedly, inspired by rising modern technology and its requirements.

*The founder of modern surgery, Ambroise Paré (chronologically between Tartaglia and Stevin, 1509-1590) had a somewhat similar social position to Stevin. In sixteenth-century Paris there were four groups of medical men: the academically trained medical doctors who wrote in Latin and did not do manual work like operating and dissecting; the "surgeons of the long robe", organized in the Collège de Saint-Côme; the corporation of the "barber-surgeons"; and the quacks who practiced illegally without belonging to any guild. The Collège of Saint-Côme had been founded in the fourteenth century as a corporation of artisans but had in the course of time successfully assimilated itself to the corporation of the doctors, demanding knowledge of Latin from its members and giving up any real medical work. The barber-surgeons remained artisans; they had shops and did shaving, leeching, vevisection, and operating. Yet their corporation took care to transmit, under supervision of the doctors and the surgeons of the long robe, a certain amount of anatomical knowledge to the apprentices. Paré, the son of a maker of strongboxes, was such an apprentice as a youth. In his twenties he worked as a kind of surgical intern at the only Paris hospital, and in the field as a military surgeon. Returning to Paris, he became a master in the guild of the barber-surgeons, and with the help of aristocratic protectors whose favor he had gained in the field, he was also admitted to the corporation of the surgeons of the long robe. He died as a surgeon to the King. In his works, all written in French, his connection with handicraft is conspicuous. He frequently scoffs at doctors "cackling in chairs" and "turning over the leaves of book"; he is proud of the manual work in surgery and emphasizes that he has not learned Latin. Yet he has gained from translations a very considerable knowledge of the classical medical literature. Socially he can be compared to Stevin or the map-maker Ortelius: he was a superior artisan who through his thirst for knowledge ascended to scholarship and entered the service of a prince. Paré invented entirely new methods in the treatment of fractures and in obstetrics, and introduced the binding of arteries to stop bleeding in operations; previously wounds had been burned with boiling oil. Paré's emphasis on experience and originality, his opposition to classical authorities and the overestimation of words deserve an extensive discussion. We shall, however, restrict ourselves to his conception of scientific progress.

Paré wrote all his works for his young colleagues. In his first publication, *La méthode de traicter les playes* (1545), he addresses the preface "to the young surgeons of good will". He has published the treatise, as he states like Dürer seventeen years earlier, "to stimulate superior minds to write on this subject so that we may all have greater knowledge"; and he concludes with the wish that God to whose honor he is writing may ordain "that some fruit and benefit to the support of the weakness of human life" issue from his labor. His *Collected Works*, frequently reprinted, appeared first in 1575. In the preface Paré promises that he will give case histories, "so that young surgeons may take courage to proceed as or, if they can, better than I do (for it is they to whom I address these writings rather than to the scholars)". And he affirms that he has not spent his life in idleness but working "for the republic, always seeking the advancement of the young apprentices of surgery to whom my writings are addressed". His interest in the training of his colleagues obviously derives from medieval craft ideals. Paré had, however, overcome any ideas of guild secrecy. His colleagues had accused him of having with his publications given everybody the means of practicing surgery. To this reproach he replies that he "is extending the gifts given him by God liberally to everybody", wishing "that there may be no one who will not become through my writings much more skillful than I am"; he does not belong to those "who make a cabala of the arts". Thus he can widen his enthusiasm for his craft to the ideal of scientific progress. "The arts", he says, "are not yet so perfected that one cannot make any addition: they are perfected and policed in the course of time. It is sloth deserving blame to stop with the inventions of the first discoverers, only imitating them in the manner of lazy people without adding anything and without increasing the legacy left to us... More things are left to be sought after than have been found". Paré explains this idea at length, cautiously but decidedly assailing the overestimation of classical medicine. God, he states, "did not give us judgement to let it rot and to stop the first outlines of the art drawn by our ancestors". Usually Paré gives the public benefit, the benefit of France, the benefit of the patients, and the honor of God as his goals. He never speaks of glory and literary immortality, although as a good craftsman he is rather proud of his reputation and even of his earnings. He is the most outspoken advocate of scientific progress before Francis Bacon. His progressive statements, however, although they speak of "the arts" in general, relate only to surgery. Once he even complains that men are too much interested in astronomy, which in contrast to anatomy will never overcome the stage of mere conjectures.*

The learned literature on the mechanical arts 1530-1640
We have studied a number of artisans, a highly instructed barber-surgeon, and two men with extensive mathematical knowledge, Tartaglia and Stevin, who were close to the artisans. College graduates have not been discussed. In the sixteenth century, however, *under the pressure of advancing technology the wall* which, since antiquity, had separated the "liberal" from the "mechanical" arts began to gradually crumble. In many countries, particularly in England, a few academically trained scholars began to be interested in technological problems.[68] Their treatises are, for the most part, written in the vernacular since they were intended to be used by navigators, gunners, surveyors, and craftsmen, or, at least, by their employers, the merchants, generals, and princes. It is often hard to determine who in sixteenth-century technological literature gives and who receives the ideas. The mathematical knowledge stems, of course, from the scholars. Actually the artisans mentioned above, William Bourne and Robert Norman, learned mathematics from the works of the scholars Recorde and Digges. On the other hand, the scholars were in the closest contact with manual workers, and several of them even earned their livelihood, for a time or permanently as instrument makers. They had, of course, the utilitarian concept of science in common with the artisans. Otherwise they would have written neither on technological subjects nor in the vernacular. In most of these works * the humanist ideal of fame is no longer mentioned and * it is expressly stated that the author published the book to benefit the public or his country.[69]

A few *anti-individualistic and "progressive"* remarks, however deserve a more extensive discussion. In 1532 the professor of mathematics at the University of Ingolstadt, Peter Apian, a contemporary of Tartaglia, published a

---

68 A list of authors in E. Zilsel: "The Sociological Roots of Science", [this vol. p. 7-21, Eds.]
69 Luca Pacioli's *Summa de Arithmetica* (Venice, 1494), is probably the first book of a learned author to stress practical utility in the preface. The text discusses for the most part problems of commercial arithmetic and contains the first printed account of double-entry bookkeeping. Luca was a monk and mathematics professor at various Italian universities, but must have had close contacts with artist-engineers and merchants. Practical utility is also expressly emphasized: in Recorde's dedication of his textbook on algebra, *The Whetstone of Witte* (1557); in Thomas Digges' prefaces to his English treatises on applied geometry and measuring instruments (*Tectonicon*, 1556; *Pantometria*, 1571); in Jacques Besson's worlds on measuring instruments and machines (*Le Cosmolabe*, 1567; *Theatrum Instrumentorum et Machinarum*, 1578); in John Dee's preface to Billingley's English version of Euclid (1570); in Thomas Hood's inaugural address as mathematical lecturer of the City of London (1588, edited by Francis R. Johnson, *Journal of the History of Ideas*, III, 1942, 94 ff.). All these authors were academically trained. Most of them combine utilitarian with humanistic, Pacioli and Recorde also with scholastic ideas.

Latin treatise on a measuring instrument for astronomers and surveyors of his invention. In the booklet is a defence of invention which somewhat anticipates Baconian ideas. *Those, Apian points out, "who reject the best things because they are new, err, for without new inventions life would return to the state of the ancients who lived lawless and uncivilized like beasts".* Countless still hidden astronomical facts could be brought to light if only we did not fail in our zeal for investigation. Apian, though an admirer of classical antiquity, does not despise the present and is convinced that nature has not grown old and tired and is still able to produce praiseworthy things. In the following year Apian published an enlarged version of the booklet in German since, as he states, he often found better mathematical understanding among the laymen than among the scholars. There he affirms that he has invented his quadrant "to benefit the whole of Christianity and almost the whole world". He also promises further treatises on measuring instruments since, "as the proverb states, I am not born to myself alone but also to those with and after me".[70]

Professor Apian (1495-1522) was a skillful mechanic. After his graduation from the University of Leipzig he made his living for several years as a maker of globes and measuring instruments. As a young and badly paid mathematics professor at Ingolstadt at the same time he ran a printer's shop. Later he became astronomy teacher to Charles V and was knighted for a mechanical planetarium.

In 1570 the learned Dutch map-maker Abraham Ortelius made a not unimportant contribution to the development of scientific method in his atlas, *Theatrum Orbis Terrarum*[71]. Ortelius wanted, as he states in the preface, to benefit the students of geography rather than to aspire to fame through the works of others. He, therefore, gave a list of about eighty cartographers and maps used in his book. *The writings of the scholastics and humanists abound with references to previous authors*. This is the first extensive bibliography in modern scientific literature. Bibliographical lists are not the cornerstone in the building of science. Yet they too manifest the modern idea of scientific co-operation.

In 1595 Ortelius' friend, Gerard Mercator, stated in his *Atlas* that in the composition and arrangement of his work he kept his eye on the benefit of the republic; "we are not born to ourselves alone but the Creator ordered us to live

---

70 *Quadrans Astronomicus* (Ingolstadt, 1532), dedication. *Instrument Buch* (*ibid.*, 1533), dedication and part 5. Apian's polemic against overestimating antiquity is manifestly inspired by Pico della Mirandola. The passage on Nature which is not yet effete is an almost literal allusion to Pliny the Younger, *ep.* VI, 2. This passage was often quoted by Giovanni Pico della Mirandola, his nephew Giovanni Francesco, and other precursors of the *querelle des anciens et des modernes* that developed in the following century. Cf. E. Zilsel, *Entstehung des Geniebegriffes*, 215, 302, 305. On Apian's life, cf. *Allgemeine Deutsche Biographie*, and "Die Apianus Druckerei in Ingolstadt" in *Veröffentlichungen der Gutenberggesellschaft*, Mainz, XXI (1930), 59-82.

71 Ortelius (1527-98) had not academic training but was a good classical scholar. As a map maker he was, together with the engravers, painters, and the renowned Antwerp piano makers, a member of the St. Luke guild in Antwerp. He made his livelihood as a map-maker, dealer in maps and antiquities and became imperial geographer to Charles V (Cf. *Biogr. Nat. Belg.*).

for the common weal". Also the editor of the later editions, the cartographer Hond, emphasized the utilitarian viewpoint. He expressly demands that authors endeavor to benefit the republic and posterity, and blames those who "sing only for themselves and the Muses".[72]

Progress of knowledge was proclaimed as a scientific and philosophical program by Francis Bacon (*Advancement of Learning*, 1605; *Novum Organum*, 1620; *De Augmentis Scientiarum*, 1623). Bacon also emphasizes the importance of scientific co-operation. In his *New Atlantis* (1627), published after his death in 1627, he describes an ideal state ruled by a body of scientists organized, according to the principle of division of labor, in nine groups. Technological and physical laboratories and agricultural stations are at the disposal of these scientists. Such institutes for research were unknown in Bacon's lifetime. There existed only observatories that, of course, were not affiliated to the universities but were establishments of rich scholars (Tycho Brahe) or princes (William IV of Hessen, emperor Rudolph II). All of Bacon's scientific ideas strictly contradicted the ideals of the "seven liberal arts" as taught at the universities. As is generally known, he himself set up the "mechanical arts" as a model to the scientists.

Bacon's ideas have been so often discussed that another discussion would be rather superfluous. Yet it was necessary to show that his idea of progress through co-operation and his utilitarian concept of science appeared before him in the sixteenth-century literature on applied mathematics, navigation, and cartography. It may not be very important which authors were Bacon's literary sources. Since he was not interested in mathematics and quantitative investigation he probably had not read treatises on measuring instruments, though it seems impossible that a scholar at the court of Queen Elizabeth where so much was done for the promotion of navigation should never have got in touch with the scientific tradition of Recorde, Digges, and John Dee.

These, however, are merely biographical questions. What really matters is the sociological origin of Bacon's ideas. Bacon and Stevin did not know each other. Yet Stevin's plea for scientific co-operation appeared a few years before the *New Atlantis*. And Tartaglia and Robert Norman, Apian, and Mercator held utilitarian views on science and deliberately worked for the gradual advancement of knowledge, whether Bacon knew their books or not. Manifestly, the idea of science which is usually regard as "Baconian" is rooted in the requirements of early capitalistic technology; its rudiments appear first in treatises of fifteenth-century craftsmen because the first engineers were artisans whereas the universities had grown out of scholasticism. However, it makes a

---

[72] Mercator's remark in the introduction to the maps of France (4th edition of the *Atlas* [Amsterdam, 1630]), 121. Hond's remark in the preface, *ibid*. The editor of the smaller edition, Petrus Montanus, also stressed the public benefit (*Atlas Minor*, Amsterdam, 1607, pref.). Mercator (1512-1594) studied at the University of Louvain, but entered a geographer's office in Antwerp after his graduation. These offices composed itineraries for merchants since reliable maps were rare, and procured heads of caravans. Often they also made maps and nautical instruments. Later Mercator established his own workshop and became the best instrument maker, surveyor, and map maker of the period (cf. *Biogr. Nat. Belg.*)

considerable difference whether notions are advanced in prefaces and casual remarks or whether they are presented as a philosophical platform to revolutionize the whole of science. Bacon used the ideas of his "predecessors" as a battering ram against scholasticism and humanism and was the first to develop their philosophical and cultural implications. The concept of scientific progress was known before him, the idea of the progress of civilization begins only with Bacon. On the other hand, Bacon fell considerably behind the military engineers and cartographers in the understanding of scientific particulars.[73]

Perhaps the clearest statement of the ideas, the genesis of which we have studied, is found *eleven years after Bacon's death* in the Cartesian *Discourse on Method* (1637). In his matter-of-fact style Descartes here explains why he published "the little that he has found". He wants, as he puts it, "to induce intelligent men to try to advance farther by contributing, each according to his inclination and ability, to the necessary experiences and by also publishing all their findings. Thus the last would start where their predecessors had stopped and, by joining the lives and the work of many people, we would all proceed much farther together than each would have done by himself". Manifestly this procedure*, advocated by Albrecht Dürer a hundred and fifty years before,* was not yet a matter of course in the Cartesian period. It is noteworthy that the quotation is not a casual remark. It concludes a lengthy explanation of the author's reluctance to communicate his ideas to the public. Descartes states that he would never have published his ideas so long as they concerned only the "speculative sciences"; when he, however, saw that his new method would revolutionize physics also he considered its concealment a violation of the law enjoining furtherance of the common weal upon us. With the help of this method, he adds, we shall understand the actions of fire, water, and all other bodies "as distinctly as we understand the various trades of our artisans, and by application of this knowledge to any use to which it is adapted we could make ourselves masters and possessors of nature". Wherever in the seventeenth century the idea of scientific progress and co-operation appears, the application of science to technology is also emphasized. The modern textbooks on the history of philosophy often disregard these Baconian traits in Descartes.

The emergence of scientific organizations 1500-1666

There remains nothing to add but a few remarks on the genesis of the first scientific organizations. The origins and the history of the first learned societies

---

73 In Campanella's utopia, *Civitas Solis* (composed in 1602, published in 1623) there are public museums but no institutions for research. Scientific progress and co-operation are not mentioned. They are not mentioned either in Campanella's interesting account of his scientific aims, *De libris propriis et recta ratione studendi syntagma* (composed in 1632, published in 1642; ed. Spampanato, Florence, 1927), where they would have been discussed if Campanella had known these ideas. In Johann Valentin Andreae's utopia *Christianopolis* (Strassburg, 1619, tr. F.E. Held, New York, 1916) there is a library, an anatomical theatre, a chemical and a physical laboratory, and an observatory. Andreae is, however, not interested in research but in education and does not mention scientific progress. He was a German protestant minister, a friend of Comenius, and a Rosicrucian.

have already often been described, but an exhaustive sociological analysis of the origins still is outstanding. The Florentine Platonic Academy, founded in 1495 by the humanist Marsilio Ficino, had little to do with scientific co-operation in the modern sense.[74] It was not devoted to research but its work, in imitation of Plato's banquet rather than his Academy, consisted of philosophical debates and speeches. Similar meetings of humanistic scholars with educated laymen had taken place since the beginning of the fifteenth century in Florence and other Italian cities. In sixteenth century Italy debating clubs and literary societies of various characters, calling themselves "academies" became suddenly very frequent.[75] Most of them occupied themselves with Italian poetry and literature, some with military science, some of a more scholarly character with the revival of classical antiquity, and a few (the academies of Telesio in Naples and of Porta in Rome)[76] even with speculative natural philosophy. Many of the Italian academies had ludicrous names - the Thirsty, the Idle, the Unstable - and in most of them the members took up equally strange surnames. Most of the academies were decidedly aristocratic associations but just the first literary society to have statute regulations and to introduce surnames - the *Rozzi* in Siena, founded about 1500 - was an artisans' association which gave dramatic performances in the style of the performance in Shakespeare's *Midsummer Night's Dream*.[77] Obviously it was an offspring of the medieval guild tradition. Even inveterate opponents of the sociological method in intellectual history might admit that the mushroom growth of countless associations in sixteenth century Italy, later in Germany, is a sociological phenomenon. It can not be analyzed here in detail. Probably the "academies" were, on the one hand, a reaction to the dissolution through capitalism of the medieval group mindedness. In the feudal period with its great variety of group traditions, clubs, i.e. rational associations, were unnecessary; only the modern era needs them as a supplement to its individualism. On the other hand, the non-scholars, particularly the nobility, had reached a sufficiently high level of education in the sixteenth century to demand intellectual conversation and literary lectures. However, the sociological roots of the literary are even more complex. At any rate the sixteenth century academies are the first clubs rather than the first scientific organizations. We have mentioned them because they manifest a social want that also plays a part in the foundation of the real scientific societies and because the first of these, the Academy of the Lynx, differs from the literary clubs only in degree.

Although several academies published, with or without the authors' names, the poems and literary treatises of their members, in only two of them did the members really work conjointly. The first was the *Aldine Academy* in Venice,

---

74 Cf. Tiraboschi: *Storia della Letteratura Italiana*, new edition, Florence 1807, VI, 103 ff. and Michele Maylender: *Storia delle Accademie d'Italia*, Bologna 1926-30, IV. 294 ff. Ficino's report on the foundation of his academy also reprinted in Plotin, *Opera*, XI ed. Creutzer, vol. I, p. XVII.
75 To the following cf. Maylender, Passim and Tiraboschi, *loc.cit*. VII/1, 139 ff.
76 Maylender, *loc.cit*. V, 295 (Telesiana) and V, 150 (Segreti). The only experiments performed by the *Segreti* were alchemistic.
77 Maylender II, 47 ff.

founded about 1500 by the humanist printer Aldo Manuzio.[78] Its members were not aristocrats but classical scholars and were bound to talk only in Greek at the meetings. Apart from the debates they prepared the editions of classical authors published by Aldo's printing office. Here the co-operative tradition of the workshop manifests itself: printing is manual work and running a publishing house a practical business. This, however, is a side branch in the genealogical tree of the modern scientific organizations. The *Accademia della Crusca* (Academy of Bran) founded in 1582 in Florence, is perhaps even more remarkable.[79] Originally an aristocratic club of the usual type with ludicrous surnames of the members, in the first years the *Crusca* only arranged lectures and carnivals and published witty and paradoxical literary treatises. However, one of the members, the cavaliere Salviati, was not satisfied with mere fun. Upon his instigation the *Crusca* began its famous dictionary of the Italian language, published in 1612. This *Vocabulari degli Academici della Crusca* is, after the astronomical encyclopedia of the Alphonsine Commission, the first scientific publication composed by several authors in collaboration. The individual authors are not given; a committee appointed by the academy devised the plan and supervised its execution. From the collective work, as it was later done in the laboratories of the *Accademia del Cimento* and the *Royal Society*, the dictionary is distinguished not only through the difference in subject matters. These later scientific societies were inspired by the idea of scientific progress and of progress in general; to the members of the *Crusca* these ideas were unknown. They did not compose their dictionary to facilitate philological research to scholars to come but to purify the Italian language; and they chose the past, the language of the fourteenth century, as the model of a correct Italian. If we take the term "practical", however, in a sufficiently wide sense, purification of language is also a practical aim. It is remarkable that in the field of knowledge the merely theoretical desire for insight was not sufficient to produce co-operation. In the humanistic studies co-operation arose rather from the requirements of a publishing house and the desire to influence language; in the natural sciences from the manual work in the laboratories and the utilitarian concept of science.[80]

When the first "archconsul" of the Crusca entered on his duties he had proposed to his fellow members that their academy would gain eternal fame in the whole of Europe. In the records of the later, and real, scientific societies similar announcements are not found. The *Accademia del Lincei* (Academy of the Lynx, Rome 1603-30) was originally an aristocratic club, or alliance, occupying itself with the study of nature, i.e. with Platonizing speculation and

---

78 Cf. Maylender IV, 294-315 and Tiraboschi VI/1 p. 115 ff.
79 Cf. Maylender I, 122 ff. and G.B. Zannoni in *Atti delle Imp. e Reale Accademia della Crusca* I (1819) pp. I ff.
80 The next dictionary to be composed by several authors was the French dictionary of the Académie Française, published in 1694. Its aims are analogous to those of the dictionary of the Crusca. To the following cf. Maylender, *loc.cit*. III, 431 ff.

astronomical observation. It was founded by three young noblemen and a somewhat neurotic medical doctor who later had to be expelled. The four young men, all of them less than twenty-six years old, vowed chastity and scientific fraternity and took up symbolic surnames. Through seven years they were the only members of the academy. When, after 1609, the invention of the telescope created a sensation in Italy the Lincei turned to more empirical studies. Their academy revived and a number of scholars, the greatest of them Galileo, and a few prelates and aristocrats joined as members. Since the foundation the Marques, later Prince, Cesi was the head of the Lincei. All his life he was writing a voluminous book, called Lynceographus, that was intended to be the charter of his academy but was never completed.[81] It manifests enthusiasm for the study of nature but nowhere provides for real co-operative research, and mentions advancement of science only in vague terms. On the other hand, the moral aims of the society and obedience to the church and the secular authorities are emphasized. Cesi's manuscript also provides for branches in all parts of the world, even in America and India, and for four ranks among the members the duties of which are extensively described. All these plans were never realized since the academy ceased to exist after Cesi's death. Cesi's sense of organization was inspired by the aristocratic clubs. When he once was for a time absent from Rome he founded two aristocratic societies of the usual type in the country.[82] Manifestly, however, he was also impressed by the organization of the religious orders. The Lincei had no laboratories and did not experiment but held scientific debates. The four founders had originally obliged themselves once a week to propose "conclusions" to a debate;[83] possibly the young men still had in mind the model of the disputations at the universities. Yet the Lincei later supported the new ideas of Galileo. Publication of a collective work on the plants, animals, and minerals of Mexico was the peak of their scientific activity. The *Mexican Thesaurus* consisted of a posthumous manuscript of the Spanish medical doctor Hernandez which had come into the possession of the Lincei and additions of several members. The scientific and financial preparation of the publication took not less than forty years.[84] Also a few other scientific works were edited by the Lincei, among them two works of Galileo.

The *Accademia del Cimento* (Academy of Trial, Florence 1657-1667) was no longer patterned after the model of the literary clubs. Experimentation was the purpose of the association, it had no charter, the members had no surnames, and

---

81 Maylender, *loc.cit.* III 453-459. An abstract, composed by an unknown member, was published in 1624 under the title *Praescriptiones Lynceae* (reprinted *ibid*.459-503). The freemason-like formula of admission reprinted in Giuseppe Gabrielli: *Le Schede Fogeliane. Acc. naz. dei Lincei. Scienze Morali. Rendiconti* VI (1939), p. 145. It is noteworthy that all these papers are composed in Latin. After 1610 Galileo wrote Italian. Also the publications of the *Accademia del Cimento* and of the *Royal Society* are composed in the vernacular.
82 Convivanti and Stabili, *ibid*. III, 466.
83 *Ibid*. III, 438.
84 It appeared in 1651 when only four Lincei were still alive. It was written in Latin, the title was changed several times, cf. *ibid*. 448-451. *Ibid*. II, 7-16 and Martha Ornstein: *The Role of Scientific Societies in the Seventeenth Century*, 3rd. ed., Chicago 1938, 76-90.

they even overdid co-operation. The scientific tasks were alloted to the members and the experiments were published collectively without giving the names of the experimenters. The *Cimento* grew out of informal meetings of naturalists and medical doctors in the palace of the Grand Duke of Florence, who in his youth had been Galileo's pupil and later, as a hobby, conducted physical experiments and established meteorological stations. Most of the members were pupils or pupils' pupils of Galileo. In 1666 the findings of the academy were published under the title *Saggi di naturali esperienzi fatti nell' Accademia del Cimento* (Trials of Natural Experiments made in the A.d.C.). The members carefully sought to avoid any similarity with the scholastics and natural philosophers and their speculations. Their publication, therefore, refrains from all theoretical explanations and may be regarded as the first manual of laboratory physics. The selection of the experiments, however, is not determined by the needs of contemporary technology, and practical utilization of the findings is mentioned only as an exception. Though the *Accademia del Cimento* existed for only ten years, its scientific work is the first example of real scientific collaboration. It is significant that this work was done in a laboratory: considering also the performance of the *Royal Society* one may call the laboratory the real birthplace of scientific cooperation.[85] Most of the members of the *Accademia del Cimento* were college graduates. The academy, however, had no connection with any university: in the seventeenth century universities laboratories were still unknown.

All of the later scientific societies of the seventeenth century were substantially influenced by the writings and utilitarian ideas of Bacon. The importance and the genesis of the *Royal Society* (established in 1660) are generally known. It developed from informal gatherings of a few college professors and many medical doctors and laymen interested in the natural sciences and their practical applications. Progress of science through systematic co-operation was the aim of the Society. This aim is mentioned in a draft of the statutes, composed by the physicist Hooke, the later Curator of the Society, who was in charge of the supervision of its apparatus and experiments.[86] *It is explained with particular clearness in the introduction to the first issue of the *Philosophical Transactions* (1666), edited by the *Royal Society*. Here Oldenburg, the secretary of the Society, states that the new periodical is being published "to the end that such Productions being clearly and truly communicated, desires after solid and useful knowledge may be further entertained . . . and those, addicted to . . . such matters may be invited and encouraged to search, try, and find out new things, impart their knowledge to one another, and contribute what they can to the Grand design on improving Natural knowledge . . . All for the Glory of God, the Honour and Advantage of these Kingdoms, and

---

[85] The origins of the modern scientific laboratory are too complex to be analyzed here. Alchemists' kitchens and observatories excepted, the first laboratories were established by scholars after the pattern of the workshops of artisans and apothecaries.

[86] Reprinted Ornstein, *loc.cit.* 108 f.

the Universal Good of Mankind". Like Descartes, Oldenburg invites explorers to "contribute" their part to a common design. Knowledge is no longer the business of literati greedy for personal fame and disputing schoolmen: the modern Western concept of scientific research has been reached.[87]* In the first decades application of science to the "mechanical arts" played a major part. The *Academie des Sciences* in Paris was founded in 1666 upon instigation of Colbert who, according to the principles of mercantilism, wanted to further the application of science to industry. As in the *Accademia del Cimento* co-operation in the Paris academy was overdone. The members, who received salaries from the king, were bound to do their scientific work in the laboratory of the Academy and to comply with a prearranged plan; only after reorganization of the Academy in 1699 was greater liberty granted to them.[88] Also in seventeenth century Germany a few scientific societies came into existence,[89] but the Prussian Academy of Science was not founded until 1700. All scientific societies of the seventeenth century were connected by many personal and literary relations. Even the tradition of the *Lincei* was not quite lost in the development of their more mature successors.[90] Although the academies did not arise independently of each other, the establishment of so many similar organizations reveals the existence of an urgent need for scientific co-operation in the seventeenth century. Neither can the mercantilistic policy of the seventeenth century princes who supported the academies of science as means to further industry be overlooked. It is the great merit of Martha Ornstein's book to have verified the fact that the modern scientific organizations - and the modern scientific procedure in general - have developed outside the universities and in a certain opposition to their tradition. Also the close connection of modern experimental science with technology is brought clearly to light in her book. The later development of the idea of progress need not be described here.

---

87 On the first scientific societies and periodicals, cf. primarily Martha Ornstein: *The Role of Scientific Societies in the Seventeenth Century*, 3rd. ed. (Chicago, 1938). Recent additions in: Harcourt Brown: *Scientific Organizations in 17th Century France* (Baltimore, 1934), and "Martin Fogel e l'idea Accademica Lincea", *Reale Accad. Naz. dei Lincei. Scienze Morali, Rendiconti*, VI/11, (1935), 814 ff.; Giuseppe Gabrielli: "Il Carteggio Linceo", *ibid. Mem.*, VI/1 (1925), 137 ff. and VI/7 (1938/39), 1 ff., and "Le Schede Fogeliane", *ibid., Rendiconti*, VI-15 (1939) 141 ff.; Francis R. Johnson: "Gresham College, Precursor of the Royal Society", *Journal of the History of Ideas*, I (1940), 413 ff. Michelle Maylender: *Storia delle Accademie d'Italia*, 5 vols. (Bologna, 1926-30); Robert Merton: "Science and Technology in the 17th Century", *Osiris*, IV (1938), 360 ff. (on the Royal Society). A sociological analysis of the precursors of the modern scientific organizations and of the first books composed by several authors in collaboration will be published elsewhere.
88 Ornstein *loc. cit.*, 139-164. On the numerous informal gatherings of scientists and laymen and the scientific clubs preceding the Academy cf. Harcourt Brown: *Scientific Organizations in Seventeenth Century France* (1620-1680). Baltimore 1934.
89 Cf. Ornstein, *loc. cit.* 165-197.
90 Cf. Harcourt Brown: *Martin Fogel e l'idea Accademica Lincea. Reale Acc. Naz. dei Lincei, Scienze Morali, Rendiconti* VI/11 (1935) 814-833.

CONCLUSION

The comparative analysis of classical antiquity and the early modern era shows that the ideas of scientific progress and co-operation are offsprings of modern capitalistic society insofar as they presuppose the existence of free labor and rapidly advancing technology. This correlation is not self-evident. Also studies like this historical analysis could be made more easily and would turn out better if they could be based on scientific co-operation. Viewed logically, the historians and philologists also could have advanced the idea that knowledge steadily progresses if, and only if, the scholars systematically used the work of their predecessors and co-operated with their colleagues. Actually they did not develop this insight and as far as progress and learned organizations exist in the humanistic studies today they came into existence only because the natural sciences took the lead. The ideal of scientific progress could also have arisen in pure mathematics, but it did not. Man is dominated by group tradition and seems to be a practical animal. Ever since artisans and scholars have existed, apprentices helped their masters in workshops and scholars were proud of their erudition. But the mere desire for knowledge did not suffice to overcome the professional or individualist pride in scholarship. Practical needs were required to change scholars into scientists and to induce them to adopt procedures and ideals from manual workers. Today one can certainly advocate scientific progress and scientific co-operation without thinking of any practical application of the theory. But in their beginnings these ideas appear again and again combined with a utilitarian concept of science. The great number of authors shows that this is not mere coincidence but a real correlation.

We have used sociological methods in our analysis. In the field of intellectual history these methods are so frequently opposed and so seldom employed that it seems expedient to conclude this study with a few methodological remarks.

(1) Sociological statements are statements on the average behavior of sociological groups and, consequently, have only statistical validity. Individuals can always take over ideas from sociological groups to which they do not belong. Benedetti was a court mechanic but took over the fame ideal from the humanists.[91] Sociological statements on individual authors are permissible only if it is verified that their ideas are characteristic of the members of some sociological group or act upon a public consisting of members of some sociological group.

(2) The various sociological subgroups are always embedded in the whole of contemporary society. Artisans differ from literati but the sixteenth century artisans belong to the early capitalistic society, the classical literati to a society based on slavery. Ideological analysis considering only the most general features of the society in which the ideas appear are amateurish: analyses studying only

---

91 Cf. above p. 150 footnote 46.

the professions of the authors superficial. Both procedures will not result in reliable statements.

(3) Since experiments are not feasible in intellectual history, comparison of analogous phenomena is virtually the only way to verify sociological statements. The comparison will yield the more reliable results the less the phenomena compared can have influenced each other. The first beginnings of the idea of progress appear in treatises of the classical military engineer, Philo of Byzantium, and of the fifteenth century master-builder, Roriczer.[92] Both were superior artisans with some mathematical knowledge, both advocated advancement of craftsmanship. This idea remained dormant in classical, and flourished in early capitalist society. *In antiquity, however, they were not taken over by scholars; in early modern capitalism they were. It would be interesting to investigate the Arabic, Persian, Indian, and Chinese literatures. It is not impossible that in these cultures also superior artisans advocated in technological treatises the gradual progress of craftsmanship - if they wrote any. Such ideas were certainly not adopted and developed by the oriental scholars, theologians, and literati. The absence of slavery, the existence of machinery, the capitalist spirit of enterprise and economic rationality seem to be prerequisites without which the ideal of scientific progress cannot unfold.*

(4) Economic facts and needs influence ideas but very indirectly. Philo built catapults; Roriczer, cathedrals; neither war machines nor churches serve economic ends. Large scale scientific co-operation took place for the first time in the laboratory of the *Accademia del Cimento*. But the members of the Academy did not select technologically important problems for their experiments. Apparently, however, economic facts play a greater part when the analysis considers the more general features of society. Both slavery and capitalism are economic concepts.

(5) In intellectual history, as in all history, all processes are influenced by tradition. Any explanation of the ideas of a period through the sociological condition only of this period is deficient since it disregards the past. We surmised that the custom of classical astronomers to compose star catalogues goes back to the astronomical records of the Babylonian priests; if this surmise is correct, the behavior of priest-scholars in an ancient oriental agrarian society had aftereffects in sixteenth century science. However, which traditions continue to exist, which revive, and which fall into oblivion depends on the conditions and needs of the present.

(6) Intellectual history is complex and our knowledge of ideologico-sociological correlations very limited. Progress will be achieved only by making hypotheses and trying to verify them philologically. Methodological bias is of no avail.

---

92 Cf. above p. 142 footnote 33, and p. 134 footnote 9.

PART II

PHYSICAL LAW AND SOCIO-HISTORICAL LAW

9

PROBLEMS OF EMPIRICISM*

I. THE RISE OF EXPERIMENTAL SCIENCE

1. Experiment and Manual Labor

Modern science is accustomed to decide all questions about reality by experience and experiment. This remarkable attitude is not at all self-evident. It is rather a late achievement in the history of mankind, and its import cannot be fully understood as long as that fact is not realized. We shall start our exposition, therefore, with a retrospective view of the rise of empirical thinking and the experimental method. Though at the beginning of the modern era empirical research proceeded from certain empirical achievements of antiquity, classical empiricism can be omitted in this brief survey. In antiquity the empirical sciences were considerably surpassed in intellectual influence by metaphysics and rhetoric, and empiricists always were but a small minority among ancient philosophers and scientists. It will be sufficient, therefore, to begin with the Middle Ages.

The seats of medieval civilization were not towns, which in the early centuries were rare, but monasteries and castles in the country. The castles and the cultural accomplishments of the knights have little bearing on theoretical thinking. The monasteries, being the centers of medieval scholarship, are more important for the present study. Monks, by the conditions of their lives, are not much disposed to look at the world with open eyes. Inclosed within walls, intrusted with the task of transmitting established doctrines to successors by scholastic instruction, they were compelled to indulge in abstract reasoning and to develop their sagacity. This attitude of mind was later taken over by the universities of the late medieval cities. Up to the thirteenth century the method of investigating that appears to be the most natural one to modern science was practically unknown. When medieval theorists, or theologians, intended to solve a problem, they looked first for relevant passages in the Holy Scripture, the patristic writings, and certain works of Aristotle. Then they compared affirmative and negative statements of colleagues and predecessors and, finally, drew conclusions by means of logical deduction from the premises collected. It

---

* [Zilsel, as in his essay 'The Methods of Humanism', makes use of 'supporting evidence paragraphs' which are given in smaller print. The original endnotes have been replaced with footnotes; Eds.]

cannot be a surprise, therefore, that the scholastic theory of syllogism is the chief contribution of medieval science.

By the end of the Middle Ages, however, a few scholars, among them Roger Bacon in England and Albertus Magnus in Germany, had begun to understand the importance of experience.[1] Their contemporary, the French nobleman Petrus de Maricourt, even experimented successfully upon magnets. This rise of new scientific methods in the thirteenth century is connected with a fundamental change in society. Towns had gradually grown up, the rural monasteries and castles were losing their social importance, and a new social class - the townsmen - entered upon the stage of history. Money and profit began to rule the lives of nations.

The period of transition from feudalism to early capitalism lasted until the fifteenth century. At its beginning, the craftsmen of the towns, to prevent economic competition, organized themselves in guilds which took care lest the working traditions of the past be broken. But when, with the strengthening of capitalism, competition proved stronger and destroyed the guilds, guild constraints upon working began to crumble. The craftsman who worked exactly as his master and his master's master had done, was surpassed by less conservative competitors; one's own spirit of enterprise, one's own experience and inventive genius, made the successful man. The age of inventions had begun. Some great inventions of this period - the manufacturing of paper, gunpowder, and guns, the mariner's compass, the printing press - are generally known. The invention of the blast furnace and the stamping-mill, the introduction of ventilators and hauling-engines in mining, numerous improvements in construction of weaving looms, ships, canals, and fortresses are scarcely less important. With the inventions of the fifteenth and sixteenth centuries the technology of the Middle Ages was completely revolutionized. Similar effects were produced by the great geographical discoveries. On new shores animals, plants, and things never seen before were found, which even the most acute monk would not have been able to deduce from his authorities. Authorities and syllogisms had been beaten by experience; a new, empirically minded type of man went out to conquer the world.

In virtue of their occupations these men were craftsmen and navigators belonging to the lower ranks of society. They were not esteemed too highly either by the noblemen or by the bankers and rich merchants of plebeian origin. Since the literati, the humanists, who wrote for upper-class publicity were not interested in such plebeian people as craftsmen, the biographies and even the names of most of the inventors are seldom known. There are few detailed and reliable literary reports even on the great discoverers.[2] The craftsmen and sailors themselves were not literary men but rather uneducated people. Their

---

1 The empirical achievements of Bacon and Albertus are often overestimated (cf. L. Thorndike, *A History of Magic and Experimental Science* [New York, 1923], Vol. II).
2 On the contemporary writings on inventors and discoverers cf. E. Zilsel, *Die Entstehung des Geniebegriffes* (Tübingen, 1926), pp. 130-143.

empiricism, very far from being science, was a matter of practical life and casual observation and, for the most part, was lacking in systematic method.

But gradually, especially in Italy, a more systematic technique of empirical research developed among certain groups of superior craftsmen whose professions required more knowledge than did their colleagues'. We are speaking of the artists, the makers of nautical and of musical instruments, and the surgeons. In the Middle Ages the painters, sculptors, and architects had not been distinguished from the whitewashers, stonedressers, and masons. With the decay of their guilds they slowly separated from handicraft and eventually, about the middle of the sixteenth century, became free artists. During the course of this evolution a remarkable group developed within their ranks. We may call them artist-engineers, for they did not only paint pictures and build cathedrals but also constructed lifting-engines, earthworks, canals and sluices, guns and fortresses, discovered new pigments, detected the geometrical laws of perspective, and invented new measuring-tools for engineering and gunnery. Many of them composed diaries and treatises on their experiences and inventions for the use of their colleagues. As they belonged neither to the Latin-speaking university-scholars nor to the humanistic literati, but were artisans, they wrote their papers in the vernacular. All of them were already accustomed to making experiments. They were the true forerunners of modern experimental science.[3]

Brunelleschi (1377-1446), the constructor of the cupola of the cathedral of Florence, was the first of these artist-engineers. Among his successors are the bronze founder Ghiberti (d. 1455), the architect and painter Lione Battista Alberti (1407-1472), who - as an exception - had had a classical education, the architect and military engineer Giorgio Martini (1425-1506), Leonardo da Vinci (1452-1519), and, finally, the goldsmith, sculptor, and gun constructor Cellini (1500-1571). The architect and gunnery expert Biringucci (d. 1538) may be mentioned because of his book on metallurgy, *Della pirotechnia*. His paper is the first treatise on chemistry based on sound experience and avoiding any alchemistic superstition. The inventors and constructors of the new clavicembali and harpsichords, and the compass-makers also belonged to the experimenting superior craftsmen. A third group was formed by the surgeons. They dissected animals and often human bodies, whereas the learned physicians seldom dared to do such untidy and sinister work. By their common interest in anatomy there were often connections between artists and surgeons.

Experiment requires manual work, and, therefore, in both antiquity and the modern era its use began in handicraft. In antiquity scientific experimentalists were extremely rare. Since rough work was generally done by slaves, contempt of manual labor formed an obstacle that only the boldest of ancient scholars dared to overcome. A similar obstacle, though by scarcity of slaves less unsurmountable, obstructed the rise of experimental science in the modern era.

---

3 On the artist-engineers and their papers cf. Leonardo Olschki, *Geschichte der neusprachlichen wissenschaftlichen Literatur,* I (Heidelberg, 1918), 45-447, and Zilsel, *op. cit.*, pp. 144-157.

The educational system under early capitalism took over from the Middle Ages the distinction between liberal and mechanical arts. In the seven liberal arts (grammar, dialectic, rhetoric, arithmetic, geometry, astronomy, and music) thinking and disputing were alone required; on them alone was the education of well-bred men based. All other arts, as requiring manual work, were considered to be more or less plebeian. There are numerous instances to indicate that up to the sixteenth century even the greatest artists of the Renaissance had to fight against social prejudice. And by the same reason the two components of modern scientific method were kept apart: methodical training of intellect was reserved for university-scholars and humanistic writers who belonged, or at least addressed themselves, to the upper class; experiment, and to a certain extent observation, was left to lower-class manual workers. Even the great Leonardo, therefore, was not a true scientist. As he had never learned how to inquire systematically, his results form but a collection of isolated, though sometimes splendid, discoveries. In his diaries he several times discusses problems erroneously which he had solved correctly years before. Gradually, however, the technological revolution transformed society and thinking to such a degree that the social barrier between liberal and mechanical arts began to crumble, and the experimental techniques of the craftsmen were admitted to the ranks of the university-scholars. Rational training and manual work were united at last: experimental science was born. This was accomplished about 1600 with Galileo Galilei, Francis Bacon, and William Gilbert. One of the greatest events in the history of mankind had taken place.

The scientific work of Copernicus (d. 1543), being rational not experimental, may be omitted here.

Galileo (1564-1642)[4] got his education at the University of Pisa and for more than twenty years he was professor at the universities of Pisa and Padua. His relations to handicraft and technology, however, are often underrated. During his student days there was no mathematical instruction at Pisa.[5] He learned mathematics privately, his tutor, Ostilio Ricci, being an architect and teacher at the Accademia del disegno which had been founded in 1562 by the painter Vasari as something between a modern academy of arts and a technical college.[6] Thus Galileo's first mathematical education was directed by persons who were artist-engineers. As a young professor in Padua he lectured at the university in mathematics and astronomy and gave private instruction on engineering in his home. For his experimental studies he established working-rooms in his house and hired craftsmen as assistants.[7] This was the first university laboratory in history. His scientific research started with studies on pumps, the regulation of rivers, and the construction of fortresses. Ever since his student days he had liked to visit dockyards and arsenals. His first printed publication (1606) described a new measuring-tool for military

---

4 On Galileo cf. Olschki, *op. cit.*, III (Halle, 1927), pp. 117-467.
5 Galileo, *Opere* (ed. nazionale), XIX, 32 ff.
6 Before the foundation of the Academia del disegno, young artists had always received their education at the workshop, like all apprentices. The new school clearly shows how engineering gradually penetrated the province of academic instruction.
7 Galileo, *op. cit.*, XIX, 130 ff.

purposes. Even in his last work of 1638, which initiated modern mechanics, the setting of the dialogue is the arsenal of Venice. His greatest achievement, the discovery of the law of falling bodies, also originated in connection with the contemporary technology. Among the gunnery experts of the period there were many discussions on the shape of the trajectory. Galileo realized that the question could not be answered before the problem of falling bodies was solved. Free falling, however, was too fast to be measured exactly. In order to slow down the movement, Galileo took brass balls, made them roll down an inclined groove, and measured the spaces, times, and velocities. He succeeded in correlating his results by means of a mathematical formula and finally determined the shape of the curve of projection. In Galileo's classical inquiry the two pillars on which modern science is based stand out: experiment and mathematical analysis. The experiments of the craftsmen, alone, would never have issued in science.

Francis Bacon (1561-1626) did not make any important discovery in the natural sciences. His writings abound with magical survivals and errors; he did not understand well the achievements of Copernicus, Galileo, Gilbert, and Harvey; the methodological prescriptions he gave in order to promote empirical research were too pedantic to be of great use to scientists. But he was the very first writer who fully realized the importance of science for human civilization. The scholastics and humanists, he explained, have only repeated sayings of the past. Only in the mechanical arts has knowledge been furthered since antiquity. Bacon, therefore, spoke of craftsmen and mariners with enthusiasm and proclaimed their works as models for the scholars. He is an enthusiastic advocate both of scientific induction and of the ideal of progress. Both ideas are closely connected: they are nothing else than the working method of early capitalist handicraft seen with the eyes of a philosopher. His whole philosophy is one great attack against the ideals of the seven liberal arts. Bacon himself performed numerous experiments: he died from a cold which he caught while stuffing a dead chicken with snow. Most of his learned contemporaries probably considered that experiment more fitting for a cook or flayer than for a former lord chancellor of England.

The first learned book of the modern era on experimental physics had appeared before Galileo's and Bacon's publications. It was William Gilbert's *De magnete* (1600). Gilbert was physician to Queen Elizabeth I. His experimental method originated partly in contemporary metallurgy and mining, partly in the experiments of the retired mariner and compass-maker, Robert Norman.[8]

Why is experiment so essential to empirical science? Mere observation is a passive affair. It means but "wait and see" and often depends on chance. Experiment, on the other hand, is an active method of investigation. The experimenter does not wait until events begin, as it were, to speak for themselves; he systematically asks questions. Moreover, he uses artificial means of producing conditions such that clear answers are likely to be obtained. Such preparations are indispensable in most cases. Natural events are usually compounds of numerous effects produced by different causes, and these can hardly be separately investigated until most of them are eliminated by artificial means. There is, therefore, in all empirical sciences a distinct trend toward experimentation. Sciences in which experiment is not feasible are handicapped. They try to solve their problems by referring to other sciences in which

---

8 Cf. "The Origins of William Gilbert's Scientific Method" [this volume pp. 71-95, Eds.].

experiments can be performed. Thus meteorology, geology, seismology, and astrophysics make use of laboratory physics and laboratory chemistry. Sociologists and economists attempt to utilize results of psychology. A few modern geologists have even attempted to imitate formation of mountains on a diminished scale in laboratories. To a large extent the poor results of the social sciences might be explained by lack of experiments. The only substitute which, under favourable conditions, can to a certain degree compensate for the lack of laboratory experiments is the use of a great number of observations when carefully compared and worked up by means of statistical methods. Until now, however, experiment has been by far the most efficient empirical method.

It is noteworthy that one of the oldest empirical sciences, astronomy of the solar system, was highly successful without any experiment. This is due to the extraordinary fact that in our solar system superimposed effects belong to very different orders of magnitude and therefore can be separated comparatively easily. The solar system is exceptionally well isolated and the sun surpasses by far all planets in mass. Were the solar system continually bombarded by heavy meteorites or, what is the same, were it passing through a dense star-cluster, and were Jupiter's mass of the same order of magnitude as that of the sun, Copernicus, Kepler, and Newton would not have achieved much.

2. Causal Research: The Mechanical Conception of Nature
The young experimental science was forced to fight hard battles with pre-scientific thinking. Primitive man does not distinguish exactly between animate and inanimate objects; he apprehends all natural events as if they were manifestations of striving, loving, and hating beings. This animistic conception of nature is predominant in all civilizations without money economy and dominated medieval thinking too. When a comet appeared in the sky or a monster was born, medieval man questioned rather the meaning, the aim, and the purpose of these events than their causes. The scholars did not think very differently either. Certainly "entelechies" and "substantial forms" of Aristotle and the Scholastics are not primitive ideas; they are complicated and highly rational constructs. Yet their animistic kernel has, as it were, only dried up; something like a soul, striving to reach its aims, still glimmers through the rational hull. The same holds of the "occult qualities" that were liked so well by the Scholastics. These could never be observed but were supposed to adhere to most objects and to produce effects by sympathy and antipathy, as if they were little ghosts. Animistic survivals like these were of no use to modern technology and had to be cleared away. Teleological explanations were gradually replaced by causal ones. Purposes of inanimate things, the meaning of natural events, and soul-like powers of physical objects cannot be ascertained by observations. On the other hand, the regular connection of cause and effect is testable by experience and experiment. Moreover, engineers are able to produce the effects they want if they know the cause. Causal explanation, therefore, became the chief aim of experimental science.

The discarding of teleological explanations may be illustrated by two well-known instances. The Scholastics, with Aristotle, explained the falling of bodies by the theory of natural places. Each body was supposed to have its correct place to which it moved when it had been brought to a wrong one. Obviously inanimate bodies were conceived as though they were cattle striving to the accustomed stable. As the theory of natural places did not give any information on the empirical details of falling, it was of no use to the artillery men of the modern era who wished to level their guns correctly. It had to be replaced by Galileo's law of falling bodies and his calculation of the parabola of projection.

The working of suction pumps was explained in the late Middle Ages by the doctrine of *horror vacui*. Water was supposed to rise in pump barrels because nature had an antipathy to empty space. Since the well-diggers of the new era could not calculate from this theory how long they might make their pipes, two pupils of Galileo, Viviani and Toricelli, experimented on pipes filled with mercury and discovered and measured atmospheric pressure.

The causal mode of investigation gave rise to a basic, and previously unknown, conception. In a well-governed state there are laws which are prescribed by the government and, for the most part, observed by the citizens. Lawbreaking is rare and is punished when detected. Let us now suppose the government to be omnipotent and the police to be omniscient. In this case laws would always be observed. The seventeenth century began to compare nature with such a perfect state, ruled by an almighty and omniscient king.[9] Thus the recurrent associations of natural processes were named natural "laws" by the scientists who investigated them - especially if they had succeeded in expressing the regularities by mathematical formulas. The term "law" became so common that people soon began to forget that it originated in a metaphor; the idea arose that *all* events, without exception, were subject to natural laws. This deterministic conviction dominated philosophy and science from Descartes, Hobbes, and Spinoza, and from Galileo, Huyghens, and Newton almost up to the present time. It impelled all naturalists to look for more and more natural laws and thus proved to be extremely fruitful.

Nevertheless, we must distinguish the empirical and the metaphysical and theological components in these ideas. The statement that there are regular connections between certain events or qualities undoubtedly is an empirical one, for such connections are observable. The case is different with the assertion that events are connected not only as a matter of fact but that, moreover, some necessity subsists, forcing one event to follow the other. Necessity never is observable; it transcends the province of experience. Metaphysical and

---

9 In fact, the metaphor is not older. In scholasticism the term "natural law" had an entirely different and merely juristical and ethical meaning. The Middle Ages could not produce the modern concept of natural law for the simple reason that the feudal state was governed not by statute but by unwritten and loose traditional law. As far as medieval princes issued regulations at all, they were for the most part privileges given to single noblemen, monasteries, and towns. They compare, consequently, rather with exceptions than with regularities in nature. Also with the ancient authors the metaphor of natural law appears but seldom and in a rather vague form. Antiquity, however, was familiar with the mythological idea of "fate".

theological additions become the more marked when necessity is interpreted as the order of a personal deity or of impersonal nature. But even without drawing in necessity, difficulties emerge and experience is transcended when it is asserted that an observed regularity will *always* hold. Obviously no statement speaking of more than a finite number of objects and containing the term "all" can be completely verified by observation. All these unempirical components were ascertained and criticized by Hume. Before Hume no scientist and no philosopher conceived natural laws merely as empirically ascertained regularities, and even nowadays the idea of lawful necessity has not entirely vanished.

The logical difficulty just mentioned is repeated on a higher level in the thesis of general determinism. What exactly does it mean to say that regularities hold "everywhere" or, which is the same thing, that there is "no" fact that is not subject to some regularity? In mathematics, if any finite number of cases is given, an equation covering the given cases can always be found. Does general determinism maintain only this analytic proposition or does it maintain more? When the nineteenth-century determinists spoke of regularities and laws, they had in mind the equations of classical mechanics. Yet those equations have since then proved inadequate. Plenty of physical facts became known in the last decades which are covered neither by mechanical equations nor by equations of a similar type. Nevertheless,it is instructive to observe how well determinism turned out - up to the twentieth century at least. Even vague and dubious assertions can render good services to empirical research as a heuristic stimulus.

As we have indicated, determinism for the most part was conceived in a special form. Up to the late nineteenth century nature was interpreted as a gigantic but lawfully functioning mechanism. Since movements, pressing, pushing, and pulling are the only essential factors in mechanical devices, in nature, as well, all processes and qualities were reduced to such movements. All other qualities, though comprising the greater portion of everyday experience - as colors, sounds, and smells - were not regarded as "real" ones. They were interpreted as "illusions", and no scientific explanation and no natural law was considered to be definitive until it was reduced to laws of mechanics.

The mechanical conception of nature began as early as Galileo. It appeared in almost all philosophers from Descartes to Kant and dominated physics from the beginning of the seventeenth to the end of the nineteenth century. As is generally known, mechanical theories of sound, heat, light, and electromagnetism were constructed in complete detail. Such theories were not completed in chemistry and remained only programs in the provinces of smell and taste.

How can the rise of this remarkable conception be explained? Man himself is, in a certain respect, a mechanical being. All actions by which he influences the world around him consist in movements, in pushing and pulling. From the days when he learns as a baby to control his limbs, he regards that way of reacting as the only natural one. Unless he happens to be a physiologist, the more subtle chemical and electrical processes within his muscles, nerves, and bowels are unknown to him. Small wonder, therefore, that, when he looks at the external world, he takes for granted the movements, pushing, and pulling that he finds there. On the other hand, he feels that all processes differing from his own actions require further explanation and tries to reduce them to mechanical ones. An electric eel, gifted with intelligence, might behave in an analogous manner. Imagine an animal of this kind, not only defending itself and attacking by electric shocks but also attracting prey by electromagnetism, splitting and absorbing its food by electrolysis! If it were a physicist, such an animal would probably look for electromagnetic explanations everywhere. To summarize, then: the mechanical conception of nature is anthropomorphic and interprets natural processes after the pattern of human actions.

Unfortunately, more facts seem to be accounted for by this biological explanation than actually prove to be correct in history. Though men have always had a like biological organization, mechanistic theories are met with only very seldom and in very few civilizations. Mechanistic physics appeared only with ancient atomists and Epicureans and in the period we are just discussing. Obviously, some additional conditions are required if theories of this type are to develop. Apparently, the state of the contemporary technology can give the additional explanation we need. Man of the seventeenth century had outgrown primitive animism and begun to see the world with the eyes of an engineer. But since the only machines with which the period was acquainted were, without exception, mechanisms such as printing-presses, weaving-looms, pulleys, and clock-work, it was rather natural to think that the whole world was a mere mechanism as well. This prejudice was shaken only when technology had entirely changed, and nonmechanical engines had become more frequent than mechanical ones. This happened in the late nineteenth century.

The scientific value of the mechanist conception of nature cannot be overlooked. Distances and shapes, pressure and movement, can be measured comparatively easily: that is to say, can be co-ordinated to numbers without great difficulty. On the other hand, qualities such as blue and cold never, themselves, appear among the results of any calculation, and their measurement is considerably more complicated. So reduction of all qualities to mechanical ones enabled the physicists to co-ordinate numbers to qualities, and this again made it possible to grasp the physical world by mathematics. By the mathematization of physics, the building-up of deductive theories was immensely furthered. We need not point out how philosophical rationalism was stimulated by this development, as these achievements belong with an account of the rational side of knowledge. The heuristic value of the mechanist conception, however, was not slight. Everyone who knows the history of physics also knows how many new empirical facts were discovered and causally explained for the first time by means of mechanistic models. Both the rational and the empirical value of the mechanical conception of nature cannot be doubted. Yet, being based on a prejudice, that conception had its dangers as well. Distinction between a "real" world of mechanics and an "illusory" one of qualities has produced a confusion of concepts that has interfered with the analysis of knowledge and philosophy for almost three centuries. This will be discussed in the next section.

II. THE PHILOSOPHY OF MODERN CLASSICAL EMPIRICISM

3. The Opposition of Outer and Inner World
After natural science had adopted the experimental method from the craftsmen, it developed rapidly during the seventeenth century. It was engaged in its special questions and, consequently, does not offer many problems to epistemological investigation. The case of contemporary philosophy was different. The most remarkable fact in empiricist philosophy of the classical periods might be said to be the tendency gradually to transfer its interest from objects to the subject. By this remarkable tendency it soon became entangled in pseudo-problems. Medieval animism and the Aristotelianism of the Scholastics were scarcely overcome when a new metaphysics developed, originating in the contrast of object and subject, of outer and inner world.

With Francis Bacon (1561-1626), interest in objects still prevailed; subjective components of knowledge are only mentioned in his doctrine of idols. Bacon is convinced that knowledge gives a rather accurate picture of nature if only we take care in avoiding a few errors and prejudices. These fallacies are induced partly by social, partly by individual, conditions by which the judgment of the observer is disturbed. They can, however, be eliminated if attention is called to them. As is generally known, such biases are called "idols" by Bacon and are discussed by him in an entirely empirical way without involving metaphysics. The beginnings of the new subject-object metaphysics appeared in empiricism with Hobbes (1588-1679).

Hobbes was a radical mechanist. Since in his opinion all processes consist in movement, he was faced with the task of explaining the origin of all other qualities that are perceived by us. He tried to master the difficulty by his terministic theory of sensation. In his epistemology sensations are distinguished from objective qualities. Sensations are not at all copies of the physical world but only correspond to physical qualities as symbols or terms do to their objects. Plato has already duplicated the world by opposing the realm of Ideas to the world of phenomena. Platonic Ideas, however, are extremely vague constructs, originating chiefly in certain ethical considerations and in the philosophy of mathematics; they were scarcely designed to help naturalists in interpretation of their scientific observations: the metaphysical background of the Platonic Ideas is obvious. Into natural science and empirical philosophy the two-worlds theory was introduced by Hobbes. Although his terministic theory of sensation contains valuable elements (which were not, however, developed until the time of the modern logic of relations), his distinction between the "real" world of movements and the subjective world of qualities has entangled the philosophy of nature in pseudo-problems for more than two centuries. They are the more serious as they are inevitable implications of the mechanistic interpretation of phenomena. As long as the latter was considered the only possible philosophy of nature, the pseudo-problems connected with the contrast of external world and subjective awareness could scarcely be eliminated.

The decisive step that changed empiricist epistemology into an introspective psychological theory of knowledge was taken by Locke (1632-1704). As generally known, Locke's theory of knowledge reduced all statements, both in everyday life and in science, either to sensation or to reflection. It belongs to the character of his philosophy that he spoke rather of ideas and sensations than of facts, observations, and statements. Modern logical empiricism restricts itself to analyzing the methods by which statements are tested and verified, whereas empiricism of the seventeenth century proceeded psychologically and investigated how ideas are obtained. Locke's polemic against rationalism became a psychological analysis of abstract ideas, on the one hand, and an attack upon innate ideas, on the other. Since innate ideas played a rather prominent part in Descartes's epistemology, this attack is of considerable historical importance in the further development of empiricism. But Locke soon turned to a discussion of the mind of the newborn child and revealed the concept of soul as the metaphysical background of all his psychological analyses. Mechanistic philosophy was considered so self-evident by Locke that its physical implications were not even pointed out by him; it appears only as his well-known distinction of primary and secondary qualities. The metaphysical background of his psychology of knowledge becomes most obvious in his exposition of the idea of substance. Substance to Locke was the "unknown support" of qualities. Although in his opinion that support is, and ever must stay, "unknown to us", he did not hesitate to distinguish material and spiritual substances and to raise the problems of their interdependence. Locke, the empiricist, accepted the metaphysical dualism of Descartes without question and discussed an interdependence, never testable by experience, between constructs that could not be tested either. The outstanding scientific and historical importance of Locke's analyses is generally known; but the metaphysical pseudo-problems, which are connected with them, must not be overlooked.

The contradiction between Locke's empiricist principle, and his theory of substance was noticed by Berkeley (1685-1753). Berkeley adopted the principles and the introspective method from his predecessor but rejected Locke's theory of substance. He started with a criticism of Locke's theory of abstract ideas. Since he found by psychological introspection that he was not able to form, for instance, the idea of a color which is neither red nor blue nor green, he rejected Locke's analysis and flatly denied the subsistence of abstract ideas. By leaving differences out of consideration and by marking common properties only, one word can obviously represent quite a group of different objects. Thus abstract ideas may and must be replaced by abstract words. Applying this analysis to the concept of substance, Berkeley concluded that substance is a mere word, void of content. The unknown "support" of qualities, which had been assumed by

Locke, although it could never be perceived, Berkeley thought could, and should, be eliminated: matter, being an abstract construct, does not exist at all. Actually there are perceivable qualities only, and these are nothing else than perceptions. Thus Berkeley's well-known equation, *esse = percipi*, resulted.

The only point that strikes us in Berkeley's remarkably consistent argumentation is that he failed to apply his analysis to the concept of mind or soul. Can souls ever be experienced? Are they not abstract supports of perceptions or qualities just as matter is? No doubt it was Bishop Berkeley's religious attitude that prevented him from drawing this dangerous consequence. Avoiding the term "idea", he always spoke of the "notion" of soul and would not admit that it is no less abstract than the idea of matter. Thus he built up his odd metaphysical system, constructing the world out of souls, ideas, and God. The tendency in empiricism both of turning to subjectivism and of getting involved in pseudo-problems has reached its peak in Berkeley.

Why did English empiricism tend to introspective psychology and how could it be invaded by subject-object metaphysics? Both in ancient and in medieval philosophy only slight beginnings of analogous ideas are to be found. The modern dualism of inner and outer world might be correlated with the dualism of soul and body and probably can be explained by the influence on mechanistic physics of theology. Belief in immortal souls and the mechanical conception of nature had never been united in one philosophical system before the seventeenth century. In antiquity the atomists and Epicureans were mechanists, but they did not believe in spiritual substances; Platonists and Stoics, on the other hand, distinguished souls and bodies but were not mechanists. Medieval theologians were not compelled to stress a chasm between spiritual and physical substances, since in their philosophy all physical objects were imbued with more or less soul-like powers; they could content themselves with Aristotelian entelechies which, being forms, were rather connected with than opposed to matter. Actual and radical dualism was introduced into philosophy by Descartes, mechanist and devout Catholic. It is comprehensible that, apart from direct Cartesian influence, similar tendencies invaded English empiricism as soon as it turned to mechanism. In their business of investigating phenomena, natural scientists are faced with the task of separating constant relations, on which all observers can agree, from the variable and unstable aspects which are offered under different conditions or to different observers in a different way. This is the sound basis for the distinction between objects and subjects. Bacon, for instance, did not misuse it. But this separation was misinterpreted as soon as the mechanistic philosophy of physical phenomena separated "real" qualities, such as motions, from merely "apparent" ones, such as colors. The rationalistic mechanist Descartes became guilty of the misinterpretation as did the empiristic mechanist Hobbes. Obviously, it was the rise of mechanistic physics that turned the harmless distinction between subjective and objective components of observation into a dualism of inner and outer world. And it is rather comprehensible that, under the influence of religious tradition, this dualism was more or less identified with the contrast of soul and matter. Thus analysis of experience turned with Locke to introspective psychology - that is, to investigation of soul. Experience became a psychical duplicate of the "real" external world, and pseudo-questions arose: whether both halves of the world actually exist and how it happens that they correspond.

4. David Hume

Locke's and Berkeley's principles were consistently applied to the concept of spiritual substance by Hume (1711-1776). As is generally known, he eliminated substantial souls and replaced them by heaps or bundles of impressions. Berkeley's philosophy, which had denied the existence of physical objects but approved the existence of souls, seemed paradoxical to common sense. To critics entangled in the problems and language of the period, Hume's position of rejecting spiritual substances as well as physical ones seemed, of course, even more paradoxical. Actually the paradox was rather diminished by him, for he treated objects and subjects in the same way and thus approached common-sense realism again. His analysis is a most important step toward overcoming the pseudo-problems of a subject-object metaphysics. In Hume's time, however, theological ideas and the mechanistic interpretation of physical phenomena still blocked the way to a complete understanding and elimination of such pseudo-problems. Moreover, his analysis was impaired by his dealing with impressions and ideas rather than with statements. This predilection for elements of knowledge which are too minute to be very useful in analysis had originated with Locke and can be traced in the development of empiricism up to Mach and the nineteenth century. The same phenomenon occurs when Berkeley, for instance, as he often does, turns philosophical criticism to criticism of language. In this case he is discussing words rather than statements. This noteworthy but prejudicial predilection apparently also originated in the psychological attitude of classical empiricism, for statements are too logical to attract the attention of psychologists. Analysis of experience shifted from ideas to statements only when axiomatic methods in mathematics of the late nineteenth century had roused interest in statements and their concatenation, and when non-Euclidean geometry and modern symbolic logic began to influence empiricism. With rationalists and with Kant, on the other hand, statements always had played an important part.

Among the most important achievements in the history of philosophy is Hume's analysis of the concept of causality. With surpassing clarity he showed that we speak of cause and effect if phenomena are connected regularly with one another. Since cause and effect are not linked by logical necessity, even the most common causal statements of everyday life depend entirely on past experiences and never can be inferred a priori. Induction, therefore, differs radically from deductive logic and is based psychologically on custom and belief. Even today the man in the street is inclined to conceive cause as a thing that by its activity produces another thing. Hume's criticism destroyed this idea of active production - that last remainder of primitive animism. Therewith it is implied that not things but processes, qualities, and relations alone are causes. And, by stressing regularity of connection, he adapted the concept of cause to the concept of natural law, the very form in which it is used in any mature empirical science. His theory of causality would not have been possible in a period in which laws were not yet known, and it was obviously inspired by Newton's and his followers' successful investigation of physical laws. Moreover, it made feasible the future development of physics. Without Hume's concept of cause, nonmechanistic physics of the late nineteenth and the twentieth centuries scarcely could have come into existence. Among later philosophers, however, his analysis has met with considerable criticism, opposition, and - even worse - complete neglect. As to induction, later philosophy has made little progress beyond Hume's remarks.

During the eighteenth century, empiricism, by emphasizing sensations, turned in France to sensationalism and, with many philosophers, to materialism. On the other hand, empiricist philosophy of the seventeenth and eighteenth centuries developed important methods and results of psychological investigation. The laws of association were discussed again and again; contemporary reports on customs of primitive peoples were used by Locke in his polemics against innate ideas, and the anthropological method was successfully developed and applied to the investigation of religion by Hume; the psychology of people with defective senses was occasionally referred to in Berkeley's *Theory of Vision* and was considerably furthered in Diderot's *Letters on the Blind* (1749) and *Letters on the Deaf-Mute* (1751). For the first time perception of space was investigated psychologically, and the visible, tactual, and muscular components of the experience of space were more or less exactly distinguished in Berkeley's *Theory of Vision*.

III. THE ADVANCE OF EMPIRICAL SCIENCE

5. Religious Problems

The last remarks have brought us to that field in the evolution of empiricism which is the most fertile one - the special sciences. Only the general expansion of the domain of empirical research can, however, be discussed here.

The empirical spirit of the modern era is entirely contrary to the spirit of medieval scholasticism and, consequently, was compelled at first to overcome the resistance of theological tradition. The rise of empiricism, therefore, is connected, though not identical, with the spreading of the Enlightenment. The beginnings of religious Enlightenment, that is, the ideas of natural religion (Herbert of Cherbury, 1624), were based rather on rational construction than on empirical investigation of the various systems. Little attention was paid by Herbert and his followers to the variety of modes of worship; priests and ceremonies were rather disliked by them. As is generally known, the main theses of natural religion stated the existence of God, immortality of the soul, and reward and punishment in the other world. They were primarily products of abstract reasoning and were based on empirical comparison only in so far as they expressed the components common to the three great monotheistic religions - the only religions well known in this period. Obviously, the articles of natural religion indicate the common content of Judaism, Christianity, and Mohammedanism. Yet the articles were void of emotional content and do not occur, as they were formulated by Herbert and his followers, in any living religious system. The religious attitude of most of the empiricist thinkers of the seventeenth and eighteenth centuries was, however, more or less determined by these rather lean ideas.

It is remarkable that the first empirical contribution to scientific investigation of religion was made in the modern era by one of the most consistent rationalists - by Spinoza. In analyzing the worldly literature of antiquity, the humanists of the Renaissance had already created and developed the methods of philological criticism. Hobbes, in his *Leviathan* (1651), had advanced a few critical remarks on the composition of the Old Testament. Spinoza, however, was the first who, in his *Tractatus theologico-politicus* (1670), dared to apply, consistently, philological methods to the Old Testament. If painstaking philological criticism is considered an empirical achievement, Spinoza's successful attempt to determine the time of composition of parts of the Old Testament may be reckoned with the major advances of empiricist thinking. The next step in this field was taken by Hume. Hume's dissertation on the *Natural History of Religion* (1753) tried for the first time to give an empirical theory of general religious development and started comparative investigation of primitive religions. Hume, however, knew for the most part only the religious ideas of a few primitive nations, of ancient Greece and Rome, and, of course, the three great monotheistic systems. Actual knowledge of the great religious systems of India and China came to Europe not before the early nineteenth century. In the later nineteenth century the empirical science of religion rose considerably, using the philological and anthropological methods of Spinoza and Hume as well as modern psychological and sociological ones. Obviously, empirical thinking could not have spread in the field of religious and anthropological research were it not for the expansion of world-trade and the colonial system that made the white race acquainted with exotic civilizations. In Locke and Hume these connections had already become visible.

6. The Natural Sciences
The natural sciences are more than two centuries older than the science of religion and comparative anthropology. In the field of astronomy and physics empirical research had reached its first overwhelming successes as early as the seventeenth century, in the period from Galileo to Newton. Chemistry joined the advance in the eighteenth century and later was followed by mineralogy, geology, and meteorology. From the methodological point of view success was obtained in the physical sciences by three means: they did not restrict themselves to mere observation but experimented wherever physical processes could be influenced by technological devices; they investigated the quantitative relations of phenomena; and they considered discovery of natural laws as the most

important goal of research. Application of mathematics to the empirical findings and the construction of deductive theories cannot be separated neatly from those empirical methods and scarcely are less important; but, as they belong with the rational side of science, they need not be discussed here.

The biological sciences developed considerably more slowly. Among them, medicine is the most ancient one. In the modern era medical investigation, as well as physical research, was hampered at first by the social prejudice against manual work. This prejudice was overcome in the late sixteenth century, and dissection came in use in anatomy about the same period as experiment did in physics. As early as the beginning of the seventeenth century Harvey experimented with animals in his embryological research and occasionally used quantitative methods in his investigation of the circulation of the blood. Nevertheless, the practical tasks of medicine affect the emotions of men so intensely that metaphysical, teleological, and even superstitious ideas and traditions were eliminated from medicine more slowly than from any other natural science. In zoology and botany quantitative and experimental methods did not come into general use before the nineteenth century. Up to that time, biologists restricted themselves chiefly to observation, non-causal description, and classification. Classification of the material must precede the investigation of causes and laws in all empirical sciences, as in business the stock of goods must be ordered and inventoried before actual trading starts. In the field of physics, as well, solid and liquid bodies, acoustical and optical phenomena, for instance, were distinguished before they were investigated scientifically. Whereas in physics, however, most classifications are elementary and, therefore, can be followed rather soon by causal research, the vaster variety of objects in the fields of zoology and botany is much harder to survey. Biological research, as a result, has, during almost three centuries, scarcely passed beyond classification and non-causal description. Moreover, it is far more difficult with plants and animals to produce alterations artificially than it is with physical and chemical objects. For all these reasons experimental methods and causal explanations were adopted by biology rather late. Instead of the concept of cause, prescientific teleological interpretation ruled the biological sciences until the nineteenth century.

The eighteenth century, however, was already aware of the difference between "artificial" and "natural" classification. Linnaeus, the most eminent biologist of the period, gave both artificial and natural classifications of plants and animals. Since plants are extremely numerous and varied, it is often not easy to determine the species to which some individual plant belongs. In order to facilitate that task, Linnaeus classified the plants by the number of their stamens; he was well aware, however, that this classification was merely artificial and hardly more than a technical device of nomenclature. His natural classifications, on the other hand, aimed at putting together plants or animals in one group which "actually" are connected with one another. Thus natural classifications includes the viewpoint of empirically ascertained relationships and, therefore, is the first step in the direction of causal research and the theory of evolution.

Lamarck's, and especially Darwin's, achievements opened entirely new ways to biological investigation. Empirical thinking was furthered by the theory of evolution in four respects. Linnaeus had still considered animal and plant species as absolutely rigid; there are as many species, he explained, as have initially been created by God. By the theory of evolution this rigid immutability was liquified, as it were: "to be" was replaced by "to become". Certainly, the concept of temporal process is not necessarily involved in the concept of natural law. There are in physics non-temporal laws of coexistence as well as laws of temporal change, though the former might be less numerous. Moreover, the latter seem to interest men more intensely; human actions are processes themselves and always aim at influencing the future. At any rate, the precursor of the concept of law - the concept of cause - contains the element of temporal change as an essential component. For that reason interpretation of living beings as subjects of a permanent temporal process was indispensable if description and classification were to be supplied by causal explanation and later by investigation of natural laws. Second, Darwin's theory of natural selection (1859) made the first successful, though yet incomplete, attempt to explain causally the obvious fact that organisms are well adapted to their environments. Since the decline of Aristotelian scholasticism, prescientific teleological traditions had been removed from the field of physics. With Darwin they suffered the first serious blow in biology. Third, Darwin's exposition of the descent of man (1871) destroyed traditional anthropocentric philosophy and thus decidedly helped eliminate obstacles to the empirical investigation of mankind. In anthropology, sociology, psychology, and ethics this influence became conspicuous very soon, even if the animal ancestors of man and natural selection were not discussed at all. Darwin's ideas throughout fitted in with the trend of modern empiricist philosophy. Already, Hobbes, Hume, and a few representatives of the French Enlightenment had, without even thinking of animal ancestors of man, consistently treated all human problems as natural, merely empirical ones. This interpretation was confirmed and enormously spread by Darwinism. And, finally, Darwin's theory called scientific attention to a few characteristic traits which can be found in any evolution - even beyond the province of biology. As it was pointed out by Herbert Spencer, for instance, specialization, or family-tree-like ramification, appears in the historical development of occupations and sciences and in the intellectual development of the human individual as well as in the phylogenetic evolution of organisms. Thus, empirical investigation of society, civilization, and mind was also considerably furthered by Darwin's ideas. On the other hand, however, the scientific value of the concept of evolution was sometimes overestimated. The merely descriptive statement of some "evolutionary" process in society or culture is only a preliminary step by which the solution of the scientific problem is prepared but not given. A final statement of causes and laws cannot be supplied simply by description of some one evolutionary process.

Darwin's theory was based on the practice of cattle-breeders and on comparative observation. He did not perform experiments. This most successful and most exact method of empirical research was applied to biological problems in the investigation of heredity and physiology in the late nineteenth century. Mendel discussed his experiments on heredity (1865) in statistical terms, whereas some physiologists of the nineteenth century transferred the ordinary methods and concepts of physics and chemistry to biology and adapted them to the new problems. Moreover, both modern genetics and modern physiology investigate quantitative relations and are seeking natural laws.

At the end of the nineteenth century teleology seemed to revive again in biology. In the final analysis Darwin's causal explanation of the fitness and adaptation of organisms deals with a statistical phenomenon only. It discusses the survival of the fittest within large groups of plants or animals, but it is not interested in the problem as to how the processes which are characteristic of life are accomplished in individuals. Such processes as the development of the fecundated egg, regeneration of injured organs, and others, were interpreted teleologically by Driesch. Even the Aristotelian concept of entelechy was reintroduced by him - a concept which two centuries before had been eliminated from science after long and arduous intellectual conflicts. The strangest phenomenon in this resurrection is that it is connected with a considerable advance of experimental method. For the aforesaid processes were investigated by Driesch by means of experiments that proved to be most fruitful. They showed that processes in one part of an organism are often influenced not only by conditions in that part but also by other, if not all, parts of the organism. Driesch and the neovitalists assume the influence of *all* parts and speak, therefore, of the efficacy of "wholeness"; wholeness, however, is interpreted by them as a nonspatial, soul-like entelechy which aims at ends. This neovitalistic "explanation" is highly contestable. Non-spatial entelechies are in no way observable, and, consequently, the entelechies of a frog and a hydra cannot be distinguished. It is not clear, therefore, what help they can give in explaining, predicting, and controlling the rather different processes occurring in these different animals. Obviously, entelechies are additional metaphysical ingredients and do not contribute anything to a solution of the empirical problems. The phenomena of regeneration and formation are now actively investigated. They cannot yet be explained satisfactorily, though many partial successes have been achieved. Nevertheless, there is no reason to assume that, in this field, solution of the scientific problems will be achieved by other means than by experiments, causal explanation, and investigation of laws.

7. Psychology

The beginnings of psychology in the modern era appear rather remote from empirical research. Descartes and Spinoza, when discussing human passions, gave theoretical constructions which were intended primarily to support the ethical ideals and theories of each philosopher. Yet it cannot be doubted that their "psychological" studies are based on a good deal of introspection into their own minds and comparative observation of fellow-men. The same methods were used more consistently by Locke, Berkeley, and Hume in their analysis of knowledge. Most psychologists of the seventeenth and eighteenth centuries were highly impressed by the scientific success of Newtonian physics; they were, therefore, seeking psychological laws, and considered the laws of association to be analogous to, and as important as, the laws of mechanics. Such physical analogies become rather conspicuous in the psychological treatises of both the physician Hartley and the chemist Priestley. Morals, also, were again and again investigated psychologically. By application of psychological methods to the problems of ethics, certainly a major advance in empirical thinking was made. On the other hand, in the eighteenth century, psychology still dealt chiefly with knowledge and rather neglected emotions, volitions, and actions.

Psychology in the nineteenth century developed considerably, both as to its subjects and as to its methods. In the psychology of knowledge the constructs of the deductive sciences began to be investigated. It was even assumed that logic could be reduced to psychology. Quite a number of epistemologists were convinced that all problems of logic would be solved if the psychological origins of logical concepts and logical propositions were found out. By their critics their line of thought was called "psychologism". Psychologism in the nineteenth-century epistemology distinctly mirrors the empirical tendencies of the period and even exaggerates them. It obviously committed the same error a chess player would make if he thought that knowledge of the historical and psychological origin of all chess rules could answer the question of which chess problems are solvable and which are not. On the other hand, psychologists with a leaning toward the natural sciences investigated sensations in close connection with psychological research. Thus experimental and quantitative methods were introduced into at least one branch of psychology.

With the decline of the Enlightenment, the irrational sides of the human mind also began to attract attention. In the early nineteenth century German romanticists were already highly interested in passion and emotion and, moreover, paid a great deal of attention to hypnotism, somnambulism, and insanity. Philosophers and psychologists belonging, as did Schelling and Schubert, to the Romantic current of thought, began to deal also with the irrational and abnormal components of mind. These were interpreted, both by writers and by theorists, metaphysically, and even more or less magical ideas frequently occur in the expositions. Again a new and fruitful field of research was discussed at first in prescientific termc, as often happens in the history of science. This initial stage, however, was overcome about the middle of the nineteenth century. Scientific psychology divested itself of magic and metaphysical ideas without resuming the eighteenth century's overestimation of intellect. It turned to voluntarism and investigated emotions, appetites, and instincts empirically, frequently in close connection with biology, physiology, and animal psychology. Psychiatrical research arose, and the comparison of normal and abnormal mind contributed fruitful data to psychology. Investigation of neuroses, and of hysteria especially, added unconscious processes to the objects of physiological research.

The new conception that mind is composed of unconscious as well as conscious elements apparently is exposed to methodological objections.[10] Viewed empirically, mind seems to be equivalent to awareness. Unconscious mental processes, therefore, seem to be a contradiction in terms and to be metaphysical constructs which can never be tested by experience. Yet, these objections do not hold. Psychology of the nineteenth century is, for the most part, based either on introspection or on empathic interpretation of the behavior of fellow-men. In this behavior, however, speaking and intentional communications form but a small part; other, and unintentional, reactions are far more numerous. No psychologist, therefore, bases his scientific assertions only on the words by which his fellow-men describes their awareness but makes use of their physiological reactions - for instance their blushing, their actions, and their omissions of action - as well. Is there any reason to abstain from investigating processes which do produce such reactions, actions, and omissions, though they cannot be described by the individual in whom they happen? Since the individual in whom they occur is not aware of them, it might be objected that they ought to be called physiological processes. Actually, we may hope that in the future they will be explored by physiological means, and certainly physiological investigation would furnish more exact results than the psychological investigations of the present time. Nevertheless, the above procedure can be justified. When an individual acts in a certain way because of some past experience which is perfectly well remembered by him but is so disagreeable that he does not like to speak of it, and when another individual has "repressed" an extremely painful experience so that it has become entirely "unconscious", both kinds of behavior can greatly resemble each other. The similarity becomes manifest in the fact that the observer can put himself psychologically in the place of both persons by means of empathy. The very possibility of empathy in both cases is the link by which conscious and unconscious processes are connected. This is the empirical reason why we are justified in using psychological terms and psychological methods in investigation of the unconscious. Physiological processes, on the other hand, work quite differently from unconscious ideas, wishes, and purposes and are not accessible to empathy.

We have made use of empathy as of an empirical symptom of the relationship between conscious and unconscious mental phenomena. It must be remarked, however, that it cannot be used as a definitive method of research. When we interpret the behavior of fellow-men by means of empathy, our interpretation mirrors experience which we happen to be familiar with. Empathy, therefore, is subject to major errors and always needs farther examination by means of more reliable methods. We may strongly feel that a man is angry, and yet all inferences based on our feeling may later prove to be wrong. When, on the other hand, his further behavior conforms with our expectations, then and only then our opinions of him are verified. Since empathy works much faster than careful scientific examinations, it supplies most of the psychological judgments in everyday life.

---

10 We need not distinguish here between unconscious and subconscious processes.

Yet scientific predictions, based on empathy, may be relied on only in so far as they are confirmed by observable actions and reactions of the individuals concerned. The method of empathic interpretation may be used in scientific psychology as a preliminary heuristic tool. Certainly, it is fruitful if its results are tested later by observation of the perceivable behavior. But it is highly fallible, and the scientific content of all assertions obtained in this way consists solely in those components which can be confirmed by observation. The precariousness of empathy has led to the rise of behaviorism, which is discussed at another place in this *Encyclopedia*. At any rate, it is an empirical fact that man can experience empathy in certain cases and is unable or less able to experience it in other ones. The cases in which it can be experienced - that is, the conscious and the unconscious mental processes - obviously have certain empirical features in common.

To summarize: if and only if empathy is more or less feasible, may an unconscious process be named "psychological" and be reckoned among mental phenomena. Nevertheless, empathy is a symptom only. With it a certain type of functioning, that is, of causal connection, becomes manifest that, apparently, is common to conscious and unconscious mental phenomena. Or, to put in a different way: conscious and unconscious components of mind are subject to kindred laws; physiological processes, to entirely different ones. The unconscious elements of mind have been introduced into psychology in order to fill the gaps of its causal explanations and to complete the domain of validity of psychological laws. This method of completing the scientific domain is entirely legitimate and is used in the physical sciences as well. Astronomers, for example, do not hesitate to discuss multiple stars with partly bright and partly dark components. Psychology of unconscious mental phenomena is not less empirical than astronomy of invisible stars. Another point, which need not be discussed here, is that, in its present state, the method of exploring unconscious phenomena needs improvement. Psychology of the unconscious is young and fruitful; but it is as inexact as all young sciences have been initially. Greater exactness may be obtained in the future by experiments and by comparison with animal psychology. In this article, however, we are not concerned with empirical details and special questions. It has only been necessary to discuss the basic question of empirical testability.

8. The social sciences

Among empirical sciences, the social sciences are youngest. They have developed gradually from at least three different sources. Jurists, political writers, and philosophers of the seventeenth century dealt with public law and political philosophy. They were, however, more concerned with rationally establishing their political aims and theories than with careful and unbiased observation of empirical facts. Hobbes, for example, was more a rationalist than an empiricist in his theory of society. He was among the most consistent representatives of the doctrine of social contract, which dominated political philosophy until the end of the eighteenth century; and this doctrine's disregard of experience is obvious today. Primitive man subscribes to rational agreements just as little as he believes in the rational articles of "natural religion". With the advance in overseas trade, savage nations gradually became better known in eighteenth-century Europe, and an increasing number of authors began to criticize, with empirical arguments, the hypothesis of a social contract.

About the same period, French writers, among them Voltaire, Raynal, and Condorcet, started investigating the history of human civilization and the development of social institutions. In political historiography, also, the advance of empirical thinking and the gradual disappearance of theological ideas might be traced. Yet, the political historians still dealt with single events and individuals and hardly belong among the predecessors of the sociologists. On the other hand, the writers who cultivated "philosophical" history, as it was called in the period of the Enlightenment, helped to prepare the way for the social sciences. They discussed general processes in the development of civilization and the interdependence of these processes. The ideas of such writers were mixed with unexamined assumptions concerning human progress and contained a good deal of wishful thinking; but, in the last analysis, their philosophy was based on empirical observation and comparison: the "philosophical" components of their expositions were but vague - and often incorrect- formulations of sociological causes and laws. Reports on exotic peoples were also published more frequently in the eighteenth century; they greatly influenced the writings of the "philosophical" historians and, therefore, may be counted among them.

A third source of the social sciences springs from the practical needs of economy. In contrast to more primitive forms of economy, capitalism requires rational regulation of economic activities. This holds true of private as well as of political economy. Since the beginning of the modern era, the princes - or rather their secretaries and counsellors - were compelled to form opinions on the question of how taxes and duties influence commerce and manufacture and are, in turn, influenced by the prosperity of the latter; their revenues depended on these questions. When, in the middle of the eighteenth century in France, problems of agriculture also became urgent, public officials and private writers began to study economies more systematically. And, finally, Adam Smith, the friend of Hume, published his comprehensive theory of economic processes (1776). He thus created the science of political economy and started a scientific development which has continued up to the present day. Political economy is the oldest among the social sciences. As it originated in the needs of an eminently rational and antitraditionalist form of economy, it was from its very beginnings based on reasoning and experience. It never went through an animistic stage and was spared controversy with Aristotelian entelechies.[11] Control and prediction of economic processes were among its aims from the beginning. Consequently, theoretical economists immediately began looking for causal explanations and very soon for economic laws. They obtained, and still obtain, the empirical material for their theories by comparative observation; in the nineteenth century

---

11 In the twentieth century, however, a few German economists attempted to introduce entelechies in political economy.

statistical methods were added and proved to be highly successful. Experiments have not yet been performed in the field of economics. The hazardous steps, in politics and business life, that are often called economic experiments are done neither for scientific purposes nor with the necessary scientific precautions.

Rational deduction and mathematics do play a large part in certain economic theories. Political economy might be the only empirical science in which, for the reasons mentioned above, the empirical elements are likely to be impaired rather by excess of rational deduction than by prescientific tradition. It is still a rather imperfect science. It is about two centuries younger than physics, and its subject matter is more complex; economic experiments are not performed, and economic research is exposed to more selfish interests, political pressure, and wishful thinking than is the case in any other science. It is comprehensible, therefore, that in political economy scientific agreement could be reached only on comparatively unimportant questions; in fact, there are separate schools which do not even recognize each other. Some of them cling to experience; the results of their inquiries are collections of material rather than theories in which facts are causally explained. Others deal with nothing but laws of economy; they investigate them by means of rational analysis of a few basic concepts and construct large deductive systems based upon scanty observations. And yet, after all, political economy might be considered the most advanced among the social sciences.

Sociology originated in the beginning of the nineteenth century. Two rather different thinkers, Hegel (1770-1831) and Comte (1798-1857), may be called the first sociologists. Hegel comes from German Romanticism. His *Philosophy of History* employs the dialectical method, which is claimed to be rational and forms the backbone of his whole philosophical system. This method can be traced back to Plato. The metaphysical mill of thesis, antithesis, and synthesis, however, is as empty as inexact; it could yield almost any result. In Hegel's system it begins its work with the concept of being. It cannot be doubted that the method's rich output has, unconsciously, been supplied by and adapted to experience, for it would be sheer magic if reasoning alone were able to derive an abundance of concrete details from a concept which is devoid of content. However, while Hegel's exposition of philosophy of nature is rather sterile and sometimes strangely contradicts later experience, his philosophy of history, law, and religion abounds with fruitful results. Apparently, the artificial mechanism of dialectic fits, approximately, certain processes which often occur in history, society, and civilization. The exact and empirical description of those processes, and the demarcation of provinces in which they occur and in which they do not, has not yet been accomplished. Hegel discussed the general lines of development in law and ethics, in morals and political institutions, in art, religion, and philosophy. He did not describe isolated events but comprehended numerous single facts in one construct and saw relationships which no one before him had noticed. These relationships always refer to temporal succession of cultural phenomena. As they are repeated in an analogous way in various fields, they may be taken as preliminary intimations of laws of sociological succession.

Regularlity, however, was always interpreted as necessity by Hegel. Fruitful empirical knowledge often appears at first in vague and metaphysical expressions; Hegel's philosophy of history is possibly the most impressive case in point.

Comte, the disciple of Saint-Simon, stemmed from the philosophical historians of the French Enlightenment. His attitude was entirely empirical and antimetaphysical. He was well trained in the natural sciences and regarded prediction as the main goal of knowledge. For him, *to know* meant *to foresee*, and science, in his system, is identified with the investigation of laws. Like Hegel, he investigated only the main lines of historical development and comprehended numerous single facts in very general statements. Utilizing the ideas of predecessors, he affirmed the sequence of a theoretical, a metaphysical, and a positive stage in the history of civilization. This succession is supposed to appear in different fields in the same way. His theory of civilization, therefore, may be taken as a preliminary formulation of a sociological law of succession. Comte coined the names of both sociology and positivism. Compared with Hegel, his expositions contain less speculation, but their empirical results are probably poorer.

It is not necessary at this point to survey the development of sociology after Comte. And even achievements so important as the introduction of the concept of evolution by Spencer and the emphasis on economic facts and economic groups by Marx need not be discussed here. But a few methodological aspects should be pointed out. Among sociologists there are today various schools and many controversies; some schools even disregard the investigations of most of the others. It might be generally agreed that sociology is an empirical science. It is based on observation and comparison, if not yet on experiment. As it does not deal with individuals but investigates groups and mass phenomena, the general sociological statements which appear in still rather uncritical forms in Hegel and Comte must be based on careful and complete collection of material if reliable results are to be achieved. It is here that statistical and sometimes even quantitative methods were successfully introduced in sociology. They were, however, largely applied to quite elementary phenomena and their use frequently resulted in mere collections of material. Causal and comprehensive sociological theories, based on statistics, are still lacking. Apart from the difficulty of putting them into practice, there is no reason why statistical methods might not be employed in the investigation of Hegel's and Comte's problems as well, that is, in investigation of cultural processes. Sociological processes are usually interpreted psychologically, that is, by empathy. It has already been pointed out that empathy proves to be fruitful as a heuristic method but that its results must always be tested by observable facts.

As to sociological laws, modern sociology does not restrict itself to the investigation of laws of succession but seeks laws of coexistence as well. There are, however, sociological schools which deny the possibility of any sociological laws. These maintain that in social research causes and laws have to be replaced by "types", by "understanding" - that is, empathy, by "wholeness", entelechies, and values. The preliminary methods of non-causal description and classification and, moreover, all pre-scientific and teleological concepts of the past are revived as ultimate goals of science in this rebellion against causality. No convincing reasons, however, have yet been given proving that causes and laws must be discarded. In everyday life we are wont to predict, more or less successfully, such social phenomena as overcrowdedness of railroad trains, the outcome of elections, and the trend of public opinion. Predictions like these presuppose that social phenomena are connected more or less regularly. It may be that in any society regularities practically always have exceptions and, consequently, hold only approximately. Social groups are seldom isolated and usually interact with one another; the number of their members is always comparatively small; the members are very different, and some of them exert disproportionate influence. These conditions do not favor group laws. With a gas inclosed in a vessel with permeable walls and consisting of only a million molecules, a few of them being extremely large, rather inexact gas laws could be ascertained. It is possible that in sociology, also, only very inexact regularities can be discovered. Yet, no physicist or astronomer would entirely disregard a regularity on the ground that it did not always hold.

One more point must be considered. The classical gas laws deal with the interdependence of temperature, pressure, and volume of the gas. Nevertheless, the single molecules of which the gas consists have neither temperature nor pressure; they whirl at random, have kinetic energy and impulse, and are, in classical mechanics, subject to laws entirely different from gas laws. In an analogous way sociological laws might connect quite different variables than psychological laws do, though social groups consist of human individuals. Thus we have come back once more to the question of empathy. If we look for social regularities by means of empathy, we may never find them, since ideas, wishes, and actions might not appear in them at all. That is, social regularities may belong to a type of connection entirely different from psychological ones. At any rate, if there are sociological laws, they can be discovered only by actually looking for them and not by discussing their possibility. And what methods are to be employed in this investigation? The empirical methods of causal research have, in all sciences, proved to be so fruitful that we shall not rashly give up hope of finding them successful in the field of sociology too.

We have surveyed the advance of empirical thinking in the domain of the special sciences. We have found that, though different means of inquiry were adapted to the various subjects of the various sciences, in the final analysis success was achieved by the same methods everywhere. These methods are: collection of the material, observation, and comparison; experiment wherever objects can be influenced by technological means; counting and measuring, if possible; causal investigation and investigation of laws. As to methods, there are no basic differences among empirical sciences. The empirical methods are best developed in physics, and physical patterns, consequently, have influenced the other sciences in an increasing degree, despite the remarkable countercurrents in sociology and biology of the last decades. As to methodological maturity and scientific success, biology is today next to physics. Sociology, on the other hand, is the least mature among the empirical sciences. Whether psychology or political economy is better developed is an open question. When such questions are to be decided, the deductive theories in the various sciences and the application of mathematics to their problems must be taken into account as well; the rational methods, however, have not been discussed here. As to historical sequence, the physical sciences are the oldest. It is noteworthy that causal psychology, which originated in the middle of the eighteenth century, and even causal political economy, which originated in the late eighteenth century, are considerably older than causal biology, which originated with Darwin and the physiology of the late nineteenth century. Yet biology has since then outstripped by far its older competitors. The historical sequence and historical development of the sciences do not exactly conform to the various degrees of complexity of their problems. They are greatly influenced by economic needs and the great struggle of ideas and can be explained only in connection with the general process of history.

IV. THE DECLINE OF THE MECHANICAL CONCEPTION OF NATURE

9. The elimination of mechanical models
In the second half of the nineteenth century important physical discoveries resulted in the breakdown of the mechanical theories of light, electricity, and magnetism. As philosophy, since the period of Galileo, had been influenced by physics to a higher degree than by any other empirical science, this physical revolution also reshaped philosophical thinking and the analysis of knowledge.

The process began with Maxwell's inquiries on electricity (1865). Using the experiments and ideas of Faraday (1791-1867), Maxwell succeeded in comprehending all laws of the spread of electromagnetic actions in two fundamental equations. Faraday, on the other hand, had interpreted electromagnetic processes mechanically and pictured electric and magnetic lines of forces as invisible ropes, acting in accordance with mechanical laws. Maxwell also obtained his equations by means of a mechanical (namely, a hydrodynamic) analogy, but the mechanical model turned out to be extremely complicated this time. Maxwell was forced to imagine a model consisting of rotating, invisible ether whirls and interspersed invisible particles pushed on by the rotations. In addition, his equations connected electrodynamics with optics, since they contained the velocity of light as an essential constant and covered all laws of propagation of light-waves as well. When H. Hertz twenty years later (1888), by his famous experiments, ascertained, for the first time, the existence of electromagnetic waves, the wave theory of light definitely became a part of Maxwell's theory of electromagnetism. Still, electromagnetic waves were supposed to spread as a kind of movement in the ether. The ether, however, had on the one hand, to penetrate everything and, on the other, to behave mechanically as a solid body, if observable optical phenomena were to be represented correctly. Thus, mechanical models gradually came to be regarded as tools with which science might possibly dispense. Why should one derive both electromagnetic and optical from mechanical laws by means of highly complex models if all laws of optics can be derived directly from simply constructed electromagnetic equations? Possibly mechanisms do not have any preferred place over other physical phenomena. If equations, wherever they are derived from, present all the observable facts in the simplest way possible, they may very well fulfil the task of science better than theories attempting to reveal a "real" world behind the phenomena.

At first, ideas like these ventured forth hesitatingly, and their implications were either not realized or met with considerable resistance from physicists. Even before the experiments of Hertz, Kirchhoff, the founder of modern astrophysics, advocated, in his *Lectures on Mathematical Physics* (1874), the opinion that physics had to "describe" rather than explain phenomena. Similar ideas were supported by Helmholtz four years later (*Tatsachen der Wahrnehmung*, 1879). The physical and epistemological implications of such ideas were developed, with greater radicalism, by the philosophical physicist Ernst Mach (*Mechanik in ihrer Entwicklung*, 1883). Mach analyzed and consistently refuted the belief in the priority of mechanics to the other branches of physics. In his conception of science, scientific explanation is equivalent to "economical description" of the observed facts; science has to represent, as he pointed out, as many facts as possible by as few concepts as possible, and it is irrelevant whether the concepts used are taken from mechanics or elsewhere. Moreover, Mach disclosed the philosophical pseudo-problems originating from the mechanical conception of nature. The opposition of an objective world of quantities and a subjective world of qualities, which had confused philosophy for three centuries, has been overcome in his analysis of scientific knowledge. Starting from different problems, the historian of physics, Pierre Duhem, and the chemist Wilhelm Ostwald helped to destroy the mechanical prejudice. Duhem (*La Théorie physique, son objet, sa structure*, 1906) pointed to the different styles of thinking favored by the English and the French, respectively, in their physical theories; he ascertained that the theories of the French, representing the facts solely by means of mathematical equations, were not less efficient than the English theories based on mechanical models. On the other hand, general energetics, advocated by Ostwald, was also suited to weaken the overestimation of mechanics; for energetics restricted itself to discussing transformations of energy in all fields equally and disregarded mechanical models.

Mechanism suffered its decisive defeat, however, in the theory of relativity and modern atomic physics. Influenced by Mach's ideas, Einstein published his fundamental papers on the special theory of relativity in 1905 and on the general theory in 1916. As is generally known, the theory of relativity pointed out the connection between certain paradoxes of light-propagation and of all spatial and temporal measurements and gravitation. Those highly general connections could no longer be represented by mechanical actions of one ether. A mechanical model of all relativistic laws has not yet been mathematically constructed in every detail. If, after the theory of relativity, a physicist still intended to be a mechanist, he would have to take refuge in a "great-" if not a "great-great-ether" behind the ether, in order to represent all relativistic facts. Evidently overcomplicated and patched up, a mechanism like this could not stand competition with the mathematical conciseness of the relativistic equations. And, finally, in 1913 Niels Bohr's paper appeared, which succeeded in representing the hydrogen atom by a planetary system of particles of electricity. From this day modern atomic physics has, with increasing success, derived all properties of matter itself - among them pushing and pulling - from actions of particles of electricity. These actions follow entirely non-mechanical laws. It can be explained only by history why modern microphysics has been named quantum *mechanics*. As is generally known, Bohr's original model of an electrical planetary system has since been given up and only equations are left. In Heisenberg's, Dirac's, and Schroedinger's quantum mechanics there is, for the most part, no question of even spatial movements of particles, let alone their pushing and pulling. Thus we have come to realize that it amounts to the same thing whether attraction of electric charges is explained by the pulling of invisible cords or the pulling of cords by the behavior of electrons and protons. We accept that theory which covers the observable phenomena and which, at the same time, is the more comprehensive, the more consistent, and the simpler. The mechanical prejudice has definitely been overcome. It needs not be mentioned that the breakdown of the mechanistic conception does not mean at all a return to premechanical animistic or teleological ideas. The statistical laws of quantum mechanics are as rational and mathematical as the laws of classical mechanics and not a bit nearer to the behavior of living beings or spirit.[12]

The breakdown of mechanistic physics took place during a period of complete revolution in technology. The levers and pulleys of the seventeenth century had receded to the background long ago. The steam engine and, in the second half of the nineteenth century, the electromotor, the dynamo, and the internal-combustion engine had taken their place. The working of all of these is based on non-mechanical processes. Merely mechanical machines, as bicycles, typewriters, and foot-driven sewing machines, have been comparatively unimportant in the economy of the last sixty years. The textile industry, in which mechanisms played a comparatively greater part, had been the leading industry up to the early nineteenth century but was fifty years later surpassed by far in economic importance by the electrical and then by the chemical industry. Men, and even the children, of the twentieth century do not feel at all strange about the working of an electric bulb, a telephone, a photographic camera, a radio set; what is even more important, they do not feel any different toward the working of these implements than toward the working of a typewriter. The movements by which man influences the world around him have, with the technology of the twentieth century, shriveled down in many cases to turning on a switch; the elec-

---

12 Cf. Philipp Frank, *Interpretations and Misinterpretations of Modern Physics*.

tromagnetic, chemical, and caloric processes which are then initiated do the rest. If man is inclined to conceive natural processes after the pattern of how he himself influences nature, it is scarcely a surprise that the priority of mechanics has completely dwindled. Certainly, modern technology and economy have not only influenced modern physical theories but have been influenced by them as well. The electrical engineer, Marconi, did succeed the theorists Hertz, Maxwell, and Mach. On the other hand, however, Mach succeeded the steam-driven factories, the dynamo, and the electromotor, and, certainly, electro-engineering and radiobroadcasting helped immensely to make the nonmechanical conception of nature less paradoxical.

10. Final remarks

The breakdown of mechanistic physics could not fail to give a new impetus to empirical thinking. With the failure of mechanistic physics, the assumption of a second world behind experience had lost its scientific support. Now the subject-object metaphysics, the pride of all philosophers, who looked down on the naïve laymen, was badly shaken; its problems began to appear as pseudo-problems. Since causes and laws were employed in the new physics as functional connections and mere regularities, the unempirical components of those concepts, already criticized philosophically by Hume, became suspect for scientists as well. All these implications were consistently developed by Mach. On the other hand, physical hypotheses and models had suddenly turned out to be unsuitable, though having proved fruitful for three centuries. Necessarily, general methodological questions arose as a result of that fact, and, for the first time in the history of modern physics, the whole internal construction of science became problematic. What part is played by simplicity in scientific theories? Which natural laws should one consider as fundamental, and which as derived, when constructing a theory? What service can be rendered to science by working hypotheses, by fictions, and by conventions? Most of these problems deal rather with the deductive side of theoretical knowledge than with its empirical components. They were raised by Mach, by fictionalism and conventionalism of the late nineteenth century, and were more or less suggested by the physical revolution. Poincaré's conventionalism, however, was influenced by modern mathematics as well as by the new physics. In the early twentieth century those mathematical and logical influences increased, united with the empiricist tradition, and resulted finally in logical empiricism - a subject which must be reserved for later treatment.

10

PHYSICS AND THE PROBLEM OF HISTORICO-SOCIOLOGICAL LAWS

The question as to the existence of laws in history has frequently been discussed. A new discussion may yet be useful, since some misconceptions based on incorrect comparisons with the natural sciences have been brought forward by both advocates and opponents of historical laws. We shall try to clarify the problem by applying a few ideas familiar to physicists and astronomers to the conditions peculiar to history. Physics is the most mature of all empirical sciences as to method. In physics the law-concept has been used for three hundred years. It may be assumed, therefore, that most of the difficulties in its application to other fields have their physical counterpart and can be clarified most easily with the help of physical concepts. A few preliminary examples of historical laws will be given towards the end of the article.

I

The relationship between historical laws and historical prophecies has sometimes been misrepresented. Astronomers can not predict from Newton's law what the position of the planet Mars will be on next New Year's Eve. In addition to the law they need the knowledge of the positions, velocities, and masses of a few celestial bodies at some given time: they need knowledge of "initial conditions" as the physicist puts it. Knowledge of a law, therefore, is not a sufficient but only a necessary condition of prediction. Evidently the same holds for history. Even if laws according to which wars between industrialized countries proceed were known, it might still be impossible to predict the outcome of the present war. Among other more intricate things we do not know is e.g. the number of airplanes on both sides. This knowledge will not be achieved before the war is ended, when it will be too late for prediction. As far as the past is concerned many analogous deficiencies, probably, never will be removed. Probably, we shall never know how large the population was in ancient Babylonia and how great the percentage of priests, noblemen, merchants, farmers, and slaves. This ignorance not only impairs our knowledge of the initial conditions but, unfortunately, also of the laws. Every scientific law asserts subsistence of current association or regular connection of certain conditions and events. When and where we are completely ignorant of essential initial conditions, we shall never find recurrent associations. Or to put it more exactly:

the more incomplete our knowledge of the initial conditions is, the more difficult is discovery of laws. We shall be very modest, therefore, in our expectations regarding historical laws.

In a few natural sciences laws appear in formulations which, beyond mere statements of recurrent associations, seem to include logical necessity. This necessity, however, springs from the deductive form only attained by those sciences. In physics e.g. the three laws of Kepler can be deduced from Newton's law; likewise virtually all laws of electromagnetism are deducible from Maxwell's equations. Yet this is a mere matter of logical form that does not affect the empirical content. When Galileo discovered the law of falling bodies by experimentation and measurement, he did not deduce it from more general mechanical laws, since such were not yet known. Deductive connection with other laws was not achieved before Newton, for deductive theories are almost always constructed a considerable time after discovery of single empirical laws. This is relevant to our problem. Investigation of historical laws still is in an embryonic stage. For a long time to come these laws must not be compared to the laws of nineteenth century mechanics or electromagnetics but to the laws of young and still undeveloped sciences such as stellar physics. Based on numerous observations e.g. the law of Leavitt-Shapley asserts the existence of a functional relation between period and luminosity of variable stars of a certain type. This and many similar isolated empirical regularities in fixed star astronomy are well verified laws[1] without any regard to the hypothetical attempts of deducing them theoretically from more general physical principles. In history where investigation of laws has hardly begun construction of deductive theories would only impair empirical research. At any rate, however, there is no basic difference between isolated empirical laws and laws connected by deduction.

There is one more reason why historical laws on the one hand and mechanical and electromagnetic laws on the other do not compare. The latter are found in laboratories. Laboratories contain artificial apparatus built for specific ends. All physical apparatus are carefully safeguarded against mechanical shock; conforming to their objectives they are isolated electrically or thermally, they are constructed airtight or protected by leads against X-rays etc. Therefore creation of "isolated systems", i.e. of systems not affected by undesirable interference from without, is among the chief aims of laboratory physics. In nature on the other hand there are, except for astronomy, practically no isolated systems. Particularly in history all systems are very incompletely isolated: cultures, countries, states always interfere with each other spiritually, politically, economically. Ancient Greece was influenced by the Orient; the Roman Empire was invaded by Germanic tribes, the civilization of these tribes was changed by classical culture and christianity; modern China is influenced by Western capitalism, modern Western philosophy by Chinese and Indian ideas etc. There

---

1 E.g. Kohlschuetter-Adams: intensity of certain spectral lines and luminosity of the star; Adams-Joy: precision of spectral lines and luminosity; Lindblad: intensity of the continuous spectrum and luminosity.

are no airtight compartments, no isolated systems in history. History, therefore, must never be compared to laboratory physics. It compares only to geophysics, i.e. to physics of earthquakes and sea-currents, to volcanology and meteorology. This is a triviality but, strangely enough, has been forgotten in most of the analyses. Moreover, the aid of laboratory physics which forms the background of meteorology and geophysics and supplies laws from which deductions can be attempted is lacking in history. All this taken into account, historical phenomena are scarcely more difficult to predict than the weather and certainly not more difficult than volcanic eruptions and earthquakes. What would scientists think of a geophysicist who gives up the search for geophysical laws because of their inexactness?

II

Psychological laws deal with the behavior of human individuals, historical laws with large groups of individuals, namely with cultures, states, nations, occupations, classes, etc. They do not correspond, therefore, to the laws of impact by which in classical gas-theory the behaviour of the single molecules is regulated ("micro-laws"), but to the gas-laws ("macro-laws"). Historical and sociological groups, however, compare to gases considerably differing from those studied in our laboratories. They are contained in vessels with permeable walls; they consist of comparatively few molecules that do not move at random but in a partially orderly way. Moreover, the impulse of some of them is considerably greater than that of others. Little wonder that under such conditions the "gas-laws" do not hold very exactly. These difficulties must be analyzed one by one.[2]

1. The "permeable walls", i.e. the incomplete isolation of the historical systems has already been discussed.

2. An ordinary gas vessel in a laboratory contains about 10^{23} molecules. Since at present the whole population of the earth amounts to about 10^9 individuals, all historical groups are by many orders of magnitude smaller than statistical systems in physics. Although the accuracy of macro-laws is dependent not only on the numbers of individuals involved, in general statistical laws of smaller groups are less exact. Certain historical groups are particularly small. E.g., a law of some intellectual development might be concerned with not more than a hundred or even as few as ten philosophers or authors. Provided all authors concerned are taken into account, investigation of macro-laws is still justified,

---

2 On the other hand the problems of modern quantum-mechanics have no bearing on our problem. Historical laws are macro-laws. Heisenberg's principle of indeterminacy questions the existence of physical micro-laws; the validity of macro-laws is not affected by quantum-mechanics. The same holds for the so often discussed problem of determinism. Even individuals with "free" will could follow statistical macro-laws.

since the authors have been singled out from considerably larger groups. Several years ago the statistician Bortkiewicz obtained his "law of small numbers" by studying soldiers of the German army who had died from having been kicked by a horse. We do not insist on the correctness of this special law of Bortkiewicz[3], but the method in general is justified. E.g. a law of intellectual history would be fairly well founded, though it is based on the observation of only fifty French philosophers, these having been singled out from forty million Frenchmen. Astronomers also investigate laws of the behavior of supernovas and "white dwarfs" though only exceedingly few of such stars are known at present. Astronomers, however, do not pick out or leave out objects according to value concepts and personal predilections.

3. Boltzmann based his theory of gases on the assumption that the molecules move "at random" ("hypothesis of disorder"), a hypothesis formulated more exactly in modern statistical physics (Von Mises). For historico-sociological groups analogous hypotheses are valid to a limited degree only. E.g. states are "organized", i.e. they consist of orderly hierarchies of subgroups. Yet, even in states a residuum of "disorderly" behavior remains. There is, however, a statistical theory of crystals too in modern physics. In crystals the atoms are arranged in three dimensional "lattices". Since over this order random oscillations are superimposed, a statistical theory of e.g. electric and thermal conductivity in crystals is possible. The same holds for historico-sociological groups: modern statistical investigation of public opinion is not prevented by the fact that states are "organized".

4. The historical influence of human individuals shows a much greater variety than the physical effectiveness of atoms. Here two kinds of human inequalities must be distinguished. The effectiveness of persons can differ because they hold different positions in an organization. If in an army a general commits treason the effects are incomparably greater than in the case of a private doing the same. Therefore, counting generals and privates, heads of states or public officials and plain civilians with the same weight can become absurd in certain statistical investigations. In states, armies, and similar organizations inequality of position is of decisive importance. For this reason the validity of historical laws certainly is smallest in political history. In this field "chance", i.e. psychological, biological, and other individual circumstances, might greatly impair macro-regularities. Macro-laws might play a greater part in the history of civilization and ideas, of art, science, philosophy, and religion where differentiated organizations are less important.

On the other hand human individuals influence history to different degrees because they differ in their personal gifts and abilities: there are good and bad artists, good and bad scientists and philosophers. Whether these differences are great enough to make statistical investigation and macro-laws impossible can not

---

3 It asserts that "dispersion" is nearly "normal" in such exceedingly small groups. Cf. L. v. Bortkiewicz: *Das Gesetz der kleinen Zahlen*, Leipzig 1898.

be decided by a priori arguments, but only by the results of empirical search after such laws. At any rate a "statistical" history, aiming at laws of intellectual, artistic, and religious developments, would greatly differ from traditional historiography. It can neither dwell on masterpieces nor disregard the mass of mediocrities. The question as to whether an artistic or theoretical work is the product of a genius or a bungler would not even enter its investigation. Certain components of these value concepts, however, would reappear in other and more objective shapes. The distinction between long- and short-lived ideas must appear also in "statistical" history, if it should represent the facts. And the same holds for both the difference between new creations and mere repetitions and the stratification of literary and artistic taste according to social ranks and various levels of education and sophistication.

III

Time is contained as variable in some physical laws but not in others. E.g. the second law of thermodynamics asserts on a temporal process, stating that the entropy of an isolated system increases with time. The law of Wiedemann-Franz, on the other hand, asserts that, without regard to time, electric and thermal conductivity of metals are proportional to each other, good electric conductors being good thermal ones and vice versa. These two kinds of regularity may be called temporal laws and simultaneity-laws.[4]

Both types of laws may be expected in history. We shall give a few examples of the first type (in preliminary and vague formulations) and indicate the empirical evidence in brackets.

1. In isolated historical systems tribal organization precedes the beginnings of the state (empirical evidence: ancient China, ancient Jews, ancient Greece, Germanic states of the early Middle Ages et al. Under external influence, i.e. in non-isolated systems, the development can proceed in a different way: British dominions, U.S.A.).

2. Individualized art and poetry are preceded by anonymous folk-art and poetry, signed paintings and sculptures by non-signed works. Or to put it more generally: the collective mindedness of the period of self sufficient domestic economy and barter economy precedes the individualistic spirit of the period of money economy and economic competition (rise of individualism in Greece in the sixth and fifth, in Rome in the third and second centuries B.C.; the Middle Ages and the Renaissance). This law goes back to Hegel and Jacob Burckhardt and has been generalized by Carl Lamprecht.

---

[4] In physics the terms "dynamic" and "static" laws are often used. This terminology has been avoided here, since it is too narrow and since, in philosophy and the social sciences, use of the word "dynamic" often covers deficiencies in scientific analysis. The magic and animistic connotations of this term, discarded in physics three hundred years ago, have not quite disappeared from the social sciences.

3. Free artists e.g. sculptors, painters, and architects, gradually develop from craftsmen, such as stonedressers, whitewashers, and masons (China, ancient Greece, the Renaissance).

4. Worship of eminent individuals first takes sides and is partial, i.e. persons with opposite or divergent aims are not admired or worshipped by the same people; impartial admiration and worship of divergent great personalities develop later: biased hero-worship precedes impartial genius-worship (classical antiquity, the pre-romantic and the post-romantic period in Europe; cf. the study of the genesis of the concept of genius mentioned in footnote 5).

All these "laws" are yet incomplete in so far as only necessary but not sufficient conditions are given. They describe temporal processes in yet rather vague formulations and may be called "historical" laws in a narrower sense. On the other hand there are in history also simultaneity-laws. Though they are usually called "sociological" laws, there is no basic difference between sociological and historical regular connections. The separation is entirely artificial, since sociology by no means disregards investigation of temporal processes. We give a few examples of simultaneity-laws.

5. Wherever learned priests are entrusted with the task of teaching priest-candidates, they systematize the vague and contradictory mythological traditions of the past and develop rational distinction, classification, and enumeration as scientific methods. Even if they turn to worldly subject-matters they develop causal investigation in a very small degree and never investigation of physical laws (Medieval Arabic and Catholic Scholasticism; Jewish Talmudism; the five orthodox Indian "philosophical" systems, especially Sankhya; Buddhistic Scholasticism in Japan).

6. If under favorable circumstances a past culture is revived after centuries, the initiators and bearers of this intellectual movement are characterized by the following traits: they do not belong to the clergy; they are scribes and secretaries in political services and develop under favorable circumstances to free literati; exceedingly proud of their ability of writing and reading they disdain the illiterate; in their idea the educated is chiefly distinguished by perfection of his literary style; as patterns of style serve literary documents of the past; the authors of these documents are being looked upon as authorities superior to all representatives of subsequent periods (the European Renaissance; the two Renaissances of Confucianism under the Han and Tang dynasties in China; probably also the Renaissance movement in the New Empire of ancient Egypt and the Neo-Parthian Renaissance).

All these historical "laws" have to be considered as preliminary and more or less probable assertions only. They are meant as examples to illustrate what form, approximately, historical laws would take and also on what kind of empirical evidence they have to be based.[5] Since experiments are not feasible in

---

[5] After collection of the historical material about twenty other hypothetical laws have been given by the author of this article at the end of his *Die Entstehung des Geniebegriffes. Ein Beitrag zur Ideengeschichte der Antike und des Frühkapitalismus*, Tübingen 1926, pp. 324-326. With the necessary

history, comparison of various countries and cultures is the only way of finding historical laws.[6] The cultures compared ought to be as independent from each other and as numerous as possible. E.g. Europe and America, the modern era and antiquity, the civilization of India, China, Egypt, and Persia must be compared and investigated from identical points of view if historical laws are to be established. In this program the only real difficulty in the application of the law-concept to history is revealed. It originates in the comparatively small number of independent historical systems. If in the natural sciences hypothetical laws are to be tested, it is in most cases easy to find suitable instances that do not influence each other. In history it is only among the primitive cultures that the number of independent instances is considerable. Higher civilizations are in most cases connected by spatial interaction or the temporal links of tradition. This is a serious obstacle to the verification of laws. E.g. analogous intellectual developments in eighteenth century France and England may be due to mutual influence and consequently must not be extrapolated and generalized. Practically the same holds for classical antiquity and modern Europe. Yet analogies e.g. in medieval Catholic and in Japanese Buddhistic Scholasticism may offer a sufficiently reliable basis for extrapolation. Far remote cultures, therefore, are particularly illuminating and convincing in the investigation of historical laws. But also in the other cases painstaking scientific criticism may often succeed in separating direct influence from actual recurrence of the same phenomenon under analogous conditions. However, prediction will always stand as the ultimate test for the correctness of a law. Evidently prediction must not be taken in a temporal meaning only. If a historical regularity, obtained by comparing certain historical systems, is confirmed in other and independent systems, it may be considered, at least for the present, as verified.

On the other hand lack of perfect analogies neither speaks against the possibility of historical laws nor does it form a basic difference between history and the natural sciences. This had been assumed erroneously by the Windelband-Rickert-school and has often been repeated since. True, no two historical individuals are completely alike and history never repeats itself. However, the repetitions in natural science are overestimated by those only who are rather remote from this field of research. He who has ever worked in a laboratory knows that even every apparatus, if it is somewhat more complex, has its individual characteristics and has to be handled with its own special tricks. No two reflectors of the same brand are perfectly alike and even less two

---

scientific accuracy a historical law has been given for the first time in Frederick J. Teggart: *Rome and China. A Study of Correlations in Historical Events*, Berkeley 1939. Professor Teggart verifies statistcally that there is a correlation between political disturbances in Western China and the Asiatic frontier regions of the Roman empire on the one hand and barbaric invasions in the Danube and Rhine region on the other. Teggart's law, however, belongs to a more special type than the "laws" indicated above. It holds, although for many events, for late antiquity only. Evidently this does not impair its scientific value. Mechanics of rigid bodies is also valid in a limited period only. There were no rigid bodies 10^{10} years ago. In empirical science all laws hold as long as the objects and systems they are concerned with exist.

6 Cf. F. J. Teggart, *loc. cit.*, p. 245.

planets. In natural science the variety of objects is mastered by the method of gradual approximation: objects may be handled as analogous in the first approximation; their differences are taken into account later in second and third approximations, when they are put together and compared in new groups. Variety of historico-sociological phenomena surpasses variety of other objects in degree only. There is no reason, therefore, why the method of gradual approximation should not be applied in comparative history too. The conformities in the cultural ideals of Renaissance humanists and Chinese literati-officials can be established first; the differences may be taken into account later.

On the other hand the method of "understanding" ("insight") [Verstehen, Eds.] which has often been recommended for social science is not sufficient when investigating historical laws. "Understanding" means psychological empathy; psychologically a historical process is "understood" if it is evident or plausible. The main objection to this criterion of the correctness of a historical assertion is that virtually always opposite historical processes are equally plausible.[7] When a city is bombed it is plausible that intimidation and defeatism of the population result. But it is plausible as well that the determination to resist increases. It would not be plausible, on the other hand, if the bombing changed the pronunciation of consonants in the bombed city. Which process actually takes place can not be decided by psychological empathy but by statistical observation only. In the final analysis the method of understanding is equivalent to the attempt to deduce historico-sociological laws from laws of introspective psychology. However, before regularities are established it is premature to attempt to deduce them. In construction of new empirical sciences the predeductive stage can not be skipped.

There is even the possibility that certain historical laws may never become psychologically evident. Certainly history is based on the behavior of human individuals reacting psychologically. Yet there are in the natural sciences macro-laws that are similar to micro-laws and others that are not. Maxwell's equations hold for the macro-phenomena caused by electric currents as well as for the behavior of (free) electrons: in this case macro- and micro-laws not only assert the same functional relations but contain also the same variables (electric and magnetic field, etc.). On the other hand the gas laws entirely differ from the laws on elastic impact regulating the behavior of molecules in classical gas-theory. In this case the macro-laws connect variables, such as pressure, volume, and temperature of the gas, which have no application at all to single molecules.[8] Likewise there may be historical macro-laws connecting parameters only that are meaningless when applied to human individuals; no term that fits emotional or intellectual processes might enter them. In this case the historical law would be deprived of the possibility of psychological understanding and empathy. Yet it

---

[7] This originates in the fact that the patterns *consent-refusal* and *affirmation-negation* predominate in the province of emotional and intellectual processes.

[8] The correspondence of one macro-parameter to a set of quite different micro-parameters is an instance of the relationship that, with a rather vague term, has been called *emergence*.

might be well verified empirically, since observable facts can be put into a correspondence also with very complex and abstract logical constructs. The law of the shifting of consonants in the Indo-European languages approaches to this type of "non-understandable" historical regularities. The number of such laws may be considerable e.g. in economic history.

History has grown out of other roots than the natural sciences. A man who, desiring to lift a load, is interested in the principle of the lever is the prototype of a natural scientist. A father who takes pictures of his son every year and collects and keeps them may serve as an analogy to the origin of historiography. He is not interested in regular connections and predictions but in the gradual development of his object because he likes it. Nobody will argue against this kind of historiography. It will always persist as long as men love their countries, their communities, and their culture. On the other hand, however, investigation of historical laws should not be obstructed by methodological objections. The greatest danger in this field is the danger of dilettantism and superficiality. The investigator of historical laws must collect, interpret, and compare an immense and highly complex material. Which scholar is an expert on modern and classical, Egyptian and Chinese history simultaneously? Astronomers have mastered analogous difficulties by division of labor and cooperation. The observatories have divided and distributed the problems, have collected, each in its field, the immense material according to identical principles and have thus produced the star catalogues and maps which form the basis of their laws. There is yet no analogous cooperation and division of labor in comparative history. Yet its problems are extremely interesting and may become important in a practical way in the future. Many scientists must establish a common program of research and cooperate according to it. By collecting and comparing the material with philological accuracy historical laws will be discovered at last not by general methodological discussions like ours.

11

PHENOMENOLOGY AND NATURAL SCIENCE*

When phenomenology was introduced as a new science by Husserl its methods were applied first to objects of logic. Later phenomenological investigation expanded gradually to the fields of psychology, ethics, esthetics, and sociology (Scheler, Pfänder, Hildebrand, a.o.). More rarely, objects of the natural sciences have been treated phenomenologically. Scattered indications of this kind are to be found in authors who do not belong to the most intimate circle of Husserl's school (Helmut Plessner, Kurt Goldstein, Walter Frost, E. Buenning). Extensively, however, the phenomenological method has been applied to objects of the natural sciences once only, namely by Hedwig Conrad-Martius, a favourite pupil of Husserl's, in her *Realontologie* (Ontology of Reality) and *Farben* (Colors).[1] Yet this less known branch of phenomenology is particularly interesting. Husserl stressed the basic difference between phenomenological ideation (*Wesensschau*) on the one hand and psychological introspection and description of the immediate data of awareness on the other. The peculiarity and scientific productivity of phenomenological method, therefore, can be studied best in a field which is as far removed from psychology as possible. We shall try to analyze the papers of Conrad-Martius more fully and shall refer to other authors occasionally as illustrations.

Hedwig Conrad-Martius investigates matter and its qualities for more than two hundred pages by means of a priori ideation. She distinguishes three possible forms of reality, the forms of body, soul and mind. Body belongs to nature, mind is supernatural, soul subnatural (§131). God belongs to all of the three realms, whereas angels are minds only (ibid.). The three states of matter, solid, liquid, and gaseous, correspond to and are "symbols" (§133) or "analogies" (§134) of body, soul, and mind respectively. This is, as the author puts it (§134), "not a construction but is based on conception of pure phenomena from the intuitive power from which we can not withdraw". Although the three states of matter can be investigated in detail by experiments only, their general characteristics can be given phenomenologically. To wit, there are three possible ways in which the particles of matter can combine. There is the "anarchic"

---

* [W. H. Cerf responded to this essay by way of his short „Review of 'Phenomenology and Natural Science' " , *Philosophy and Phenomenological Research*, 1941, 1, p. 513. Zilsel's reply to Cerf's response, 'Concerning Phenomenology and Natural Science' is reproduced in this volume pp. 214-2150; Eds.]

1 *Jahrbuch für Philosophie u. phänomenologische Forschung* vol. 6 (1923) pp. 159-333 and Ergänzungsband (1929), pp. 339-370.

structure in which everyone is against every other one. Thus results general repulsion and the gaseous state. There is the "atomistic" structure, everyone being for himself, with mutual indifference and the liquid state resulting. And finally, in the "communist" structure everyone is for everyone, and unity and the solid state result (§162). This is the "a priori characterization" of the three states of matter. There follows an a priori discussion of hardness, brittleness, softness etc. and of glass, crystals, and metals. The essence of metallicity e.g. is suppleness i.e. "the combination of valiant power, compliance, and self-preservation" (§175).

The following sections deal with sound, temperature, and light. In sounds the essence of material objects manifests itself sensually (§189 ff.). The essence of temperature is immediate affection of the human body: the state of the thing directly acts upon the state of the body (§200), warmth corresponding to life, cold to death (§201). Light is "the most wonderful phenomenon": the author refers to God's words "Let there be Light" (§212). When a thing is made red-hot it reaches a peak of activity, an eruption occurs, and it begins to glow. "Primary luminous state is ecstasy of matter.... These are not metaphorical phrases but must be taken most literally and verbally" (§217). Later this statement is amended somewhat: light *is* not ecstasy of matter but ecstasy of matter "posits" light (§237), or even better "implies" light (§239). The last thirty pages deal with colors referring to Goethe's theory, which however is partially criticized. Blue is the color of Mary's coat and of the magic flower, it is the color of humility. Purple is the color of majesty, brown of monks' cowls, wood and plainness etc. (§279 ff.).

Historically the exposition is influenced by German Romanticism, by Novalis, Schelling, and Baader, a contemporary of Schelling with theosophical leanings, who is quoted several times (§§118, 131). The trichotomy of matter, soul, and mind can be traced back even to the Gnostics (Hippolytus, Valentinus). As to method the difference from the natural sciences is stressed (§§145, 159, 162, 185). Yet some results are based on experiments, though on experiments performed by means of quite elementary physical instruments. E.g. in discussing brittleness the definition of a textbook on mineralogy is quoted and agreed with (§171). Apparently the use of steel points and chisels is compatible with a priori Wesensschau, the use of galvanometers and X-ray tubes is not. Several statements which are expressly said to originate in a priori Wesensschau obviously are based on experience. E.g. it is emphasized (§182) that we hear immediately in the sound of an object whether the object is hard, liquid, or elastic; it would be "delusion" to ascribe this to association of ideas. Likewise it is maintained (§278) that we immediately see whether a piece of butter is hard or soft. It would be instructive to test these phenomenological statements with individuals who never had touched objects of the kind in question before. The experiences which, unconsciously, are utilized again and again are, however, rather defective. E.g. it is deduced a priori, as it is claimed, from the mere essence of fire (§229) that water is opposite to it and, consequently, extinguishes it; the fact that fire is extinguished e.g. by carbon dioxide as well is omitted, presumably because this occurs more rarely in everyday life. Or it is explained a priori that light comes into existence by glowing (§§213-219). The fact that there is also cold light (phosphorescence, luminescence) is omitted. All these a priori expositions, as they are claimed to be, are in fact primitive and, consequently, defective and rather often incorrect natural science. Some statements, being based on vague preconceived ideas, are entirely arbitrary and quite insusceptible of confirmation. E.g. genuine metals, such as steel, are distinguished from spurious metals, such as aluminum (§175); genuine liquids, such as water, from spurious liquids, such as molten gold (§159). This distinction is supported by linguistic considerations: gold melts and becomes solid, whereas ice thaws and freezes; "on this difference much could be said phenomenologically" the author says (ibid.). It may be mentioned that in several other articles also of the *Jahrbuch für Philosophie*[2] linguistic and merely verbal investigations play a considerable part. Certain phenomenological subtleties are so closely linked with the peculiarities of German syntax and vocabulary that they cannot be translated at all into other languages. We need not point out that our author makes abundant use of vague associations of ideas particularly in her symbolic interpretations and analogies.

The investigation never provides a method of testing and verifying the results obtained and never seeks after the causes of the phenomena discussed. This is the most essential difference from the method of the natural sciences. Yet the ends, values, and meaning of the phenomena are discussed again and again. We shall give a few examples: presenting itself in its state of rigidity is "the peculiar end" of glass (§178); metals are "noble", gold and steel are "perfect metals" and their perfection is "marvellous" (§176); there are two kinds of flames, holy flames which are unselfish and radiant, and unholy flames, being selfish and smoky (§§225-228, 244-248). As to symbolic interpretation of the phenomena we need not repeat the numerous examples quoted before. Obviously it is assumed that *Wesensschau* is able to recognize the ends, meanings, and values of the phenomena a priori, whereas their causes can be found empirically only and, consequently, must be left to natural scientists. This is expressed rather clearly by Hans André, a philosopher who is near to phenomenology, in his book on *Archetype and Cause in Biology* (*Urbild und Ursache in der Biologie*. Muenchen 1931). There he states e.g. that (p. 94) green conforms to the essence of plants since by their destination plants are open to light. This aspect of the problem is said to be the highest. On a lower scientific level the green color of plants is investigated teleologically and on the lowest level causally. Altogether disregard of causality is even more conspicuous in phenomenological ontology than disregard of experience.

In his prefatory remarks to the first volume of the *Jahrbuch für Philosophie und phänomenologische Forschung* Husserl declares phenomenology to be "an

---

2 e.g. H. Aumann: *Zum deutschen Impersonale*, Ergänzungsbd. (1929) pp. 1 ff.; F. Neumann: *Die Sinneinheit des Satzes und das indogermanische Verbum*, ibid. pp. 297 ff.

unlimited field of strictly scientific investigation". The new periodical shall be "not an arena of vague reformatory ideas but a place of serious scientific work". It may be left to the readers to decide whether this promise has been kept in Conrad-Martinus' investigation of physical phenomena. Historically the method used by her is not so new as it seems to be. Natural phenomena were viewed and discussed teleologically and their meanings were interpreted a priori long before their causes were investigated. These pre-scientific methods were familiar to Babylonian and Indian priests as well as to Chinese mandarins and medieval scholastics. When St. Hildegard of Bingen declared in the 12th century that rivers rise in the sea and flow uphill[3] she certainly used a method of knowledge which cannot have been so different from phenomenological *Wesensschau*. Hedwig Conrad-Martius sometimes approaches ideas that are even more ancient. The statement that the gaseous state correspond to and is a symbol of mind revives primeval animistic ideas of breath-like soul. The analogy between phenomenological and prescientific methods is confirmed by the exposition of a historian of philosophy who is influenced by Husserl's ideas. Walter Frost in his book on Francis Bacon[4] emphasizes that Bacon's natural philosophy anticipates ideas of the last decades. Actually the non-causal elements in Bacon's concept of "form" which Frost has in mind are less anticipations of modern phenomenology than remnants of medieval scholasticism.

The rise of phenomenological ontology can be understood historically most adequately if it is viewed in connection with the remarkable anticausal rebellion in the philosophy of the last fifty years. Since about 1890 in European, especially in German philosophy, new methodological concepts have been advocated: values in Windelband, Rickert, and their school, ends and aims in neovitalism, ideal types in Dilthey and Max Weber, interpretation in meaning in Troeltsch. With these new methods phenomenological *Wesensschau* belongs. All of them have in common that they do not provide a method of testing and confirming the results obtained, that they are meant to supplant causal investigation and, in the final analysis, that they are familiar to prescientific civilizations. Oriental priests and scholars, medieval monks have always interpreted the meanings of phenomena, have viewed their essences, values, and ends, and distinguished and classified types. Since the beginning of the seventeenth century, since the period of Galileo, these prescientific methods gradually were displaced by causal and verifiable investigation. There is hardly another characteristic that distinguishes western civilization and the modern era as distinctly spiritually as does the predominance of causal research. The fact that the precausal methods are being revived again at present is a most remarkable phenomenon in the history of ideas, and one which probably can be explained sociologically only. The surface of a river shows countercurrents and eddies only when a rock is hidden on the bottom. Unfortunately the causal explanation

---

3 *Subtilitates* II, 5 (Migne, Patres Latini, vol. 197).
4 *Bacon und die Naturphilosophie*, Muenchen 1927, pp. 81 ff., 103 ff.

of the anticausal tendencies in modern philosophy and sociology is highly complex and cannot even be attempted here.

We have analyzed phenomenological ontology only. In this field the background of the new methods is particularly clear. The application of phenomenology in the fields of logic, psychology, and sociology have not been discussed. Certainly the relationship of logic and psychology offers subtle problems; certainly it is necessary first to ascertain and describe the data of awareness and the mere facts of society before their connections and changes are investigated. It may be doubted whether phenomenology has formulated these problems correctly. It is especially open to question, whether it has separated the real problems from pseudoproblems originating in verbal expression only. It is rather obvious, however, that in these other fields too the historical roots of the phenomenological method are to be found in the remarkable endeavor to substitute for the methods of causal research.

12

CONCERNING
'PHENOMENOLOGY AND NATURAL SCIENCE'

In his review of my article "Phenomenology and Natural Science" Mr. Walter Cerf does not represent the contents of the article quite correctly. My analysis had not arbitrarily picked out some phenomenological paper, especially exposed to criticism, and endeavored by "ridiculing" it to deal "a deadly stroke" to the "entire life work of Husserl". The article had rather carefully distinguished applications of the phenomenological method to problems of logic, of psychology and sociology, and of the natural sciences. It had expressly restricted itself to analysis of this last "less known branch of phenomenology". (Methodological objections to the other branches were indicated in the last eight lines only). It had selected the *Realontologie* of Hedwig Conrad-Martius as an instance because this paper is the only one in the Jahrbuch that deals with physical facts, although similar expositions are not rare in phenomenologically influenced contemporary philosophical literature. Mr. Cerf rejects Miss Conrad-Martius' "unfortunate" paper as "romantic revelries". For the methodological reasons pointed out in my analysis I agree with his judgment and assume it will contribute to purifications of the phenomenological method and of philosophical method in general. Before Mr. Cerf's admission, however, the *Realontologie* had not been adversely criticized in the numerous articles of the *Jahrbuch* on phenomenological methodology. On the contrary it had been at length and rather enthusiastically reviewed by a phenomenologically influenced philosopher in the official journal of the "Gesellschaft Deutscher Naturforscher and Ärzte", (*Naturwissenschaften*, vol. XIV, 1926, pp. 947 f.).

Mr. Cerf takes offense particularly at the historical remarks in the last but one paragraph of my article and at the restriction to Wesensschau in my analysis. As a matter of fact Husserl has proceeded beyond Wesensschau in his later investigations. Also my historical assumption on the origin of phenomenological ontology may be erroneous. But does it really mean "to prostitute the methods of historical research", as Mr. Cerf maintains, if one starts from the starting point? Does it contradict the spirit of historical research, if one tries to find out the historical roots of phenomenological ontology by starting from Wesensschau and comparing it with the methods of the natural sciences?

---

* [This is a reaction to W. H. Cerf 'Review of 'Phenomenology and Natural Science', *Philosophy and Phenomenological Research*, 1941, 1, p. 513. Cerf replied in his 'In Reply to Mr. Zilsel', *Philosophy and Phenomenological Research*, 1941, 2, pp. 220-2. Eds.]

My analysis has assumed that phenomenological ontology - together with several other methods in contemporary philosophy and sociology - has grown out of the "endeavor to substitute for the methods of causal research".

This assumption is denied by Mr. Cerf. Clarification of the complex methodological performance involved in the application of phenomenology to the objects of the natural sciences might be forwarded if Mr. Cerf answered the following questions:

1. Miss Conrad-Martius investigates the properties of the three states of matter, of crystals, glass, metals, flames, etc., by means of phenomenological methods. Admittedly her results are inadequate. Could such objects be investigated by means of other phenomenological methods in a more adequate way and could this be exemplified? An answer to this question would contribute to demarcation of the objects which are susceptible to phenomenological investigation and would help to clarify the possibility of phenomenological ontology.

2. The paper of Conrad-Martius always disregards causal relations between the physical facts studied. In the paper phenomenological analysis always results in a priori establishment of values, meanings, and ends of physical objects. Is this a consequence of incorrect application or a necessary result of phenomenological method? In other words, is the phenomenologist able to recognize values, meanings, and ends of non-man-made physical objects? And can causal relations be investigated by phenomenological study of the physical objects which are connected by them? An answer to these questions would help to clarify the methodological tasks of phenomenological ontology.

3. If the phenomenological publications are arranged according to the kind of objects (and facts) investigated, approximately the following sequence results, starting with the most numerous and ending with the rarest group: papers on logico-mathematical objects, on psychological and aesthetic objects, on sociological objects, on objects of the natural sciences. Does this sequence result from merely historical causes? Or is it the consequence of some intrinsic properties of the phenomenological method and, if so, of what properties? An answer to this question would shed light on the relationship of phenomenology to the various sciences in general and to the natural sciences in particular. This last relationship was the object of my analysis.

13

HISTORY AND BIOLOGICAL EVOLUTION

What is the relationship of history to the phylogenetic evolution of man? Historians, like all specialists, are wont to restrict themselves to their own problems and, therefore, do not deal with this question. Only some popular books on the history of the world cross the dividing line between social and natural science. They start with the origin of the solar system, describe the development of the crust of the earth and of life, turn to prehistoric civilization and ancient Egypt, and eventually finish with the history of present times. It is striking that those expositions become more and more detailed the nearer they come to their end. Near the beginning one page describes ten million years, near the end ten months.

Dividing lines between different sciences have barred scientific progress so often, that it certainly is useful and even necessary to consider history from a naturalist's point of view. But if this is done, it must be done correctly. Since the crust of the earth became solid 1 or 2×10^9 years have passed, whereas the whole history of mankind since the period of the first Egyptian and Sumerian kings until present times has lasted about 5000 years. So "geological" to "historical" time is as 300,000 to 1. History, therefore, even from the naturalist's point of view is scarcely one section among other sections of the evolution of life. To think e.g. that the biological rise of mammals during the tertiary period and the political rise of Germany since 1933 belong to one line of evolution is the same as to consider the transition from winter to summer a continuation of the dying away of the glacial period.

How would natural scientists view the problem? When a scientist who is investigating the interdependence of some quantities records the results of his observations in graphic form, he usually gets a curve which is not smooth, but is disturbed by convexities of various sizes. In analyzing this result he has to separate the trend of the main curve from the convexities and the large convexities from the superimposed smaller ones. Without carrying through the separation he never will be able to find the natural laws which he is looking for, for differences of orders of magnitude always give a hint that effects are superimposed the causes of which are different. To illustrate: when one studies the movement of stars, he has to separate their circular movement during a day, their tiny oscillations during a year, and the extremely slow movements by which the shapes of the constellations are changed during a thousand years. The daily circular movement is based on the daily rotation of the earth, the annual oscillations are caused by the annual revolution of the earth around the sun in connection with the aberration of light and the parallax, and the slow secular shiftings are the peculiar movements of the stars by which they wander through space. Without the separation of those movements belonging to different orders of magnitude there can be no scientific astronomy, for quite different causes and different natural laws correspond to them. It would be easy to add more examples, but the given one may be sufficient.

§ 2. Let us apply this to history. In the end the historical evolution of mankind is enclosed in the astrophysical evolution of our galaxy. If, therefore, we start from this evolution we can take out of it the evolution of our solar system and out of this the geological evolution of the earth, each process belonging to a different order of magnitude and, therefore, needing its own method of research. In the development of the earth the origin and the evolution of life is a partial process which again has its own laws and includes among others the genesis of man. So we have reached man, but events connected with man can be observed and explored more in detail. If we observe animals in intervals of about 100,000 years, we can notice their phylogenetic variations. E.g. we notice the phylogenetic variation of man by comparing the skulls of pithecanthropus, of Heidelberg man and of recent man. On the other hand distinct variations appear in human behavior within one hundred years. As to mankind, therefore, finer and very remarkable processes are superimposed upon its phylogenetic variations. Those processes are marked by the fact that they happen, vanish, and change with much greater velocity than biological changes. They belong to a special order of magnitude, therefore they are subject to special laws and the research on them and their causes requires special methods. Those very changes, joined together, form the History of Mankind. Even faster processes, lasting about minutes, occur with human individuals: these are the biological and psychological reactions of men. And we could even descend to molecular and atomic processes, if we like to take human individuals to pieces and if it is required.

We may, of course, also begin the construction the other way round with those processes which are fastest. We could begin with the reactions of human individuals. Then the secular variations of these reactions would form history. And we should get biological evolution, if we comprehend man and the other animals and if we inquire into their variations by one order of magnitude slower. To history at any rate we have assigned a special province among human occurrences. *The realm of history comprehends human occurrences and their causes which are slower by one degree than the reactions of the individuals and faster by one degree than biological evolution.* Thus we have stated, as it were, a definition of history.

We have to add a few remarks. First: our "definition" of history is behavioristic as it does not speak of the mental world, but of the reactions of men and their changes, and even more it is only quantitative. Some philosophers, especially German ones, emphasize that history is entirely different from natural science, that it is a "mental" science, a *Geisteswissenschaft*, and they infer from this that there are no historical laws. When the relationship of history to science is discussed, usually the concepts of mental phenomenon and value are introduced by these philosophers and the discussion approaches more or less metaphysics. We have tried to avoid this and to give its due to the peculiarity of history by a quantitative criterion only and without involving metaphysics.

Secondly, we have to realize that in empirical science it is neither usual nor useful to start by giving a definition of the subject that is to be treated. No physicist would begin the theory of electromagnetism by defining what electricity is: he will rather start by explaining the equations of Maxwell. Therefore, we shall have to prove that our "definition" of history is not at all scholastic and sterile, but is able to disclose essential points of the historical process. Since in history, as well as elsewhere, peculiar laws and causes correspond to the peculiar order of magnitude, we shall be able to show that it is the difference between tradition and heredity that corresponds to the different velocities of historical and biological evolutions referred to in our definition. Obviously phylogenetic evolution goes on much slower than historical evolution, because and only because heredity is a much more powerful brake than tradition is. Likewise we shall see that on the other hand the difference of velocity between historical changes and the reactions of individuals is closely connected with the fact that historical processes are social events. Certainly such secular changes of reactions and habits as form the subject of history would not occur among mere hermits; these changes need reciprocal influence between many individuals. Only in societies is there tradition and therefore evolution with historical velocity. So our quantitative definition which at first might have seemed to be superficial and sterile discloses the fact that history is a social not an individual process the velocity of which is determined by the resistance of tradition not of heredity.

Thirdly we have to notice that the characteristic velocity of history is a peculiarity of mankind that distinguishes man from animals. E.g. changes of language can be observed after one or two centuries and, therefore, belong to the historical processes. Alfred the Great, speaking old Saxon, would not be able to make himself understood to the man in the street of today's England, but our dogs certainly bark quite in the same way as the Anglo-Saxon dogs did and no doubt even many men are still alive who entirely resemble the contemporaries of King Alfred as to biological marks and reactions. The example shows that only man possesses reactions which are plastic enough to change within a century and, therefore, are studied by history and sociology. This does not depend at all on the fact that man forms societies, but it simply means that only man produces civilization. Also in societies of bees there is a technique of building as in human societies and there are dances which in some way correspond, as the biologist von Frisch has shown, to human language. No doubt, also those social reactions have changed and developed, for the bees descend from solitary hymenoptera and even today various species of hymenoptera have reached quite different stages of evolution in their social habits. But, certainly, those evolutions lasted not centuries but periods of a geological order of magnitude. We may, therefore, affirm that man is the only historical animal. Or, what means the same, only man produces civilization.

§ 3. The difference between history and biology is illustrated rather well by the difference between nation and race or, to use a term less ambiguous, variety. Except for the Chinese there is no nation today which is much older than a thousand years, for nations rise, develop, change, and vanish within a few centuries. A thousand years ago there were no English but only Celts and Anglo-Saxons and Danes and Normans. These are nations or nationlike groups and are products of history. Human varieties on the other hand, being products of biological evolution, behave quite differently. Also the various human "races" must have changed and developed for they most probably descend from the same species of ape. But within historical periods the physical types of man seem to be quite unchangeable. We can not discuss however the relationship between nation and physical type in greater detail. We wish rather to discuss the different laws which are valid in "racial" evolution on the one hand and in national evolution on the other.

The circumstances and laws by which species and varieties arise and change are not yet clear. The better known are the circumstances and laws under which physical types persevere, the better known is heredity and its laws. We know that the transmission of physical qualities is based on the fact that the genes of the germ-cells are carried over materially from the parents to the offspring and we know the laws of Mendel according to which "racial" qualities are distributed among the descendants, if the parents to different physical types. All those results of genetics, as generally known, are valid for men too.

But in human societies complicated finer processes are super-imposed upon the rough biological events. Obviously national continuity connecting generations is based not on heredity but on those finer processes. Let us e.g. suppose all Spanish documents and manifestations of language, literature, architecture, and so on destroyed, but the children by some miracle kept alive, growing up and propagating. There is no doubt that the children will transmit the physical qualities of their ancestors to their offspring according to Mendel's laws, but certainly there will be no longer any Spaniards. We do not say tradition is *sufficient* to preserve nationality, for nations fill up gaps and increase more or less preponderantly by their own biological descendants. Physical type and heredity, therefore, are not altogether unimportant for the maintenance of nationality: race-philosophy, if it says this, contains a grain of truth. But at any rate the bond of tradition is necessary if nations are to persevere, for nations are annihilated, if this bond is cut through.

Tradition differs entirely from heredity and follows different laws. If you cross red-flowering and white-flowering peas, three quarters of the offspring are red, one quarter is white and so on, according to Mendel's laws. On the other hand, when a Slovak marries a Hungarian woman, it will depend on much more complicated laws to which nation the children and the children's children will belong. Unfortunately these laws are so complicated that the scientists have not been able yet to find them. But when they will have succeeded, then they will have found a law of tradition, a sociological law, a historical law. For all historical and sociological processes depend on tradition. All objects, the changes of which are studied by historians, like e.g. customs, languages, religions, styles of arts, political aims and institutions, are transferred from one generation to the next one not by begetting. Civilization is a texture of rather complicated human reactions which are not hereditary, but are acquired many years after birth owing to the influence of example and teaching. Now we can understand why the velocity of history belongs to a different order of magnitude than the velocity of biological evolution. As we have stated, this difference is based on the fact that the laws of tradition entirely differ from the laws of heredity.

Likewise we can understand now why man is the only historical animal. Let us remember the difference between reflexes and spontaneous actions. Reflexes and instincts - for instincts are nothing else than chains of reflexes - respond to external stimulation with rigid and unchangeable reactions. On the other hand many reactions can be checked, conditioned, and modified by former experiences: they may become plastic and may even adapt themselves to circumstances, if circumstances vary. In this case we call them spontaneous actions. Only reflexes and instincts can be inherited, whereas spontaneous actions are influenced by example and teaching and can be transmitted to the next generation by and only by tradition. Physiologically reflexes and instincts are based upon nervous processes in the spinal cord and the interior parts of the brain, whereas in spontaneous actions also cortical processes are always involved. As in man the cortex of the brain is more highly developed than in any other animal, in human behavior spontaneous actions play the most important part.

The more the inherited instincts recede into the background, the more it becomes necessary that the animal develops its behavior and acquires its habits after birth. Therefore human babies and children take a much longer time than newborn animals to become able to maintain their lives by themselves. Two more circumstances contribute to this special position of man. Firstly, man, like all mammals and many other animals, takes care of his new-born and half grown-offspring. Secondly, man lives in groups and like all gregarious animals is endowed with instincts of imitation. He is not only able to, he is also inclined to learn. Thus human children are always born in communities of grown up people and are cared for and influenced by them through many years. Therefore reactions and habits of one human generation are always transmitted by example and education to the next one. Were man like ants to possess only instincts instead of choice, were they like cockchafers to leave their new-born offspring to their fate, were they like moles to live isolated, there would be no tradition, no civilization, and no history.

14

SCIENCE AND THE HUMANISTIC STUDIES*

The greatest problem in the unification of theoretical research is the problem of unifying science and the humanistic studies. We use the term "humanistic studies" to include all branches of history and the studies of languages, literature, fine arts, and religion as they are usually prosecuted. When speaking of science we have in mind the empirical, primarily the natural sciences. In denying the name "science" to history or philology we do not wish to say that they are less valuable or interesting e.g. than physics but that they are substantially different from it.

In contrast to the natural sciences the humanistic studies are concerned with human activities and their products: the two branches of research deal with different objects. Yet both kinds of objects can be reduced to observable facts. It is possible, therefore, to construct all concepts in both fields logically in such a way, that they form terms of the same vocabulary, the vocabulary of the physical language (Carnap). However, there still remains a difference. The two kinds of research are concerned with different *relations* between the objects investigated, they *connect* the facts in a different way. This difference may be interpreted as a difference of syntax.

How are facts connected by the scientists? He is primarily interested in natural laws. Universal implications are their logical expression. The concept of causality is but a subspecies of the law concept and we still always use it with this meaning. Knowledge and only the knowledge of natural laws makes control of nature possible.

In the humanistic studies single facts are very often described separately. Only in history connections between facts play a great part. The temporal succession of the states of the same single object is the connection in which the historian is interested. It must be mentioned, however, that the historical kind of connection in a few cases is studied also in the province of non-human objects (historical geology, paleontology, phylogenetics).

Connection by a law and connection by temporal succession differ essentially. Illustration: everybody knows the advertisements in which luminous figures pass over a screen of numerous electric bulbs which are turned on and

---

* [This is a previously unpublished essay. It was found among the papers Charles Morris donated to the University of Chicago Library (Department of Special Collections, file: Unity of Science Movement, Box 3 folder 4). It is either a general outline or the complete text of Zilsel's contribution to the Sixth International Congress for the Unity of Science held at the University of Chicago, 2-6 September 1941. Eds.]

off in a suitable way. The physicists studies the connection between the processes in the bulbs and in the electric leads. He wishes to control the working of the apparatus and is interested, therefore, in the causal connections but does not establish direct relations between the successive configurations of lighted bulbs. The spectator in the street, on the other hand (the historian, the humanist) is interested in the moving pattern i.e. connects one position of the luminous figure with the next, disregarding the processes in the wiring behind the screen.

Our analogy apparently exaggerated the difference between the historical and the scientific way of connecting phenomena. The analogy seems to fit only the most primitive stage of historiography (annals, chronicles). Actually a sufficiently complete description of human actions often approaches a causal description. This is a consequence of two facts: (1) Many laws are of succession; in many isolated systems the state at the time t is the cause of the state at the time $t+dt$. (2) In the field of human behaviour tradition is active: creation of an office, a law etc. causally influences the subsequent historical processes. Yet it is mere coincidence when the historical and the scientific ways of connecting phenomena occasionally agree. (a) Frequently the historian describes successive facts that are in no way causally connected. (b) He often disregards effects and discontinues the study of causal chains if they lead out of the system of his inquiry. (c) He often disregards causes and picks up causal chains coming from the outside of the system described. (d) Even if the succession described happens to be a causal one he does not investigate it in the manner appropriate to causal analysis. This difference is decisive. Scientifically causal connections are not investigated by following one single series of events but by comparing numerous analogous series and by analyzing their agreements and differences. (Instances of the cases a-d).

The prototype of a historian is a father who takes pictures of his son in short intervals and keeps and collects them because he likes him. Certainly this humanistic historiography will persist as long as men like their countries, their communities, and their culture. Yet the same facts could and should be investigated also in the way of science, i.e. by investigating laws (universal implications). This possibility is usually denied by historians.

Methodological objections have been brought forward. (1) It has been pointed out that historical objects are individual phenomena which never repeat themselves perfectly alike (Windelband, Rickert). Actually however the same holds for the objects of the natural sciences though in a lesser degree. In science this difficulty is mastered by the method of gradual approximation. The same method can be applied also to historical objects. (2) It has been maintained that establishing and distinguishing "types" is the only method appropriate to historical and sociological objects (Dilthey, Max Weber). However the typological method is but a slightly modernized modification of the enumerations, distinctions, and classifications favoured by all Schoolmen. These methods are used as preliminary expedients in natural science too. There they precede and prepare investigation of laws.

There are certain parts of the humanistic studies that compare to the natural sciences in exactness, namely the preparatory methods by which the material is established in history and especially classical philology (critique and emendations of texts and sources, textual criticism, paleography). All of them presuppose validity of universal implications. If in the field of historical phenomena further laws are to be established, comparative and statistical methods must be used, since laboratory experiments are not feasible. Analogous processes in as many different cultures as possible must be compared. And the single facts in each culture must be collected and worked up by statistical methods. Success or failure of prediction is the decisive test of the correctness of a historical and socio-logical law.

Summary

1. Even after unification of their vocabulary science and the humanistic studies would essentially differ.
2. The scientist connects facts mainly by universal implications, the historian by temporal succession.
3. The humanistic kind of historiography persists as long as men like their culture.
4. Yet historical facts could be investigated scientifically too. A scientific history must use comparative and statistical methods.

PART III

APPENDICES

APPENDIX I

THE SOCIOLOGICAL ROOTS OF SCIENCE*

Outline

1. Introduction: Problem and Method.
Problem of the investigation: which social conditions caused the occurrence of science?

This problem would be relatively easy to solve if there had been several cultures in the development of human history in which science arose and others in which it did not. Those circumstances which are common to the "scientific" cultures, and absent in the "non-scientific" cultures, would be the causes sought for the occurrence of science. In fact the problem is more difficult to solve and the result is thus more uncertain 1) because most of the various cultures are not mutually exclusive systems, but rather are dependent upon one another, and 2) because science is primarily found in the European-American culture of the Modern Period, and in a considerably smaller measure in Antiquity, in the Arabic culture of the Middle Ages, and in China. Our problem can thus not only be treated with a sociological comparison, but must be analyzed "historically" as well.

We must first clarify which ideological structures take up the place of science in the "non-scientific" cultures. Answer: such structures which correspond with one another are magic, theology, humanism, science. Note that these structures not only correspond to one another but also form a chain of development. That is: magic continues to exist in the "theological", magic and theology in the "humanistic", magic, theology, and humanism in the "scientific" cultures.

Chapter 2: The Predecessors of Science.
A) Magic (compare Lévy-Bruhl). Places of occurrence: primitive cultures. Characteristics: magic through similarity and contact, *natura naturans*, animism, pre-rational outlook. Emotional and unconscious drives, traditions influence behaviour much stronger than empirical observation and rational thinking.
B) Theology. Places of occurrence: old Egypt and Babylon, India, Chinese-Japanese Buddhism, Jewish Middle Ages ("juridical" theology), Arabic Middle Ages, Christian Middle Ages. Characteristics: rational procedures, mythology becomes rationalized and systematized. Scholarly procedures (Scholastic), mostly schools of professional

---

* [Zilsel presented this project outline, along with an outline of his project *on the concept of law* (see Appendix II), to the director of the *International Institute of Social Research,* Max Horkheimer, in April 1939.; Eds.]

priests (excepting the Jews). "Letter" knowledge. Very strict bonds to authority and tradition. Predilection for divisions and enumerations (example: Indian Samkhya). *bene docet cui bene distinguit.*[1] In contrast, the empirical and causal interest is absent. Predilection for commentaries and quotes. Interest in systematic logic (Middle Ages, India.) - Theology developed rational methods for the first time in the history of man.
C) Humanism. Places of occurrence: European Renaissance, Chinese Confucianism (compare Max Weber), Antiquity (Rhetoricians and Grammarians), perhaps the New Persian Empire as well. Characteristics: worldly and not theologically interested, rational, bound to authority (yet worldly authorities), high esteem of language and style (in its oral expression: the art of speaking), stylistic-rhetorical ability as single precondition for the occupation of secretarial professions (Renaissance, China). Ambition, glory-ideal, pride in knowledge, gentleman-ideal of the upper-class, disdain for handwork. No search for laws of nature. -Humanism developed the methods of philological and historical research.
D) Science. Compare chapter 3.

Chapter 3: Science.
Places of occurrence. Above all the European-American modern. Less developed: Antiquity, Arabic Middle Ages. In all of these cultures cities are the residence of culture; a money economy, a market economy, and competition predominate. These are the social and economic preconditions for the overcoming of magic and theology. They are already fulfilled in the humanistic cultures. Characteristics of science:
1) Science has worldly interests (Galileo occupies himself with falling stones, not with angels). 2) Science has causal interests. Repression of the miracle. "Disenchantment of the world" (Max Weber). Bound to the development of technology: Age of discoveries.

The concept of a law of nature arises as a further development of the concept of cause. It grows out of the comparison of nature and the state. It thus presupposes a rationally founded body of law, i.e. it presupposes the overcoming of feudalism. Development of the concept of a law of nature: a) gradual expansion of the areas subject to a lawful perspective. Laws are first sought in inanimate nature (astronomy, physics, and biology beginning with the 19th century), then in the behaviour of the human individual (Psychology as program as early as Descartes, Spinoza, Hobbes; association-psychology of the 17th and 18th centuries, modern psychology of the unconscious) finally the laws of human society (economics, sociology), b) the content of the concept of law becomes stripped of its metaphysics and is transformed. Laws of nature are first divine commands and the principles of nature's construction. God as constructor of the world-clock, bound to deism. Gradual elimination of the theological aspects of the concept of law. Laws of nature as functional connections (E. Mach). Statistical laws. The modern "crisis of causality", quantum mechanics. Through all the changes of the concept of law there remains: law = regular connection.

---

1 [He teaches well who distinguishes well, Eds.]

3) Science relies not upon authority and tradition but rather on independent thinking (scientific criticism, autonomous thinking). Bound to the economic change of the emerging early capitalism: breaking of the guilds and cooperatives, dissolution of the handwork and guild tradition through competition. Occupations loose their hereditary nature, each individual is the forger of its own happiness. Individualism of the early modern period (Jac. Burckhardt). The spirit of scientific criticism is nothing other than a transfusion from economic individualism to the ideological arena. Science is also individualistic and essentially without authority. Science would rupture society and would lead to anarchism if another aspect of science didn't introduce other social bonds (scientific cooperation; compare Ch. 6).
4) Science is of a rational and mathematical nature. Theology was also rational, yet it sought rational divisions, science seeks rational laws. Tendency towards mathematization in science. The mathematics of the early modern period grew out of a merchant arithmetic on the one hand, and out of the desire for calculations of military engineers and architects, representative geometry out of the perspectival needs of the painters. Still alongside numerological and Pythagorean influences. Proof of these elements in the work of Francesco di Giorgio Martini, Piero della Francesca, Luca Pacioli, Tartaglia, Cardano et al. Early capitalism is also of a rational and mathematical nature: development of bookkeeping. The rational and mathematical spirit of science is nothing other than the transfer of these economic methods into the sphere of ideology. Rationalisation also enters into politics (Theoreticians: Machiavelli, Guicciardini; the utopia literature), into the military (mercenary armies with a chain of command instead of the feudal knight armies), and into the judiciary (positive right instead of common law and the feudal privileges, introduction of Roman law).
5) Science is empirical. Observation and experiment. Age of inventions and discoveries. Empirical aspects of craftsmanship, manufacture and commerce of the early modern period.

Chapter 4: Science and Humanism.
Essential insights into the development of science can only be obtained when we study the final subtleties, i.e. the difference between science and humanism. The first representatives of worldly and rational thinking in the late Middle Ages (14th century) were not scientists, they did not seek laws of nature and didn't experiment. They were rather urban, princely, and papal secretaries like Lovato and Mussato. They grew out of the Chancellery. These are the origins of humanism. Goal: Mastery of word and letter. From this, pride in knowledge, joy in quoting, pride in memory, erudite collections.

The human pursuit of knowledge has two sides. On the one hand, pride to know more than one's fellow man. On the other, knowledge as a means to control nature. The first aspect leads to knowledge of the past, the second to knowledge of the future (prediction). Humanism is turned towards the past. Often even epigonic attitudes. "There are no longer any great men" (Leonardo Bruni, Salutati, Bessarion). The humanists are bound to authority (Cortese is proud to be known as *simia Ciceronis*).[2] Dispute between Ciceronians and Anticiceronians (Polizian, Cortese, Joh. Franc. Pico della Mirandola, Pietro Bembo, Erasmus). Strife between Platonists and Aristotelians (Gennadius, Georg of Trapezunt, Teodoro Gaza, Plethon, Bessarion). Problem: why does early capitalism ideologically lead to the Renaissance, i.e. to the (alleged) rebirth of Antiquity? Why is it apparently turned, at first, to the past?

Attempt to solve the problem: The urban society based on money economy of early capitalism is more similar to the urban economic society of Antiquity than to feudal society of the Middle Ages. Thus arises an opportunity to lean on Antiquity. The ideology of the oldest intellectually active group of early capitalism seizes this opportunity, namely the humanists. It is notable that new urban-economic societies generally tend to seek their ideals in the past. For we find similar patterns in the urban and economic development in China (ca. 500B.C.). (Confucius). Mandarins, i.e. literary officials are also the representatives of Chinese Humanism: the importance of literary style, pride of knowledge, have a strong connection to the authority of the past. Unsuccessful antihumanist revolution of the first Emperor, Qin Shi Huang Ti (the large bookburning of 217 B.C.) Two Confucian "Renaissances" in the Han dynasty (second century B.C.) and Tang dynasty (sixth century B.C.). Somewhat related events are also to be found in Antiquity: Rhetoric, Alexandrian commentators and grammarians: high esteem of style, glory-ideals. On the other hand, Antiquity lacked a connection to the past, because a Greek urban-economic past did not exist. "Renaissances" demand a certain degree of national homogeneity.

Back to the early capitalism of the Renaissance! Handwork is scorned among the humanists: *artes mechanicae*, and *artes liberales*. Mouthwork valued more highly than handwork. Literary doctors more highly valued than dissecting wound doctors, literati more highly valued than artists. In the contemporary catalogues of famous men, out of 967 famous men only 4.5% were fine artists . Artists arose out of the less esteemed handwork tradition and sought recognition through their connections to literati. Advancement not completed until the 16th century. Exactly the same process in Antiquity. Inventors and discoverers were almost always ignored as members of the lower class and the *artes mechanicae* in the Renaissance. Most of the inventors are unknown today. Analysis of the contemporary literature on inventors and discoverers. Most of the architects and engineers of Antiquity are also unknown. The exceptional role of Archimedes and its causes.

Chapter 5: The Development of Experimental Science Around 1600.
Among the humanistic literati of the Renaissance there is a lower class of craftmen, inventors, seamen, and surgeons who were not held in the highest of esteem by their contemporaries. Class of superior artisans who need a considerable degree of knowledge. These also compose written treatises (Brunelleschi, Leon Battista Alberti, Giorgio Martini, Leonardo da Vinci, Cellini, Vannoccio Biringucci, Albrecht Dürer). Related and connected to these are the constructors of musical instruments and the

2 [Cicero's ape, Eds.]

surgeons. These invent, experiment, and dissect. They aren't yet scientists, for they still proceed in a quite unsystematic fashion and are usually not educated enough. Yet experimental science grew out of their circle.

The birth of experimental science occurs around 1600 through the success of these superior artisans within the circle of systematically trained university-scholars.

Galilei (1564-1642). His social background. His educational path. Condition of the contemporary universities. His connection to the artist-engineers (his teacher Ostilio Ricci). His plebeian sympathies. His first writings deal with technical problems and inventions. The discovery of the law of falling bodies born of the need of contemporary ballistics (the large significance of military technology for the developing science is in general noteworthy). Galileo's negative disposition toward humanism.

Baco von Verulam (1581-1625). His position on Antiquity and his antihumanism. His position on handwork and on experiment. Magical remains in Bacon. He realizes first the cultural significance of the new science, yet is himself scientifically unproductive.

Other antihumanistic movements of the 16th century: Aretino, Giordano Bruno, Telesio, Patrizzi. Antihumanistic aspects in Descartes. Obviously, the distance to ancient society grew so strongly through the further development of the economy of early capitalism that the re-establishment of Antiquity is no longer seen as a sufficient ideal. Independent and new cultural ideals are established.

Chapter 6: Scientific Cooperation.
The glory ambition of the humanists. Humanists as economically dependent on patrons. The humanist as dispenser of fame, as *dispensator gloriae*: he makes his patron and himself famous through his ability to write and speak. That is, fame as the foundation of the professional ideology of the intellectual workers. Baco, on the other hand, rejects individual glory as a goal. His ideal: advancement of learning. Furtherance of human culture through the mastery of nature. That is, objectivization of the ideals of culture and a stricter organisation of societal activity.

Bacons *Nova Atlantis*: the House of Solomon as governmental research institution with a hierarchy of officers and divided labor. Cooperation of the researchers. Progress is to be achieved through the planned cooperation of the researchers: each builds upon the results of the other. These ideas are foreign to Antiquity and to the Renaissance. Related ideas by Thomas More, Campanella, and Descartes.

The development of the scientific organisation influenced by Bacon's *Nova Atlantis*. 1654 foundation of the Royal Society. 1664 Proceedings of the Royal Society. 1663 foundation of the Académie française. Development of the present state of scientific cooperation.

Conclusion.
Science in Antiquity (the Hippocratic school, Democritus, Archimedes, descriptive science). Problem: why doesn't science come to its full development in Antiquity? Attempt to solve the problem: the culture of Antiquity is a culture of a small class of independently wealthy. Occupational work is generally not held in esteem. As

a result of the slave-system, handwork is in particular so despised that the manner of thought employed by the privileged class of workers can only on rare occasion rise to the upper class. Moreover, it is for these reasons that machine-technology cannot arise. For machine-technology is bound to experiment and natural science.

Problem: Why didn't science arise in China where the slave-system fell out of use? Various attempts to solve the problem.

Attempt to solve the main problem: Science originates in urban cultures with a money economy, a market economy, and competition so that first the magical and theological mode of thinking is destroyed and then the manner of thinking of certain groups of superior artisans rises to the level of the literary educated upper class. If rough manual labor is mainly carried out by slaves, this rise is hampered.

APPENDIX II

LAWS OF NATURE AND HISTORICAL LAWS*

Outline

Problem: To clarify the claim that there are "laws" in history and sociology through an analysis of the concept of a law of nature with examples taken from modern science. It shall at the same time be shown that there isn't any fundamental difference between history and sociology on the one hand, and science on the other, which would make the search for laws in history and sociology completely hopeless.

Introduction: Birth and Transformation of the Concept of a Law of Nature.

1. Law and Introspection.
The concept of cause. "Forces" in physics, history, and sociology. Law of nature and function. Isolated ("empirical") laws and deductively connected laws. We should expect, at first, only isolated laws from history and sociology, just as we do from the still undeveloped sciences. Deductively connected laws in economics. Isolated laws are never logically necessary; only the deductive derivation of a law from another, already established law, is logically comprehensible. Psychological insight (*Einfühlung* = empathy). Historical-sociological laws may, yet need not be, psychologically understandable. Historical "understanding".

2. The Exactness of Laws and Predictions.
Predictions presuppose knowledge of the law and the given constellation. Isolated systems. As the systems found in history and sociology are not easily isolatable, history and sociology should not be compared with the physics of the laboratory, but rather with geophysics. "Violation" of laws, "exceptions", law and "empirical rule". "Peculiarities" of the constellation in physics and history. Exactness of laws and their need to be fully comprehensive.

3. Micro-laws and Macro-laws.
Statistical physics. Macro-laws frequently contain other orders of magnitude than the micro-laws which form their basis. Historical laws can only be macro-laws. Physical and historical-sociological orders of magnitude. The relationship of

---

* [Zilsel presented this project outline, along with the an outline of his project *On the Sociological Roots of Science* (see Appendix I), to the director of the *International Institute of Social Research*, Max Horkheimer, in April 1939, Eds.]

historical-sociological laws to psychological laws. The "presupposition of disorder" in physics. Historical-sociological systems are usually ordered, yet there are also physical laws concerning partially ordered macro-systems. Difficulties in history and sociology. "Mneme" and "wholeness" in biology, physics, history, and sociology. Laws and even mathematical structures may be investigated in mnemonic and whole systems. Excursion: can history be portrayed in a mathematical fashion? Although there are not any reasons in principle against this idea, the attempt is unfruitful at the moment.

4. Laws and Time.
Laws of succession and laws in which time is not contained as a variable; in physics, history, and sociology. Dollo's biological law. The irreversibility of physical macro-structures. Historical irreversibility. The distinction between dependent and independent variables; on cause and effect in physics, history, and sociology. The speed of biological and historical processes. Concerning the "biological" conception of history.

5. The Causal-crisis in Contemporary Quantum Physics.
The Heisenberg-Bohr relation. The macro-laws of physics are not affected by the causal-crisis. Relay effects in physics and history. The problem of free will doesn't have anything to do with the problem of historical-sociological laws.

6. On the "Dialectic" of History.
Thesis and antithesis. The transformation from quantity to quality. Synthesis and mneme. Why is there change in history?

APPENDIX III

BIBLIOGRAPHY OF WORKS CITED BY ZILSEL

Accademia della Crusca, Firenze Atti della Reale Accademia della Crusca, 1819-1919/20 (1921).
Accoltus, B. *Dialogus de praestantia virorum sui aevi,* Parma 1689.
Adamson, J.W. 'The Extent of Literacy in England in the 15th and 16th Centuries: Notes and Conjectures', *The Library,* 1930, 4th ser., X, 2, pp.163-93.
Agricola, G. *Bermannus sive de re metallica dialogus,* Basel 1530.
———., *De re metallica,* Basel 1556.
Agricola, R. *De inventione dialectica,* 1480, Cologne 1518.
Alberti, L.B. *Opuscoli morali,* ed. Bartoli, C., Venice 1568.
———., *Trattato della pittura,* 1435/36, De pittura (Latin), Basel 1540, Della pittura (italian),Venice 1547.
Andreae, J.V. *Christianopolis,* Strassburg 1619, ed. and transl. Held, F.E., New York 1916.
Apian, P. *Cosmographiae introductio,* Ingolstadt 1529.
———., *Instrument Buch,* Ingolstadt 1533.
———., *Quadrans astronomicus,* Ingolstadt 1532.
Aquinas, T. *Summa theologica,* Complete German-Latin edition, ed. Albertus-Magnus-Akademie Walberberg, 36 vol., Salzburg 1933.
Aristeas, *The letter of Aristeas,* ed. and transl. byThackeray, H. St. J., London 1917.
Aristoteles, *Opera, ex recensione Immanuelis Bekkeri,* ed. by the Preußische Akademie der Wissenschaften, 5 vol., Berlin 1831-1870.
Armitage, A. *Copernicus: The Founder of Modern Astronomy,* London 1938.
Arnim, H. *Stoicorum veterum fragmenta,* Stuttgart 1905.
Arnobius. *Disputationes adversus gentes,* Rome 1542.
Augustinus, A. *De civitate dei, c.* 400, Venice 1470.
———., *De libero arbitrio,* ed. Prosper, Augustinus, Ambrosius, Basel 1524.
Auzout, A. L` *éphéméride du comète,* Paris 1665.
Bachet, C.G. *Arithmetic des Diophant,* Paris 1621, reprinted in B. Botfield, *Praefationes et epistolae editionibus principis auctorum veterum praepositae,* Cantabridge 1861.
Bacon, F. *Advancement of Learning,* London 1605.
———., *Novum organon,* London 1620.
———., *De dignitate et augmentis scientiarum,* London 1623.
———., *Nova atlantis: opus imperfectum,* ed. Rawley, W., London 1627.
Bacon, R. *Fr. Rogeri Bacon opera quaedam hactenus inedita,* ed. Brewer, J.S., vol. l: *Opus tertium. Opus minus. Compendium philosophiae,* London 1859.
Barlow, W. *The Navigator's Supply,* London 1597.
———., *Magnetical Advertisements,* London 1616.
———., *A Brief Discovery of the Idle Animadversions of Mark Ridley,* London 1618.
Beck, L. *Die Geschichte des Eisens in technischer und kulturgeschichtlicher Beziehung,* 5 vol., Braunschweig 1884-1903.
Bembo, P. *Opera,* Basel 1556.
Berkeley, G. *A Treatise Concerning the Principles of Human Knowledge,* London 1710.

Bessarion, N. *In calumniatorem Platonis,* Venice 1516.
Besson, J. *Theatrum Instrumentorum et Machinarum,* Lyon 1569.
———., *Le cosmolabe. Ou instrument universel, concernant toutes observations qui se peuvent faire par les sciences mathematiques tant au ciel, en la terre, comme en la mer,* Paris 1567.
Billingsley, H. *The Elements of Geometrie of Euclide of Megara,* London 1570.
Biondo, M. *Della nobilissima pittura,* Venice 1549.
———., Von der hochedlen Malerei, *Quellenschriften für Kunstgeschichte und Kunsttechnik des Mittelalters und der Renaissance,* V, Vienna 1873.
Biondo, F. *Roma instaurata. De origine et gestis Venetorum. Italia illustrata,* 2 vol., Verona 1481.
———., *Roma triumphante,* Brescia, Verona 1482.
Biringucci, M. *De la pirotechnia,* Venice 1540.
Blundeville, T. *The Theoriques of the Seven Planets, Shewing all their Diverse Motions,* London 1602.
Boccaccio, G. *Opere volgari,* Florence 1827-34.
Bodin, J. *Advocati methodus, ad facilem historiarum cognitionem,* Paris 1566.
———., *Les six livres de la république,* Paris 1576.
———., *Universae naturae theatrum,* Lyon 1596.
Borkenau, F. *Der Übergang vom feudalen zum bürgerlichen Weltbild. Studien zur Geschichte der Philosophie der Manufakturperiode,* Paris 1934.
Borough, W. *Discourse of he Variation of the Cumpas, or Magneticall Needle,* London 1581.
Bortkiewicz, L. *Das Gesetz der kleinen Zahlen,* Leipzig 1898.
Botfield, B. *Praefationes et epistolae editionibus principibus auctorum veterum praepositae,* Cantabridge 1861.
Boume, W. *Inventions or Devices,* London 1578.
Boyle, R. A. *Defence of the Doctrine Touching the Spring and Weight of the Air,* London 1662.
———., *A free Enquiry into the Vulgarly Receiv'd Notion of Nature,* London 1686.
Breasted, J.H. *The Dawn of Conscience,* New York 1933.
Brown, H. *Scientific Organizations in Seventeenth Century France (1620-1680),* New York 1934.
———., 'Martin Fogel e l'idea Accademia Lincea', *Reale Accademia Nazionale dei Lincei. Scienze Morali, Rendiconti,* VV11, 1935.
Brunner, K. *Gründlicher Bericht des Büchsengiessens,* in Johanssen, O., 'Karl Brunners gründlicher Bericht des Büchsengiessens', *Archiv für die Geschichte der Naturwissenschaften und der Technik,* VII, 1916, pp. 165-84, 245-55, 313-23.
Bruno, G. *De la causa, principio et uno,* Venice 1584.
Burckhardt, J. *Die Kultur der Renaissance in Italien,* Basel 1860.
Bury, J..B. *The Idea of Progress; an Inquiry into its Origin and Growth,* New York 1932.
Campanella, T. *De libris propriis et recta ratione studendi syntagma,* Paris 1642.
———., *Civitas solis,* 1602, in Campanella, T.,T. Adami, *Realis philosophiae epilogisticae partes quatuor: hoc est de rerum natura, nominum moribus, politica, (cui "Civitas solis" iuncta est) & economica, cum adnotationibus physiologicis,* Frankfurt 1623.
———., *Lettere,* ed. Stampanato, V., Bari 1927.
Canter, W. *Gulielmi Canteri ultraiectini novarum lectionum libri quatuor. In quibus, praeter uariorum autorum, tam graecorum quam latinorum, explicationes et emendationes,* Basel 1564.
Cantor, M. *Vorlesungen über Geschichte der Mathematik,* Leipzig 1899.
Cardanus, H. *De subtilitate libri XXI,* Basel 1552.
———., *Opera omnia,* Lyon 1663.

Cassirer, E. *Das Erkenntnisproblem in der Philosophie und Wissenschaft der neueren Zeit*, 2nd ed., Berlin 1911.
Caxton, W. *The Dictes and Sayings of the Philosophers*, Westminster 1477.
Cennini, C. *Il libro dell' arte*, trans. Thompson D.V., New Haven 1933.
———., *Das Buch von der Kunst oder Tractat der Malerei*, c. 1390, transl. Ilg, A., *Quellenschriften für Kunstgeschichte und Kunsttechnik des Mittelalters und der Renaissance*, vol. 1, Vienna 1888.
Copernicus, N. *De revolutionibus orbium caelestium libri VI*, Thorn 1873.
———., *Dissertatio de optimae monetae cudendae ratione*, Thom 1873.
———., *Inedita Copernicana*, ed. Curtze, M., *Mitteilungen des Copernicus-Vereins für Wissenschaft und Kunst zu Thorn*, no. 1, Leipzig 1878.
———., *Three Copernican Treatises*, transl. Rosen, E., New York 1939.
Cortese, P. *Letter to Poliziano*, in Poliziano, A., *Opera*, Basel 1553.
D'Arezzo, D. *Fons memorabilium universi*, 1370.
D' Irsy, S. *Histoire des universités*, vol. I, Paris 1933.
D' Ollanda (Hollanda), F. *Da pintura antiga. Vier Gespräche über die Malerei: geführt zu Rom 1538, Quellenschriften für Kunstgeschichte und Kunsttechnik des Mittelalters und der Renaissance*, IX, Vienna 1899.
Dee, J. *Preface to Billingsley's English Version of Euclid*, London 1570, in Hood, Th., Inaugural address as mathematical lecturer of the city of London, London 1588, ed. Johnson, F.R., *Journal of the History of Ideas*, III, 1942.
Deferrari R.J., et al. *A Concordance of Ovid*, Washington 1939.
Dehio, G. *Geschichte der deutschen Kunst*, 2 vol., Berlin 1921.
Delvaille, J. *Essai sur l' histoire de l'idée de progrès*, Paris 1910.
Descartes, R. *Oeuvres de Descartes*, ed. by Adam, Ch., P. Tannery, 12 vol., Paris 1897-1910.
Diels, H. *Antike Technik*, 3rd ed., Leipzig 1924.
———., *Fragmente der Vorsokratiker*, ed. Kranz, W., 5th ed., Berlin 1934.
Digges, L. *A Booke named Tectonicon*, London 1556.
———., *A Geometrical Practise, named Pantometria*, London 1571.
Digges, Th. *A Mathematical Discourse of Geometrical Solids*, in L. Digges, *A Geometrical Practise, named Pantometria*, London 1571.
———., *An Arithmeticall Militare Treatise, named Stratioticos*, London 1579.
Diophantus, A. *Arithmetica*, Paris 1621.
Duhem, P. *La théorie physique, son objet, sa structure*, Paris 1906.
———., *Etudes sur Léonard da Vinci*, Paris 1906.
———., *Les origines de la statique*, Paris 1905.
Dürer, A. *Underweysung der Messung mit dem Zirckel unn Richtscheyt in Linien, Ebnen unnd gantzen Corporen*, Nürnberg 1525.
———., *Vier Bücher von menschlicher Proportion*, Nürnberg 1528.
———., *Etliche Underricht zu Befestigung der Stett, Schloss und Flecken*, Nürnberg 1527.
Egnazio, G.B. *De exemplis illustrium virorum*, Paris 1554.
Erasmus, D. *Opera*, Basel 1540.
Erman, A. *Ägypten und ägyptisches Leben im Altertum*, ed. Ranke, H., Tübingen 1923.
Estienne, H. *Ciceronianum lexicon graecolatinum*, Paris 1557.
———., *Lexicon graecolatinum*, Geneva 1593.
Estienne, R. *Dictionaire francoislatin: contenant les motz & manières de parler francois, tournez en latin*, Paris 1539.
Euclides, *Opera omnia*, ed. Heiberg, I., H. Menge, 8 vol., Leipzig 1883-1916.
Falero, F. *Tratado del esphera y del arte del marear*, Sevilla 1535.

Farrington, B. 'Vesalius and the Ruin of Ancient Medicine: A Neglected Point of View', *The Modern Quarterley,* 1938, 1, l, pp.23-8.
Ficino, M. *Theologia platonica de immortalitate animorum,* Florence 1482.
Filelfo, F. *Epistolarum familiarum,* Venice 1502.
————., *Cent-dix lettres grecques de Filelfe,* notes and comments by Legrand, E., Paris 1892.
Fowler, Th. *Bacon's Novum organum,* edition with introd., notes, etc., 2nd ed., Oxford 1889.
Fracastoro, G. *De sympathia et antipathia rerum,* Lyon 1554.
————., *Homocentrica: Eiusdem de causis criticorum dierum per ea quae in nobis sunt,* Venice 1538.
Francesca, P. della, *De prospectiva pingendi,* 1484, ed. Winterberg, C., 2 vol., Strassburg 1899.
Frost, W. *Bacon und die Naturphilosophie,* Munich 1927.
Gabrielli, G. 'Il carteggio Linceo', *Reale Accademia Nationale dei Lincei, Classe di Scienze Morali, Storiche e Filologiche,* Rendiconti, VI, 1, 1925, pp. 137ff. and VI, 7, 1938/39, pp. 1ff.
————., 'La schede Fogeliane', *Reale Accademia Nationale dei Lincei, Classe di Scienze Morali, Storiche e Filologiche,* Rendiconti, VI, 15, 1939.
Galilei, C. *Le opere di Galileo Galilei,* 20 vol., Florence 1890-1909.
Gechauff, T. *Archimedis syracusani opera,* Basel 1544.
Ghiberti, *I commentarii,* Florence 1477.
Gilbert, W. *De magnete, magneticisque corporibus et de magno magnete tellure, physiologia nova plurimis et argumentis demonstrata,* London 1600.
————., *De mundo nostro sublunari. Philosophia nova, opus posthumum. Ab autoris fratre collectum pridem et dispositum,* Amsterdam 1651.
Gudger, E.W. 'The Five Great Naturalists of the Sixteenth Century: Belon, Rondelet, Salviani, Gesner and Aldrovandi: a Chapter in the History of Ichthyology', *Isis,* 1934/5, XXII, pp. 21-40.
Günther, R.T. *The Astrolabes of the World,* Oxford 1932.
————., *Early Science in Oxford,* Oxford 1923-45.
Gurvitch, G. *Natural Law,* Encyclopedia of the Social Sciences, New York 1933, XI, pp. 284 ff.
Hakluyt, R. *The principall navigations, voiages, traffiques and discoveries of the english nation,* London 1589.
Harriot, Th. A *Brief and True Report of the New Found Land of Virginia,* London 1588.
Hellmann, G. *Rara magnetica 1269-1599: P. de Maricourt, F. Falero, P. Nunes, J. de Castro, G. Hartmann, M. Cortes, G. Mercator, R. Norman, W. Borrough, S. Steven,* in Hellmann, G., *Neudrucke von Schriften und Karten über Meteorologie und Erdmagnetismus,* vol. 10, Berlin 1898.
Hipparchus, *In arati et eudoxi phaenomena commentariorum,* ed. Manitius, C., Leipzig 1894.
Hobbes, Th. *Opera philosophica,* London 1839-1845.
Hood, Th. *Inaugural Address as Mathematical Lecturer of the City of London,* London 1588, ed. Johnson, F. R., *Journal of the History of Ideas,* 1942, III, pp. 94-106.
————., *The Making and Use of a Sector,* London 1598.
————., *The Mariners Guide,* London 1596.
————., *The Use of Both the Globes,* London 1592.
————., *The Use of the Celestial Globe in Plans,* London 1590.
————., *The Use of the two Mathematicall Instrumentes, the Crosse Staffe and the Iacobs Staffe,* London 1590.

Hooke, R. *Lectures de potentia restitutiva*, London 1678.
————., *Micrographia*, London 1665.
Hooker, R. Off the Lawes of Ecclesiastical Polity, *The Works*, ed. Keble, J., 3 vol., 7th ed., Oxford 1888.
Hues, R. *Tractatus de globis et eorum usu*, London 1594.
Hume, D. *The Natural History of Religion*, London 1753.
Hutten, U. v. *Epistolae obscurorum virorum*, Hagenau 1515.
Huygens, Chr. *Oevres complètes*, 22 vol., The Hague 1888-1950.
Jähns, M. *Geschichte der Kriegswissenschaften vornehmlich in Deutschland*, Munich 1889-91.
Jastrow, M. *The Civilization of Babylonia and Assyria; its remains, language, history, religion, commerce, law, art and literature*, Philadelphia 1915.
Johnson, F.R. *Astronomical Thought in Renaissance England*, Baltimore 1937.
————., 'Gresham College, Precursor of the Royal Society', *Journal of the History of Ideas*, 1940,1, pp. 413-38.
Jöcher, C. G. *Allgemeines Gelehrtenlexicon*, Leipzig 1751.
Kaibel, G. *Epigrammata graeca ex lapidibus conlecta*, Berlin 1878.
Kelsen, H. 'Die Entstehung des Kausalgesetzes aus dem Vergeltungsprinzip', *Journal of Unified Science*, 1940, 8, pp. 69-130.
————., *Vergeltung und Kausalität*, The Hague 1941.
Kepler, J. *Opera omnia*, ed. Frisch, C., 8 vol., Frankfurt 1858-1871.
Leonardo da Vinci, *Das Buch von der Malerei: nach dem Codex Vaticanus 1270*, ed. Ludwig, H.., *Quellenschriften für Kunstgeschichte und Kunsttechnik des Mittelalters und der Renaissance*, vol. 15-18, Vienna 1882.
————., *Il codice atlantico*, ed. Piumati, G., Milan 1894-1904.
————.,*Les manuscrits de Léonardo de Vinci*, ed. Ravaisson-Mollien, M., Paris 1881-1891.
————., *Trattato de la pittura*, ed.Trichet Du Fresne, R., Paris 1651.
Locke, J. *An Essay Concerning Human Understanding*, London 1690.
Lombardus, P. *Textus sententiarum*, c. 1150, ed. de Gorichem, H., Basel 1492.
Lucretius, C. *Index Lucretianus*, ed. Paulson, J., Gothenborg 1911.
————., *De rerum natura*, ed. Munro, H., 4th ed., Cambridge 1893.
Maricourt, P. *Petri Peregrini Maricourtensis de magnete, seu rota perpetui motus libellus*, Augsburg 1558, in Hellmann, G., *Rara Magnetica*, Berlin 1898.
Maylender, M. *Storia delle Accademie d'Italia*, 5 vol., Bologna 1926-30.
Medina, P. de, *Arte de navegar*, Valladolid 1545.
Mercator, G. *Atlas minor*, ed. with introd. by Montanus, P., Amsterdam 1607.
Merton, R.K. 'Science, Technology, and Society in Seventeenth Century England', *Osiris*, 1938, IV, pp. 360-632.
Michelangelo [Buonarotti], *Rime e lettere*, Florence 1903.
Montaigne, M. de. *Essais*, Bordeaux 1582.
Montesquieu, Ch.. *De l' esprit des lois*. Geneva 1748.
Muratori, L. *Rerum italicarum scriptores: ab anno aerae christianae 500 ad 1500*, ed., 28 vol., Milan 1723-1751
Neudörfer, J. *Des Johann Neudörfer Schreib- und Rechenmeisters zu Nürnberg Nachrichten von Künstlern und Werkleuten daselbst aus dem Jahr 1547: nebst der Fortsetzung des Andreas Gulden*, reprinted in Lochner, G., *Quellenschriften für Kunstgeschichte und Kunsttechnik des Mittelalters und der Renaissance*, X, Vienna 1875.
Newton, I. *Isaaci Newtoni opera quae extant omnia*, ed. Horsley, S., 5 vol., London 1779-1785.
Nizolius, M. *De veris principiis et vera ratione philosophandi contra pseudophilosophos*, Parma 1553.

Norman, R. *The New Attractive: Containing a Short Discourse of the Magnes or Lodestone*, London 1581, 2nd ed. by Borough, W., London 1596.
Nunes, P. *Tratado da sphera: com a theorica do sol e da luna e ho primeiro livro da geographia da Claudia Ptolomeo*, Lisbon 1537.
Olschki, L. *Geschichte der neusprachlichen wissenschaftlichen Literatur*, vol. l: *Die Literatur der Technik und der angewandten Wissenschaften vom Mittelalter bis zur Renaissance*, Heidelberg 1919, vol. 2: *Bildung und Wissenschaft im Zeitalter der Renaissance in Italien*, Leipzig, Rome, Florence 1922, vol. 3: *Galilei und seine Zeit*, Halle 1927.
Ornstein, M. *The Role of Scientific Societies in the Seventeenth Century*, 3rd ed., Chicago 1938.
Ortelius, A. *Additamentum Quintum, Theatri Orbis Terrarum*, Antwerp 1595.
————., *Theatrum orbis terrarum,*, Antwerp 1570.
Oswald, J.C. A *History of Printing, its Development Through Five Hundred Years*, New York 1928.
Pacioli, L. *Summa de arithmetica, geometria, proportioni e proportionalita*, Venice 1494.
Palissy, B. *Discours admirables, de la nature des eaux et fontaines: tant naturelles qu' artificielles, des metaux, des fels et salines, des pierres, des terres, du feu et des emaux*, Paris 1580.
————., *Recette véritable par laquelle tous les hommes de la France pourront apprendre à multiplier et augmenter leurs thrésors*, La Rochelle 1563.
Pare, A. *La méthode de traicter les playes*, Paris 1545.
————., *Oeuvres complètes*, ed. Malgaigne, S. F., Paris 1840-41.
Pascal, B. *Oeuvres complètes*, ed. Brunschvicg, L., et al., 14 vol., Paris 1908-1923.
Pasquali, G. *Storia della critica del testo*, Florence 1934.
Pastrengo, G. *De originibus rerum*, Venice 1547.
Patrizzi, F. *Nova de universis philosophia*, 2nd ed., Venice 1593.
Paulsen, F. *Geschichte des gelehrten Unterrichtes auf den deutschen Schulen und Universitäten: vom Anfang des Mittelalters bis zur Gegenwart*, 3rd ed., Leipzig 1919.
Pauly, A. *Paulys Realencyclopädie der classischen Altertumswissenschaft*, ed. Wissowa, G., 24 vol., Stuttgart 1894-1963.
Petrarch, F. *Francisci Petrarcae epistolae de rebus familiaribus et variae: tum quam adhuc tum quae nondum editae*, ed. Fracassetti, G., 3 vol., Florence 1859-63.
————., *De remediis utriusque c. Eiusdem de contemptu mundi*, Rotterdam 1649.
Philo of Byzantium, *Philons Belopoiika: viertes Buch der Mechanik*, ed. and transl. Diels, H., E. Schramm, Abhandlungen der Preussischen Akademie der Wissenschaften, Philosophisch-historische Klasse, 1918, no. 16, Berlin 1919.
Pico della Mirandola, G. *De hominis dignitate*, Basel 1530.
————., *De studio divinae et humanae philosophiae*, Strassburg 1507.
Pliny (the elder), *Historia naturalis*, ed. Gaza, Th., Rome 1470.
Plotin, E. *Opera omnia*, ed. Creuzer, Fr., 3 vol., Oxford 1835.
Plutarchus, *Vitae*, Stuttgart 1925.
Poggio, B. *De varietate fortunae*, ed. Dominico, G., Paris 1723.
————., *Dialogus de infelicitate principum*, Paris 1511.
Poliziano, A. *Miscellanea*, Florence 1489.
————., *Opera*, Basel 1553.
Porta, G. della, *Magia naturalis*, Naples 1589.
Ptolomaeus, C. *Handbuch der Astronomie*, ed. and German transl. by Manitius, K., 2 vol., Leipzig 1912-1913.
————., *Opera quae exstant omnia*, ed. Heiberg, J., 3 vol., Leipzig 1898-1903.
————.*Libros del saber de Astronomia*, 1276, ed. Castilla, A., 5 vol., Madrid 1863-67.
Putnam, G. H. *Books and their Makers During the Middle Ages,* 1 vol., New York 1896-97.

Ramus, P. *Dialecticae partitiones,* Paris 1543.
Ramusio, G. B. *Delle navigationi et viaggi,* Venice 1550.
Recorde, R. *The Whetstone of Witte, Whiche is the Seconde Parte of Arithmetike,* London 1557.
Rehm, A. 'Zur Rolle der Technik in der griechisch-römischen Antike', *Archiv für Kulturgeschichte,* 1938, XXVIII, pp.135-162.
Robortelli, F. *Utinensis disputatio de ante critica corrigendi antiquorum libros,* 1557, Nürnberg 1747.
Roriczer, M. *Büchlein von der Fialen Gerechtigkeit,* Regensburg 1486.
Sabellicus, M. A. *De rerum et artium inventoribus poema,* Leipzig 1511.
Saggi di naturali esperienze fatti nell' accademia del cimento sotto la proteaone del serenissimo principe Leopoldo di Toscana (Trials of Natural Experiments made in the A.d.C.), ed. Magalotti, L., Florence 1666.
Salutati, C. *Epistolario,* ed. Novati, F., 4 vol., Rome 1891.
Santi, G. *Federigo di Montefeltro, duca di Urbino: cronaca,* first ed. by Santi, G., 1305, ed. Holtzinger, H., Stuttgart 1893.
Sanuto, L. *Geografia,* Venice 1588.
Sardus, A. *De rerum inventoribus libri II,* Mainz 1577.
Sarton, G. *Introduction to the History of Science,* 3 vol., Baltimore 1931.
————., 'Simon Stevin of Bruges (1548-1620)', *Isis,* 1934, XXI, pp. 241-262.
Scaliger, J.C. *Exotericarum exercitationum liber XV,* Paris 1557.
Schlosser, J. v. *Materialien zur Quellenkunde der Kunstgeschichte* (Wiener Akademie Berichte), 10 vol., Vienna 1914-1920.
Schlund.E. 'Petrus Peregrinus de Maricourt', *Archivum Franciscanum Historicum,* 1911, IV, pp.436-55 and 1912, V, pp. 22-40.
Schmuttermayer, H. *Fialenbüchlein 1484 oder 1489,* reprinted with an introduction in *Anzeiger für Kunde der deutschen Vorzeit,* 1881, 28, Neue Folge, pp. 66-78.
Seneca, L. *L. Annaei Senecae Naturalium quaestionum libros VIII,* ed. Gercke, A., Leipzig 1907.
Spinoza, B. *Opera,* ed. Vloten-Land, 4 vol., 3rd ed., The Hague 1914.
————., *Tractatus theologico-politicus,* Hamburg 1670.
St. Germain, Chr. *Doctor and Student,* 15th ed., London 1571.
Stevin, S. *L` arithmetique,* Leyden 1585.
————., *De Havenvinding,* Leyden 1599, reprinted in Hellmann, G., *Rara Magnetica,* Berlin 1898.
————., *De thiende,* Leyden 1585.
————., *Hypomnemata mathematica,* Lyon 1608.
————., *Les oeuvres mathematiques,* ed. Girard, A., Leyden 1634.
————., *Mémoires mathematique,* 4 vol., Leyden 1605-08.
————., *Wiscontige Gedachtenissen,* Leyden 1608.
Suarez, F. *Opera omnia,* ed. Berton, Ch., M. Andre, 28 vol., Paris 1856-1878.
Symonds, J.A. *Renaissance in Italy,* 3 vol., London 1877.
Tartaglia, N. *Euclide Megarense philosopho: solo introduttore delle scientie mathematice:diligentemente reassettato,* Venice 1665.
————., *General trattato di numeri et misure,* Venice 1556-60.
————., *Opera Archimedis Syracusani philosophi et mathematici ingeniosissimi,* first Latin ed. of Archimedes, Venice 1543.
————., *Questi et inventioni diverse,* Venice 1546.
————., *Ragionamenti sopra la sua travagliata inventione,* Venice 1551.
Taylor, E.G. *Tudor Geography 1485-1583,* London 1930.

Teggart, F. *Rome and China: A Study of Correlations of Historical Events*, Berkeley 1939.
Telesio, B. *De natura iuxta propria principia*, Rome 1665.
————., *De rerum natura iuxta propria principia*, Naples 1570.
Thompson, S. P. 'Petrus Peregrinus de Maricourt and his Epistola de Magnete', *Proceedings of the British Academy*, 1905/06, II, pp. 337-408.
————., 'The Family and Arms of Gilbert of Colchester', *Transactions of the Essex Archaeological Society*, 1906, IX, pp. 197-211.
Thorne, S. E. 'St. Germain's doctor and Student', *The Library*, 4th ser., X, 1930, pp. 421-26.
Tiraboschi, G. *Storia della letteratura Italiana*, new ed., 9 vol., Florence 1805-13.
Trissino, G. *La Poetica*, Venice 1562.
————., *Tutte le opere*, Verona 1729.
Valeriano, P. *De litteratorum infelicitate*, Venice 1620.
Valla, L. *Dialectice libri tres seu eiusdem reconcinnatio totius dialectice et fundamentorum universalis philosophie: ubi multa adversus Aristotelem*, Paris 1509.
————., *Treatise on the Donation of Constantine*, reprinted and trans. by Coleman, C., New Haven 1922.
————., *Elegantiae*. c.1440, in Hutten, U.v., *Phalarismus: Dialogus Huttenicus*, Mainz 1517.
Vasari, G. *Opera*, ed. Milanesi, G., 9 vol., Florence 1878-85.
Vergilius, P. *De rerum inventoribus*, Venice 1499.
Villard de Honnecourt, *Bauhüttenbuch*, c. 1235, Critical edition of the Bibliotheque National Paris, ed. Hahnloser, H., Vienna 1935.
Voigt, G. *Die Wiederbelebung des classischen Alterthums oder das erste Jahrhundert des Humanismus*, 3rd ed., Berlin 1893.
Volateranus, R. *Commentariorum urbanorum liber I*, Rome 1506.
Voltaire, *Oeuvres de Voltaire: Avec notes, préfaces, avertissemens, remarques historiques et littéraires*, ed. Beuchot, M., Paris 1829-1840.
Vossler, K. *Poetische Theorien in der italienischen Frührenaissance*, Berlin 1900.
Wallis, J. 'A Summary Account Given by Dr. John Wallis of the General Laws of Motion', *Philosophical Transactions of the Royal Society of London*, 1668, III, pp. 864-66.
Wilamowitz-Moellendorff, U. *Geschichte der Philologie*, in Gercke, A., E. Norden, *Einleitung in die Altertumswissenschaft*, 3 vol., 3rd ed., Leipzig 1927.
————., 'Antigonos von Karystos', *Philologische Untersuchungen*, 1881, IV, pp. 263-91.
Winternitz, M. *Geschichte der indischen Literatur*, Leipzig 1920.
Wren, Ch. 'Lex natura de collisione corporum', *Philosophical Transactions of the Royal Society of London*, 1668, III, pp. 867-8.
Wright, E. *Certain Errors in Navigation*, London 1599.
Zannoni, G. B. *Atti delle Imperiale e Reale Accademia della Crusca*, 1819, I, 1ff.
Zinner, E. *Die Geschichte der Sternkunde*, Berlin 1931.

APPENDIX IV

BIBLIOGRAPHY OF EDGAR ZILSEL'S WORKS WITH A BIBLIOGRAPHY OF SECONDARY LITERATURE CONCERNING ZILSEL

Books

Zilsel, E. (1916) *Das Anwendungsproblem. Ein philosophischer Versuch über das Gesetz der grossen Zahlen und die Induktion*, Leipzig: Barth.
--- (1918) *Die Geniereligion. Ein kritischer Versuch über das moderne Persönlichkeitsideal*, Leipzig & Wien: Braumüller. Reprinted in 1990 as *Die Geniereligion. Ein kritischer Versuch über das moderne Persönlichkeitsideal, mit einer historischen Begründung*, (Herausgegeben und eingeleitet von Johann Dvorak), Frankfurt am Main: Suhrkamp.
--- (1926) *Die Entstehung des Geniebegriffes. Ein Beitrag zur Ideengeschichte der Antike und des Frühkapitalismus*, Tübingen: Mohr. [Republished in 1972 by Olms Verlag (Hildesheim & New York), with a preface by H. Maus.]
--- (1976) *Die sozialen Ursprünge der neuzeitlichen Wissenschaft*, Frankfurt am Main: Suhrkamp (Herausgegeben und übersetzt von Wolfgang Krohn mit einer biobibliographischen Notiz von Jörn Behrmann).[1]
--- (1992) *Wissenschaft und Weltanschauung: Aufsätze 1929-1933*, (Mit einem Vorwort von Karl Acham, herausgegeben und eingeleitet von Gerald Mozetic), Wien, Köln, Weimar: Böhlau Verlag.[2]

Essays

--- (1912) 'Mozart und die Zeit. Eine didaktische Phantasie', *Der Brenner* (Halbmonatsschrift, Innsbruck: Brenner-Verlag, pp. 268-71.)
--- (1913) 'Bemerkungen zur Abfassungszeit und zur Methode der Amphibolie der Reflexionsbegriffe', *Archiv für Geschichte der Philosophie*, 26:431-48.
--- (1921a) 'Versuch einer neuen Grundlegung der statistischen Mechanik', *Monatshefte für Mathematik und Physik*, 31:118-56.
--- (1921b) 'Der einführende Philosophieunterricht an den neuen Oberschulen', *Volkserziehung* (published by Österr. Unterr. Amt), 324-41.
--- (1924) 'Kant als Erzieher', *Schulreform*, 3:182-8.
--- (1927) 'Über die Asymmetrie der Kausalität und die Einsinnigkeit der Zeit', *Die Naturwissenschaften*, 15,12:280-6.

---

1 This is a translation, in order of appreance, of (1942c) [49-65], (1942a) [66-97], (1941a) [98-126], (1945) [127-150], (1940) [151-6], (1942b) [157-99], (1941c) [200-211], (1940b) [212-219]. The numbers in square brackets refer to the page numbers in the German edition.
2 This collection contains the following essays in order of appearence (1929) [31-44], (1932c) [45-57], (1933a) [58-73], (1931b) [77-87], (1931c) [88-98], (1931a) [101-44], (1932d, review) [145-9], (1933c) [153-66], (1933b) [167-178]. The numbers in square brackets refer to the pages of (1992).

--- (1928) 'Naturphilosophie', in F. Schnass (ed.), *Einführung in die Philosophie*, (pp. 107-43). Osterwieck-H: A. W. Zickfeldt.
--- (1929) 'Philosophische Bemerkungen', *Der Kampf*, 22:178-86.
--- (1930) 'Soziologische Bemerkungen zur Philosophie der Gegenwart', *Der Kampf*, 23:410-24.
--- (1931a) 'Geschichte und Biologie, Überlieferung und Vererbung', *Archiv für Sozialwissenschaft und Sozialpolitik*, 65:475-524.
--- (1931b) 'Materialismus und marxistische Geschichtsauffassung', *Der Kampf*, 24:68- 75.
--- (1931c) 'Partei, Marxismus, Materialismus, Neukantianismus', *Der Kampf*, 24:213-20.
--- (1932a) 'Das mechanistische Weltbild und seine Überwindung', *Der Atheist*, 6,9:129- 31.
--- (1932b) 'Bemerkungen zur Wissenschaftslogik', *Erkenntnis*, 3,2/3:143-61.
--- (1932c) 'Die geistige Situation der Zeit', *Der Kampf*, 25:168-76.
--- (1932d) 'Die Arbeitsschule - eine metaphysische Anstalt', *Wissenschaft und Schule*.[3]
--- (1933a) 'Die gesellschaftlichen Wurzeln der romantische Ideologie', *Der Kampf*, 26:154-64.
--- (1933b) 'Das Dritte Reich und die Wissenschaft', *Der Kampf*, 26:486-93. (Written under the name Rudolf Richter.)
--- (1933c) 'SA philosophiert', *Der Kampf*, 26:393-402. (Written under the name Rudolf Richter.)
--- (1935) 'P. Jordans Versuch, den Vitalismus quantenmechanisch zu retten', *Erkenntnis*, 5:56-65.
--- (1937) 'Moritz Schlick', *Die Naturwissenschaften*, 25,11:161-67.
--- (1940a) 'Copernicus and Mechanics', *Journal of the History of Ideas*, 1:113-8. (Reprinted in P. P. Wiener & A. Noland (eds.), *Roots of Scientific Thought: A Cultural Perspective*, [pp. 276-80]. New York: Basic Books, 1957.)
--- (1940b) 'History and Biological Evolution', *Philosophy of Science*, 7:121-8.
--- (1941a) 'The Origins of William Gilbert's Scientific Method', *Journal of the History of Ideas*, 2:1-32. (Reprinted in P. P. Wiener & A. Noland (eds.), *Roots of Scientific Thought: A Cultural Perspective*, [pp. 219-50]. New York: Basic Books, 1957.)
--- (1941b) 'Phenomenology and Natural Science', *Philosophy of Science*, 8:26-32.
--- (1941c) 'Physics and the Problem of Historico-sociological Laws', *Philosophy of Science*, 8:567-79.
--- (1941d) 'Concerning 'Phenomenology and Natural Science'', *Philosophy and Phenomenological Research*, 2:219-20.
--- (1942a) 'The Genesis of the Concept of Physical Law', *The Philosophical Review*, 51,3:245-79.
--- (1942b) 'Problems of Empiricism', in O. Neurath (ed.), *Founadations of the Unity of Science: Towards an International Encyclopedia of Unified Science* Vol. II, 8, (pp. 53-94). Chicago: University of Chicago Press.
--- (1942c) 'The Sociological Roots of Science', *The American Journal of Sociology*, 47:544-62.
--- (1945) 'The Genesis of the Concept of Scientific Progress', *Journal of the History of Ideas*, 6:325-49. (Reprinted in P. P. Wiener & A. Noland (eds.), *Roots of Scientific Thought: A Cultural Perspective*, [pp. 251-75]. New York: Basic Books, 1957.)

---

3 Source of this reference is Zilsel's own bibliography which he wrote in 1939 while in New York. We have not been able to get hold of a copy of this essay.

BIBLIOGRAPHY OF EDGAR ZILSEL'S WORKS

Reviews

--- (1924a) 'Review of G. Heymans *Die Gesetze und Elemente des wissenschaftlichen Denkens. Ein Lehrbuch der Erkenntnistheorie in Grundzügen* (Leipzig: J. A. Barth, 1923, 4. Auflage)', *Die Naturwissenschaften*, 12:33-5.

--- (1924b) 'Review of P. Hertz *Über das Denken und seine Beziehung zur Anschauung. 1. Teil: Über den funktionalen Zusammenhang zwischen auslösendem Erlebnis und Enderlebnis bei elementaren Prozessen* (Berlin: J. Springer, 1923)', *Die Naturwissenschaften*, 12:348-9.

--- (1925a) 'Review of A. C. Elsbach *Kant und Einstein. Untersuchungen über das Verhältnis der modernen Erkenntnistheorie zur Relativitätstheorie* (Berlin & Leizig: W. de Gruyter & Co., 1924)', *Die Naturwissenschaften*, 13:406-7.

--- (1925b) 'Review of H. Reichenbach *Axiomatik der relativischen Raum- und Zeitlehre* (Braunschweig: Fr. Vieweg & Sohn, 1924)', *Die Naturwissenschaften*, 13:407-9.

--- (1925c) 'Review of B. Bauch *Das Naturgesetz. Ein Beitrag zur Philosophie der exakten Wissenschaften* (Berlin & Leipzig: B. G. Teubner, 1924)', *Die Naturwissenschaften*, 13:407.

--- (1926a) 'Review of E. Bleuler *Das Psychoide Prinzip der organischen Entwicklung* (Berlin: J. Springer, 1925)', *Die Naturwissenschaften*, 14:644-6.

--- (1926b) 'Review of J. Günther *Allgemeine Ontologie der Wirklichkeit* (Halle: M. Niemeyer, 1925)', *Die Naturwissenschaften*, 14:646.

--- (1926c) 'Review of O. Hölder *Die Mathematische Methode. Logisch-erkenntnistheoretische Untersuchungen im Gebiete der Mathematik, Mechanik und Physik* (Belin: J. Springer, 1924)', *Die Naturwissenschaften*, 14:646.

--- (1926d) 'Review of E. Lohr *Atomismus und Kontinuitätsttheorie in der neuzeitlichen Physik* (Leizig & Belin: B. G. Teubner, 1926)', *Die Naturwissenschaften*, 14:12 86.

--- (1927a) 'Review of H. Weyl *Philosophie der Mathematik* (München & Belin: R. Oldenburg, 1926)', *Die Naturwissenschaften*, 15:24-7.

--- (1927b) 'Review of V. Kraft *Die Grundformen der wissenschaftlichen Methoden* (Wien & Leipzig: Hölder-Pichler-Tempsky, 1926)', *Die Naturwissenschaften*, 15:101.

--- (1927c) 'Review of A. Meyer *Logik der Morphologie im Rahmen einer Logik der gesamten Biologie* (Berlin: J. Springer, 1926)', *Die Naturwissenschaften*, 15:101-2.

--- (1927d) 'Review of H. Cornelius *Grundlagen der Erkenntnistheorie* (München: E. Reinhardt, 1926, 2. Auflage)', *Die Naturwissenschaften*, 15:103.

--- (1927e) 'Review of J. H. Tummers *Die spezielle Relativitätstheorie und die Logik* (Maeseyck: J. Denis, 1924)', *Die Naturwissenschaften*, 15:294.

--- (1927f) 'Review of J. M. Keynes *Über Wahrscheinlichkeit* (Leipzig: J. A. Barth, 1926)', *Die Naturwissenschaften*, 15:867-9.

--- (1927g) 'Review of A. Eleutheropulus *Die exakten Grundlagen der Naturphilosophie* (Stuttgart: F. Enke, 1926)', *Die Naturwissenschaften*, 15:869.

--- (1928a) 'Review of W. Gent *Die Philosophie des Raumes und der Zeit. Historische, kritische und analytische Untersuchungen* (Bonn: F. Cohn, 1926)', *Die Naturwissenschaften*, 16:60-1.

--- (1928b) 'Review of K. Lasswitz *Geschichte der Atomik vom Mittelalter bis Newton* (Leipzig: L. Voss, 1926)', *Die Naturwissenschaften*, 16:151-2.

--- (1928c) 'Review of W. Forst *Bacon und die Naturphilosophie* (München: E. Reinhardt, 1927)', *Die Naturwissenschaften*, 16:152-3.

--- (1928d) 'Review of B. Russell *Die Analyse des Geistes* (Leipzig: F. Meiner, 1927)', *Die Naturwissenschaften*, 16:1030-1.

--- (1929a) 'Review of B. Russell *Philosophie der Materie* (Leipzig & Berlin: B. G. Teubner, 1929)', *Die Naturwissenschaften*, 18:933-4.

--- (1929b) 'Review of H. Bergmann *Der Kampf um das Kasusalgesetz in der jüngsten Physik* (Braunschweig: F. Vieweg & Sohn, 1929)', *Die Naturwissenschaften*, 18:934-5.

--- (1929c) 'Review of D. Hilbert & W. Ackermann *Grundzüge der theoretischen Logik* (Berlin: J. Springer, 1928)', *Die Naturwissenschaften*, 18:935-6.

--- (1929d) 'Review of J. Reinke *Wissen und Glauben in der Naturwissenschaft* (Leipzig: J. A. Barth, 1929)', *Die Naturwissenschaften*, 18:936.

--- (1931) 'Review of A. Kranold *Vom ethischen Gehalt der sozialistischen Idee* (und *das Verhältnis des Marxismus zur Ethik* (Neurer Breslauer Verlag, 1930)', *Der Kampf*, 24:189-90.

--- (1932a) 'Review of R. v. Mises *Wahrscheinlichkeitsrechung und ihre Anwendung in der Statik und theoretischen Physik* (Leipzig & Wien: F. Deuticke, 1931)', *Die Naturwissenschaften*, 20:472-4.

--- (1932b) 'Review of K. Jaspers *Die geistige Situation der Zeit* (Berlin & Leipzig: W. de Gruyter, 1931)', *Die Naturwissenschaften*, 20:474-5.

--- (1932c) 'Review of G. P. Gonger *A World of Epitomizations. A Study in the Philosophy of the Sciences* (Princeton, N. J.: Princeton Unversity Press, 1931)', *Die Naturwissenschaften*, 20:776-7.

--- (1932d) 'Review of O. Neurath *Empirische Soziologie: Der wissenschaftliche Gehalte der Geschichte und Nationalökonomie* (Berlin: J. Springer, 1931)', *Der Kampf*, 25:91-4.

--- (1932e) 'Review of A. S. Eddington *Das Weltbild der Physik und ein Versuch seiner philosophischen Deutung* (Braunschweig: F. Vieweg u. Sohn, 1931)', *Angewandte Chemie*, 45,11:230/1.

--- (1933a) 'Review of H. Dingler *Geschichte der Naturphilosophie* (Berlin: Junker & Dünnhampt, 1932)', *Die Naturwissenschaften*, 21:224.

--- (1933b) 'Review of W. Dubislaw *Die Philosophie der Mathematik in der Gegenwart* (Berlin: Junker & Dünnhampt, 1932)', *Die Naturwissenschaften*, 21:224-5.

--- (1933c) 'Review of H. Andre *Urbild und Ursache in der Biologie* (München & Berlin: H. Oldenburg, 1931)', *Die Naturwissenschaften*, 21:721.

--- (1933d) 'Review of E. Bünning *Mechanismus, Vitalismus und Teleologie* (Göttingen: Verlag 'Öffentliches Leben', 1932)', *Die Naturwissenschaften*, 21:792.

--- (1933e) 'Review of B. Bavink *Ergebnisse und Probleme der Naturwissenschaften. Eine Einführung in die heutige Naturphilosophie* (5. Auflage, Leipzig: S. Hirzel, 1933)', *Die Naturwissenschaften*, 21:899-900.

--- (1933f) 'Review of P. W. Bridgman *Die Logik der heutigen Physik* (München: M. Huebner, 1932)', *Zentralblatt für Mathematik und ihre Grenzgebiete*, 5:147.

--- (1934) 'Review of H. Stotz *Die Welt der physikalischen Theorien. Die Formen und Prinzipien ihrer Konstruktion* (Giessen: Diss, 1932)', *Zentralblatt für Mathematik und ihre Grenzgebiete*, 8:98.

--- (1935a) 'Review of E. Wind *Das Experiment und die Metaphysik* (Tübingen: J. C. B. Mohr (Paul Siebeck), 1934)', *Die Naturwissenschaften*, 23:19-20.

--- (1935b) 'Review of K. Popper *Logik der Forschung. Zur Erkenntnistheorie der modernen Naturwissenschaften* (Wien: J. Springer, 1935)', *Die Naturwissenschaften*, 23:531-2.

--- (1935c) 'Review of F. Warrain *Essai sur les pricipes des algorithmes primitifs. Addition - soustraction - multiplication - division - puissances -racines* (Paris: Hermann & Cie, 1934)', *Zentrallblatt für Mathematik und ihre Grenzgebiete*, 11,:97/ 8.

--- (1937) 'Review of E. Meisner *Erkenntniskritische Weltanschauung auf der Grundlage der Arbeitsbedingungen des Gehirns* (Leipzig: F. Meiner, 1936)', *Die Naturwissenschaften*, 25:494-5.

Comments

--- (1932) 'Comments on Aster, E. v. und Th. Vogel: Kritische Bemerkungen Zu Hugo Dinglers Buch *Das Experiment. Erkenntnis, zugl. Ann. Philosoph.* 2, 1-20 (1931)', *Zentralblatt Für Mathematik und ihre Grenzgebiete*, 2, :177.--- (1932) 'Comments on Dingler, Hugo: *Über den Aufbau der experimentellen Physik. Erkenntnis, zugl Ann. Philosoph.* 2, 21-38 (1931)', *Zentralbaltt für Mathematik und ihre Grenzgebiete*, 2, : 177.
--- (1932) 'Comments on 'Reichenbach, Hans: *Schlußbemerkung. Erkenntnis, zugl. Ann. Philosoph.* 2, 39-41 (1931)', *Zentralblatt für Mathematik und ihre Grenzgebiete*, 2, :177.
--- (1933) 'Comments on 'Hahn, Hans: *Logik, Mathematik und Naturerkennen* (Wien: Gerold & Co, 1933); Reichenbach, Hans: Kant und die Naturwissenschaft *Naturwiss.* 21, 601-06 & 624-26 (1933); Höningswald, Richard: Kausalität und Pysik. Eine methodologische Überlegung. S-B preuss. Akad. Wiss. H 16/7, 568-78 (1933)', *Zentralblatt für Mathematik und ihre Grenzgebiete*, 5, :195.
--- (1934) 'Comments on Jordan, Pascual: *Quantenphysikalische Bemerkungen zur Biologie und Psychologie. Erkenntnis* 4, 215-52 (1934)', *Zentralblatt für Mathematik und ihre Grenzgebiete*, 9, :387.
--- (1934) 'Comments on Jensen, Paul: *Kausalität, Biologie un Psychologie Erkenntnis 4*, 165-214 (1934)', *Zentralblatt für Mathematik und ihre Grenzgebiete*, 9, :387.

Obituaries

Hollitscher, W. (1954) 'Zur Erinnerung an Professor Edgar Zilsel: Die Geburt der Modernen Wissenschaft', *Oesterreichisches Tagebuch: Wochenschrift für Kultur, Politik und Wirtschaft*, 20, Oktober 19, col. 1.
Kristeller, P. O. (1948) 'Memorial Notice of Edgar Zilsel', *Philosophical Review*, 57, p. 375.

Responses to Zilsel's work

I: Reviews of his books

a. *'Das Anwendungsproblem'* (1916)

Author's Summary (*Selbstanzeige*), *Kantstudien*, 21, 335.
Bavink, (1916), 'Review of Zilsel (1916)', *Zeitschrift für den physikalischen und chemischen Unterricht*, 29.J., 198.
Hahn, H. (1917), 'Review of Zilsel (1916)', Monatshefte für Mathematik und Physik, 27-28 year, 37-38.
Sterzinger, O. (1918), 'Review of Zilsel (1916)', *Archiv für Geschichte der Philosophie*, 31 Bd., 184-7.

b. *'Die Geniereligion'* (1918)

Author's Summary (*Selbstanzeige*), *Kantstudien*, 24B, 165.
Henning, H. (1920), 'Review of Zilsel (1918)', *Zeitschrift für Psychologie und Physiologie der Sinnesorganen*, I Abtlg., Zeitschrift für Psychologie, 84, 356.
Reimer, W. (1919), 'Review of Zilsel (1928)', Centralblatt Lit. Dtschld., Nr. 16/17, p. 179/80.

c. '*Die Entstehung des Geniebegriffes*' (1926)

Anler, Ch. (1927), *La Revue critique l'histoire et de littérature*, 244.
Baumgart, D. (1927) 'Review of Zilsel (1926)', *Zeitschrift für Ästhetik und allgemeine Kuntwissenschaft*, vol. 22, nr.4.
Baeumler, A. (1929), 'Review of Zilsel (1926)', *Logos*, 18 vol., 140-43.
Böckmann, P. (1928), 'Review of Zilsel (1926)',*Zeitschrift für deutsche Bildung*, 53, iv.
Brügel, F. (1927) 'Review of Zilsel (1926)', *Der Kampf*, 20:42-3.
Croce, B. (1926) 'Review of Zilsel (1926)', *La Critica*, 24:297-8.
Evola, J. (1928) 'Review of Zilsel (1926)', *Bilychnis*, 32 vol., p.162.
Gmelin, H. (1930) 'Review of Zilsel (1926)', *Sprachen d. neuer.*, 35 vol., 396.
Hashagen, J. (1927) 'Review of among others Zilsel (1926)', *Schmollers Jahrbuch für Gesetzgebung, Verwaltung und Volkswirtschft im Deutschen Reiche*, 51:630-3.
Hundhausen, C. (1927) 'Review of Zilsel (1926)', *Kölner Zeitschrift für Soziologie*, 6th year, p.92.
Jordan, L. (1930) 'Review of Zilsel (1926)', *Zeitschrift f. romanische Philol.*, 50 vol., 362-69.
Anonymous (1926) 'Review of Zilsel (1926)', *Revue de Metaphysique et de Morale*, 33rd year, Suppl. Juillet-Sept., p.13,II.
Lukàcs, G. (1928) 'Review of Zilsel (1926)', *Archiv für die Geschichte des Sozialismus und der Arbeiterbewegung*, 13:299-302.
Messinger, K.A. (1927) 'Review of Zilsel (1926)', *Pruess. Jahrbuecher*, 112, 201-9.
Nestle, W. (1927) 'Review of Zilsel (1926)', *Philologische Wochenschrift*, 47,48:1453-6.
Onians, R. B. (1926) 'Review of Zilsel (1926)', *The Classical Review*, 40:171.
Pitrou, R. (1927) 'Review of Zilsel (1926)', *Revue internationale de sociologie*, 35, 308-12.
Rehm, W. (1926) 'Review of Zilsel (1926)', *Deutsche Literaturzeitung*, 47, 36:1756-60.
Weinberger (1929), 'Review of Zilsel (1926)', *Wochenschrift, Berliner Philolog.*, Abt. 1921, *Philos. Wochenschrift*, 49 year, p.398.
Wundt M. (1926) 'Review of Zilsel (1926)', *Literarische Wochenschrift*, p.868.
Vossler K. (1926) 'Review of Zilsel (1926)', *Frankfurter Zeitung*, Feb. 18, nr. 131, col. 3.

d. '*Geschichte und Biologie, Überlieferung und Vererbung*'

Latten, W. (1931) 'Review of Zilsel (1931)', *Kölner Zeitschrift für Soziologie*, 10 year, p.406.

e. '*Problems of Empiricism*'

Beth, E.W. (1949) 'Review of Zilsel (1942b)', *Algemeen Nederlands Tijdschrift voor Wijsbegeerte en Psychologie*, vol. 41, p. 136.
Hempel, C.G. (1942) 'Review of Zilsel (1942b)', *Philosophical Abstracts*, received 2-42.
Margenau, H. (1943) 'Review of Zilsel (1942b)', *Philosophical Review*, Jan. 1943.
Nagel, E. (1942) 'Review of Zilsel (1942b)', *The Journal of Philosophy*, Dec. 4, 1942.

II: Direct responses to his essays

Adler, M. (1931) 'Wozu schreibt man Bücher? Melancholische Betrachtung zu einer Buchbesprechung', *Der Kampf*, 24:125-31. (Reply to [1931b].)
Carnap, R. (1932) 'Erwiderung auf die vorstehenden Aufsätze von E. Zilsel und K. Duncker', *Erkenntnis*, 3,2/3:177-88. (Reply to [1932b].)

Cerf, W. H. (1941a) 'Review of 'Phenomenology and Natural Science'', *Philosophy and Phenomenological Research*, 1:513. (Reply to [1941b].)
--- (1941b) 'In Reply to Mr. Zilsel', *Philosophy and Phenomenological Research*, 2:220-22. (Reply to [1941d].)
Keller, A. C. (1950) 'Zilsel, The Artisans, and the Idea of Progress in the Renaissance', *Journal of the History of Ideas*, 11:235-40. (Reply to [1945]; reprinted in P. P. Wiener & A. Noland (eds.), *Roots of Scientific Thought: A Cultural Perspective*, (pp. 281-6). New York: Basic Books, 1957.)
Reichenbach, H. et al. (1935) 'Zu Edgar Zilsel 'P. Jordans Versuch, den Vitalismus quantenmechanlich zu retten'', *Erkenntnis*, 5:178-84. (Reply to [1935].)
Taube, M. (1943) 'Dr. Zilsel on the Concept of Physical Law', *Philosophical Review*, 52:304-5.

III: Indirect responses to his essays

Crombie, A. C. (1975) 'Some Attitudes to Scientific Progress: Ancient. Medieval and Early Modern', *History of Science*, 13:213-30. (Reaction to [1945].)
Hall, A. R. (1959) 'The Scholar and the Craftsman in the Scientific Revolution', in M. Clagett (ed.), *Critical Problems in the History of Science*, (pp. 3-23). Madison: University of Wisconsin Press; reprinted in A. R. Hall *Science and Society* (Aldershot: Variocum, 1994.).
Milton, J. R. (1981) 'The Origin and Development of the Concept of the "laws of nature"?', *Archives Européennes de Sociologie*, 22:173-195.
Molland, A. G. (1978) 'Medieval Ideas of Scientific Progress', *Journal of the History of Ideas*, 39:561-77. (Reaction to [1945].)
Mommsen, T. E. (1951) 'St. Augustine and the Christian Idea of Progress', *Journal of the History of Ideas*, 12:346-74. (Reaction to [1945].)
Needham, J. (1951) 'Human Laws and the Laws of Nature in China and the West', *Journal of the History of Ideas*, 12:3-30, 194-230. (Independent confirmation of [1942c].)
Oakley, F. (1961) 'Christian Theology and the Newtonian Science: The Rise of the Concept of Laws of Nature', *Church History*, 30: 433-57. (Critique of [1942a])
Schramm, M. (1981) 'Roger Bacons Begriff vom Naturgesetz', in P. Weimar (ed.), *Die Renaissance der Wissenschaften im 12. Jahrhundert*, (pp. 197-209). Zürich: Artemis Verlag. (Discussion of R. Bacon's conception of law.)
Steinle, F. (1995) 'The Amalgamation of a Concept - Laws of Nature in the New Sciences', in F. Weinert (ed.) *Laws of Nature: Essays of the Philosophical, Scientific and Historical Dimensions*, (pp. 316-68). Berlin & New York: W.de Gruyter. (Discusses the differences of the concepts of law of nature among 17th century authors.)
Rossi, P. (1970) *Philosophy, Technology, and the Arts in the Early Modern Era*, New York: Harper Torch Books. (Chap. 2 is relevant for [1945].)
Ruby, J. E. (1986) 'The Origins of Scientific 'Law'', *Journal of the History of Ideas*, 47:341-359. (Rebuttal of [1942c].)

IV: Reviews of the 1976 German edition of Zilsel's essays on the Social Origins of Science, translated and edited by Wolfgang Krohn

Freudenthal, G. (1977) 'Review of E. Zilsel' *Das Argument* 103:395-7.
Hornung, W. (May 28, 1976) 'Review of E. Zilsel,' *Die Zeit*.
Kaminski, W. (1976) 'Review of E. Zilsel,' *Philosophischer Literaturanzeiger* 29(5):264-7.
Ley, H. (1976) 'Review of E. Zilsel,' *Deutsche Zeitschrift für Philosophie* 26(1):132-6.

Ludwig, K. (1976) 'Review of E. Zilsel,' *Technikgeschichte* 43(4):327-328.
Prenzel, I. and Ueding, G. (Nov. 8, 1976) 'Review of E. Zilsel,' *Hessischer Rundfunk.*
Rammert, W. (1977) 'Review of E. Zilsel,' *Kölner Zeitschrift für Soziologie und Sozialpsychologie* (4):813-5.
Rigler, E. (Dec. 1977) 'Review of E. Zilsel,' *Wiener Tagebuch.*
Schleir, H. (1977) 'Review of E. Zilsel,' *Zeitschrift für Geschichtswissenschaft.*
Thom, A. (1980) 'Review of E. Zilsel,' *Wissenschaftliche Zeitschrift* 29(4): 408-9.

Secondary literature

Books
Dvorak, J. (1981) *Edgar Zilsel und die Einheit der Erkenntnis*, Wien: Löcker Verlag.

Essays
Dahms, H. J. (1993) 'Edgar Zilsels Projekt 'The Social Roots of Science' und seine Beziehungen zur Frankfurther Schule', in R. Haller & F. Stadler (ed.), *Wien-Berlin-Prag: Der Aufstieg der Wissenschaftlichen Philosophie; Zentenarien Rudolf Carnap - Hans Reichenbach - Edgar Zilsel,* (pp. 474-500). Wien: Verlag Hölder- Pichler-Tempsky.
Dvorak, J. (1985) 'Wissenschaftliche Weltauffassung, Volkshochschule und Arbeiterbildung im Wien der Zwischenkriegszeit am Beispiel von Otto Neurath und Edgar Zilsel', in J-H. Dahms (ed.), *Philosophie, Wissenschaft, Auklärung. Beiträge zur Geschichte und Wirkung des Wiener Kreises,* (pp. 129-43). Berlin & New York: W. de Gruyter. [Related shorter version (1991) 'Otto Neurath and Adult Eduaction: Unity of Science, Materialism and Comprehensive Enlightenment', in T. E. Uebel, 1991, (pp. 265-74).
--- (1990) 'Zu Leben und Werk Edgar Zilsels und zur Soziologie des Geniekults', in E. Zilsel, (1990), pp. 7-40.
--- (1993) 'Wissenschaft als gesellschaftliche Auseinandersetzung und als kollektiver Arbeitspross - Edgar Zilsel und sein Werk', in R. Haller & F. Stadler (ed.), *Wien--Berlin-Prag: Der Aufstieg der Wissenschaftlichen Philosophie; Zentenarien Rudolf Carnap - Hans Reichenbach - Edgar Zilsel,* (pp. 424-46). Wien: Verlag Hölder--Pichler-Tempsky.
Fischer, K. R. (1993) 'Das Historische Bewußtsein bei Carnap, Reichenbach und Zilsel', in R. Haller & F. Stadler (ed.), *Wien-Berlin-Prag: Der Aufstieg der Wissenschaftlichen Philosophie; Zentenarien Rudolf Carnap - Hans Reichenbach - Edgar Zilsel,* (pp. 555-62). Wien: Verlag Hölder-Pichler-Tempsky.
Fleck, C. (1993) 'Marxistische Kausalanalyse und funktionale Wissenschaftssoziologie. Ein Fall unterbliebenen Wissenstranfers', in R. Haller & F. Stadler (ed.), *Wien-Berlin-Prag: Der Aufstieg der Wissenschaftlichen Philosophie; Zentenarien Rudolf Carnap - Hans Reichenbach - Edgar Zilsel,* (pp. 501-24). Wien: Verlag Hölder-Pichler-Tempsky.
Götz, C. M., & Pankratz, T. (1993) 'Edgar Zilsels Wirken im Rahmen der Wiener Volksbildung und Lehrerfortbildung', in R. Haller & F. Stadler (ed.), *Wien-Berlin-Prag: Der Aufstieg der Wissenschaftlichen Philosophie; Zentenarien Rudolf Carnap - Hans Reichenbach - Edgar Zilsel,* (pp. 467-73). Wien: Verlag Hölder-Pichler-Tempsky.
Krohn, W. (1985) 'Edgar Zilsel zur Methodologie einer exakten Geisteswissenschaft', in J-H. Dahms (ed.), *Philosophie, Wissenschaft, Auklärung. Beiträge zur Geschichte und Wirkung des Wiener Kreises,* (pp. 257-75). Berlin & New York: W. de Gruyter.
--- (1990) 'Edgar Zilsel und die marxistische Tradition', *Unpublished MS/ Paper presented at the Inter-University Centre Dubrovnik.*

--- (1993) 'In Search of Laws: Edgar Zilsel, the Vienna Circle, and Marxian Tradition', *Unpublished MS/ Paper presented to the 1993 annual History of Science Society Meeting*,

Nemeth, E. (1995) 'Wir Zuschauer' und das 'Ideal der Sache'. Bemerkungen zu Edgar Zilsels 'Geniereligion', *Vorträge des Instituts Wiener Kreis*, 1997, pp. 157-70.

Rutte, H. (1993) 'Zu Zilsels erkenntnistheoretischen Ansichten in der Phase des Wiener Kreises', in R. Haller & F. Stadler (ed.), *Wien-Berlin-Prag: Der Aufstieg der Wissenschaftlichen Philosophie; Zentenarien Rudolf Carnap - Hans Reichenbach - Edgar Zilsel*, (pp. 447-66). Wien: Verlag Hölder- Pichler-Tempsky.

Stadler, F. (1979) 'Aspekte des Gesellschaftlichen Hintergrunds und Standorts des Wiener Kreises am Beispiel der Universität Wien', in H. Berghel, et al. (eds.), *Wittgenstein, The Vienna Circle and Critical Rationalism*, (pp. 41-59). Wien: Hölder-Pichler-Tempsky. [English translation, and slightly shortened: (1991) 'Aspects of the Social Background and Position of the Vienna Circle at the University of Vienna', in T. E. Uebel (ed.), 1991, pp. 51- 77.

Uebel, T.E. ed. (1991), *Rediscovering the Forgotten Vienna Circle: Austrian Studies on Otto Neurath and the Vienna Circle*, Kluwer Academic Publishers.

Zilsel, P. R. (1982) 'Portrait of My Father', *Shmate: A Journal of Progressive Jewish Thought*, 1,1:12-3. [German translation: (1988) 'Über Edgar Zilsel', in F. Stadler (ed.), *Vertriebene Vernunft (Vol II): Emigration und Exil Österreichischer Wissenschaft*, (pp. 929-32). Wien & München: Jugend & Volk; also in [1990], pp. 41-7.).]

INDEX OF NAMES

Accoltus 26, 27, 42, 66, 148, 235
Achilles 43, 135
Agnolo 150
Agricola, G. 80, 81, 92, 235
Agricola, R. 55, 64, 68, 235
Aischrion 137
Alberti 13, 20, 36, 43, 53, 67, 68, 91, 173, 230, 235
Albertus Magnus 172, 235
Alceste 135
Alciati 26, 63, 66, 69, 70
Aldovrandi 51
Alphonso 147, 148, 156
Amerbach 61
Amyot 64
Anaxagoras 76, 130
Anaximander 99
André 211
Andreae 161, 235
Apian 15, 154, 158-160, 235
Apollonius 9, 49, 109, 136
Aquinas 76, 104, 106, 120, 235
Aratus 137
Archimedes 9, 14, 18, 19, 49-51, 77, 87, 91, 94, 101-103, 109-111, 130, 131, 136, 154, 230, 232, 241
Aretino 37, 40, 46, 52, 67, 231
Argyropulos 26
Aristeas 139, 235
Aristophanes 50, 101
Aristotle 38, 51, 55, 67, 76-78, 91, 100-102, 124, 125, 130-133, 135, 137, 139, 140, 151, 171, 176, 177
Arnobius 103, 104, 106, 107, 235
Ascham 64
Astruc 60
Augustine 40, 67, 104, 235, 250
Aurispa 25
Auzout 116, 235
Avicenna 77, 80
Baader 210
Bachet 43, 50, 148, 235
Bacon, F. xxix, xxxiv, 5, 6, 7, 15-17, 19, 29, 32, 38, 42, 44-47, 56, 67, 76, 78, 87, 92, 93, 107, 108, 129, 131, 146, 151, 154, 158, 160, 161, 165, 174, 175, 180, 182, 212, 231

Bacon, R. 79, 93, 172
Baif 63
Baker 88
Bandino d'Arezzo 34
Barlow 85, 92, 235
Barzizza 26, 64, 69
Bebel 64
Beccadelli-Panormita 26, 41
Bellay 63
Belleau 63
Belon 52, 238
Bembo 26, 37, 43, 45, 48, 230, 235
Benedetto xliv, 15, 26, 27, 66, 83, 148, 167
Benedictus 149
Bentley 55, 118
Berkeley 119, 181-183, 188
Bessarion 26, 36, 37, 229, 230, 236
Besson 236
Billingsley 236
Biondo 25, 43, 236
Biringucci 13, 20, 77, 173, 230, 236
Bisticci 25, 36, 40
Blundeville 84, 236
Bodin 66, 106, 236
Boëthius 139
Bohr 234
Boltzmann xli, 203
Borkenau 96, 236
Borough 20, 72, 85, 87, 88, 154, 236, 240
Bortkiewicz 203, 236
Bourne 14, 20, 153, 154, 158
Boyle lv, lviii, 10, 97, 117, 121, 236
Bracciolini 26
Brahe 77, 112, 156, 160
Brant 63
Brisson 63
Brown 129, 166, 210, 236
Brunelleschi 13, 77, 91, 173, 230
Bruni 25, 27, 36, 44, 47, 48, 63, 91, 229, 231, 236
Brunner 153, 236
Buchanan 64
Budé 58, 59, 63
Bühler xli
Burckhardt 20, 204, 229, 236
Buridan 11
Bury 129, 143, 236

Busch 64
Busleiden 59, 63
Caesar 47, 48, 64
Calvin 60
Camerarius 64
Campanella 6, 17, 78, 91, 161, 231, 236
Canter 55, 56, 236
Cantor 20, 236
Cardanus 43, 74, 77, 79, 80, 229, 236
Carnap xv, xxi, xxii, xxx, xl, xli, xlvi, xlvii, lvii, lviii, 221, 250-252
Cartesian 114, 115, 118, 121, 161, 182
Casaubonus 44, 45, 54, 56, 57, 63, 69, 149
Cassirer xliii, 96, 237
Cavendish 82, 84
Caxton 237
Cellini 13, 91, 173, 230
Celsus 139, 140
Celtes 61, 64
Cennini 43, 150, 151, 153, 237
Cerf xxxviii, 209, 214, 215, 250
Cesi 164
Cheke 64
Cherbury 184
Chrysoloras 26, 41
Clemanges 63
Cole 14
Comenius 161
Condorcet 192
Confucius 11, 230
Conrad-Martius 209, 212
Conversino 25
Copernicus 7, xxxi-xxxiii, lvii, 9, 11, 17, 52, 75, 77, 85, 108, 123-127, 174-176, 235, 237
Cortese 230, 237
Corvinus 64
Coulomb 94
Cratander 61
Croce xliv, 249
Croke 64
Crusius 64
Ctesibius 132
Cuspinianus 63
Dante 23, 24, 27, 28, 36, 41, 142
Daremberg 140
Darwin 187, 196
Decembrio 26, 66
Dee 52, 107, 160, 237
Deferrari 101, 237

Delvaille 129, 237
Democritus 18, 100, 102, 105, 130, 232
Descartes 6, 10, 17, 36, 38, 46, 47, 56, 67, 68, 97, 112-118, 143, 154, 161, 166, 177, 178, 181, 182, 188, 228, 231, 237
Digges 9, 15, 19, 52, 158, 160, 237
Dilthey xxxvi, 212, 222
Diocles 136
Diogenes 136, 140
Dionysius 135
Diophantus 9, 49-51, 136, 148, 154, 237
Dolet 61
Dollfuss xxii
Dollo xxxvi
Doneau 63
Dorat 63
Drake 84
Driesch 188
Duauf 31
Duerer 88, 90
Duhem 11, 20, 197, 237
Dürer 9, 13, 14, 20, 42, 152, 153, 157, 161, 230, 237
Dvorak 243, 251
Dyroff xliii
Egnazio 237
Einstein 197, 245
Elzevir 61
Empedocles 76, 136
Epicurus 102, 130, 139
Erasmus 36, 37, 41, 43, 53, 61, 62, 67, 230, 237
Erastus 77, 80
Eratosthenes 133, 142
Erman 31, 237
Estienne, H 44, 45, 55, 69, 149
Estienne, R 55, 69
Euclid 9, 14, 35, 49-52, 77, 91, 102, 103, 107, 109, 110, 112, 115, 136, 154, 155, 158, 237
Evans xvi, xxiv-xxvi
Faber Stapulensis 63
Facio 26
Falero 72, 237, 238
Fallopius 77
Fantis 79
Faraday 196
Farrington 90, 238
Feigl xvi, xxii
Feltre 26, 40

INDEX OF NAMES

Fichet 64
Ficino 25, 47, 66, 68, 77, 162, 149, 238
Filelfo 26, 40, 41, 44, 61, 66, 149, 238
Finaeus 15
Flavius 135
Fowler 45, 107, 238
Fracastoro 73, 76, 77, 79, 238
Francesca 9, 20, 229, 238
Frege xli
Frisch 111, 218, 239
Frischlin 64
Froben 61, 62
Frost 209, 212, 238
Gabrielli 164, 166, 238
Galen 51, 68, 76, 77, 136, 141
Galileo ix, xxxi, xxxii, xxxiv, 5, 7, 11, 15-17, 19, 20, 35, 36, 38, 46, 47, 49, 51, 67, 68, 72, 73, 76, 87, 90, 92- 95, 108-110, 114, 119, 126, 127, 148, 154, 164, 165, 174, 175, 177, 178, 185, 196, 201, 212, 228, 238
Gaza 26, 51, 230, 240
Geber 77
Gennadius 230
Germain 105, 241
Gesner 52, 238
Ghiberti 13, 20, 77, 91, 173, 238
Gilbert xiii, xxix, xxxiii, lvi, lvii, 5, 7, 15, 19, 46, 71-90, 92-94, 108, 174, 175, 238, 242
Giotto 150
Giovio 26
Godefroy 63
Goldstein 209
Gomperz ix, xxii, xli
Gonzaga 26
Grabmann 20
Grocyn 64
Grotius 83, 155
Grouchy 63
Grynaeus 51, 64
Guarino 26, 40, 69
Gudger 52, 238
Gunther 20, 117
Günther 238, 245
Gunzburg 20
Gutenberg 47
Hadley 64
Hahn xvii, xxiv, xxxix, xlvi, xlvii, lvii, 247
Hakluyt 15, 92, 238

Haller xxi, xxx, xlvi, lvii, lviii, 251, 252
Halliwell 20
Hariot 82
Hartley 189
Hartmann xxi, 64, 86, 238
Harvey xxix, xxxiv, 5, 175, 186
Hegel l, 193, 194, 204
Hegius 64
Heisenberg 234
Hellmann 72, 79, 83, 86, 238, 239, 241
Helm 153
Helmholtz 197
Heraclitus 99, 101, 102, 135, 139
Hermes 74, 78, 155
Herodotus 135
Hertz 196, 197, 199, 245
Hessen ix, 64, 156, 160
Heynlin 64
Hildebrand 209
Hildegard of Bingen 212
Hipparchus 101, 137, 138, 142, 143, 238
Hippocrates 18, 51, 76, 136, 142
Hippolytus 210
Hitler xxiii, xxvi, xxxiii
Hobbes lv, lviii, 116, 119, 177, 180, 182, 185, 187, 191, 228, 238
Homer 43, 50
Hood 15, 237, 238
Hooke 97, 116, 117, 165, 239
Hooker 106, 107, 239
Horace 43
Horkheimer xvi, xxii, xxvii, xxxiii, xxxvii, 227, 233
Hotman 63, 66
Houghton xvi, xvii, xxvii, 20
Hues 15, 82, 84, 92, 239
Hume 178, 183-185, 187, 188, 192, 199, 239
Husserl xi, 209, 211, 214
Hutten 54, 61, 68, 239, 242
Huyghens 10, 114, 116, 117, 126, 177
Irsy 237
Jähns 153, 239
Jaspers xi, xxi, 246
Jodelle 63
Joergensen xlvii
Johnson xxv, 15, 20, 21, 158, 166, 237-239
Kaibel 144, 239
Kendall 82-84

Kepler xxxi, xxxii, xxxiv, lvii, 17, 111-114, 119, 126, 147, 176, 201, 239
Kirchhoff 197
Kohn 98
Koyré xxxii, lvii, lviii
Lagos 83
Lambin 63
Lamprecht 204
Larkey 21
Latimer 64
Latini 27, 28, 34, 212
Lemaire 63
Leonardo da Vinci 4, 12, 13, 20, 36, 90, 91, 110, 173, 230, 237, 239
Leonicus 26
Leto 26
Lévy-Bruhl 227
Lily 64
Linacre 64
Linnaeus 186, 187
Lipsius 54, 64
Livy 136
Locke 119, 181-185, 188, 239
Lombardus 142, 239
Loschi 25, 66
Lovati 23, 28, 229
Lovejoy xxxiii
Lucianus 18, 100
Lucretianus 102, 239
Lucretius 51, 102, 130, 144, 239
Luder 61, 64
Lukacs ix, xliv
Luther 58, 60
Mabillon 55
Mach xi, xl, xli, xliv, 183, 197, 199, 228
Manetti 25, 69
Manuzio 25, 26, 40, 149, 163
Marconi 199
Maricourt 74, 79, 80, 93, 172, 238, 239, 242
Marsili 16, 26, 110
Marsuppino 25
Martini 173, 229, 230
Martyr 15, 68, 92
Marx 2, 194
Mauretanus 77
Maxwell 196, 199, 218
Maylender 129, 149, 162-164, 166, 239
Mazochi 25
Medici 25, 26, 41
Medina 15, 83, 92, 239

Meister xli, xlii
Melanchthon 59, 61, 64
Melville 64
Mendel 188, 219
Menodotus 136, 141
Mercator 15, 82, 159, 160, 238, 239
Merton ix, 20, 90, 166, 239
Mestlin 85
Michelangelo 43, 53, 239
Mill 94, 172, 193
Miltiades 50
Minerva 52, 53
Mises xl, 203, 246
Molyneux 58, 63
Montanus 160, 239
Montesquieu 239
Montreuil 58, 63
Mussato 23, 28, 229
Musuro 26
Muth 64
Neander 64
Nemeth x, xvii, xx, lviii, 252
Neudörfer 150, 239
Neumarkt 63
Neurath xv, xvi, xxii, xxiii, xxxix- xlii, xlv-xlvii, lviii, lix, 3, 244, 246, 251, 252
Newton xxxii, 17, 97, 114, 118, 119, 176, 177, 185, 201, 239, 245
Nicander 99
Niccoli 25, 34, 41
Nizolio 26, 55, 66, 68, 239
Norman 14, 15, 20, 72, 85-90, 92, 94, 154, 158, 160, 175, 238, 240
Novalis 210
Nuñes 15, 72, 83, 92, 238, 240
Ollanda 237
Olschki ix, 20, 90, 173, 174, 240
Oresme 11
Ornstein 20, 129, 164-166, 240
Orpheus 74, 78
Ortelius xxix, 15, 88, 157, 159, 240
Ostwald 197
Oswald 62, 240
Ovid 47, 50, 100, 101, 104, 106, 107, 237
Owen 64
Palissy 20, 240
Pamphilus 53
Panormita 26, 41, 44, 61, 149
Papiniamus 50

INDEX OF NAMES

Pappus 101
Paracelsus 77, 80, 86, 89
Paré xxix, 15, 20, 156, 157
Pascal 110, 240
Pasquali 56, 240
Passerat 63
Pastrengo 34, 240
Patrizzi 71, 74, 78, 91, 112, 121, 231, 240
Paulsen 48, 57, 61, 240
Pauly 101, 132, 143, 240
Pepi 31
Perotti 55, 56, 69
Petrarch 23-25, 27, 28, 33, 34, 36, 37, 40, 41, 45, 55, 58, 63, 66, 240
Peuerbach 51, 63
Pfänder 209
Philo of Byzantium 42, 132, 134, 141, 142, 168, 240
Piccinino 42
Piccolomini 25, 26, 48
Pico della Mirandola, F. 25, 37, 41, 43, 47, 48, 159, 230
Pico della Mirandola, G. 25, 37, 48, 67, 68, 148, 159, 240
Pirckheimer 63, 152
Pizicolli 25
Plantin 61
Plato 43, 51, 76, 78, 100, 130-135, 137, 138, 144, 151, 180, 193
Plessner 209
Plethon 230
Pliny the Elder 43, 51, 76, 80, 136, 139, 140, 240
Pliny the Younger 134, 159
Plotin 162, 240
Plutarch 18, 51, 103
Poggio 26, 27, 41, 42, 44, 66, 240
Poincaré xl, 230
Poliziano 26, 37, 54, 237, 240
Polybius 101
Pomponazzi 26, 66
Pontano 26
Porcellio 26, 42
Porta 77, 83, 162, 240
Posidonius 132
Priestley 189
Ptolemy 51, 76, 80, 85, 123, 125, 137, 139, 142, 143, 147, 156
Putnam 62, 240
Pythagoras 87, 100, 103, 130

Quintilian 37, 68
Rabelais 64
Raleigh 82
Ramus 15, 55, 63, 68, 92, 241
Ranke 31, 237
Raynal 192
Razes 77
Recorde 9, 19, 52, 158, 160, 241
Regiomontanus 50-52, 63
Rehm 143, 241, 249
Reichenbach xvi, xxi, xxii, xxvii-xxx, xl, xli, xliv-xlvii, xlix, lvii, lviii, 245, 247, 250-252
Reinhold 85
Reininger xli, xlii
Reuchlin 64
Rheticus 50
Riccboni 65
Ricci 5, 15, 174, 231
Rickert xxxvi, 206, 212, 222
Rienzi 23, 28, 65
Robortelli 26, 44, 54-56, 241
Roche 14
Rondelet 52, 238
Ronsard 63
Roriczer 151, 153, 168, 241
Rotz 20
Rousseau 53
Rutte xlvi, lviii, 252
Sabellicus 34, 67, 241
Sabinus 59
Sadoleto 26
Saeldner 48
Sagredo 92
Saint-Simon 194
Sallust 134, 135
Salutati 25, 27, 28, 33, 36, 40-43, 45, 54, 229, 241
Salviani 52, 238
Sandys 20, 25, 45, 51, 54, 56, 63, 68, 149
Santi 53, 241
Sanuto 83, 241
Sardus 34, 241
Sarton xvi, xxiv, xxvii, xxxiii, 20, 147, 155, 241
Sbrullius 42
Scaliger 56, 64, 77, 80, 155, 241
Schedel 64
Scheler 209
Schelling 189, 210

Schlick x, xvi, xl-xlii, xliv, xlvii, lviii
Schlosser xli, 20, 42, 241
Schlund 79, 241
Schmuttermayer 151, 153, 241
Scholz xliii
Schöner 50
Schubert 189
Selling 61, 64
Seneca 23, 43, 54, 101, 143, 241
Sextus Empiricus 141
Sigonio 26, 44, 54
Simoneta 41, 68
Smith xxxiv, xxxv, xxxvii, lviii, 192
Sombart xxxvi, 20
Spencer 187, 194
Spinoza xxi, 60, 114-116, 121, 177, 185, 188, 228, 241
Stadler xvii, xxi, xxiii, xxviii, xxx, xxxix, xlii, xlvi, liv, lvii, lviii, 3, 251, 252
Stapulensis 63
Stifel 9
Strabo 76, 80
Strada 25, 28
Sturm 64
Suarez 106, 107, 115, 121, 241
Symonds 20, 25, 45, 241
Tacitus 60, 80
Taddeo 150
Tartaglia 9, 14, 16, 19, 20, 35, 49, 67, 90, 110, 154-156, 158, 160, 229, 241
Taylor 20, 153, 241
Taysner 79
Teggart 206, 242
Telesio 47, 78, 91, 162, 231, 242
Thales 74, 76, 143, 145
Themistocles 50
Theophrastus 76
Thomaeus 26
Thompson 71, 79, 81, 151, 237, 242
Thorne 106, 242
Thou 63, 98, 110
Thucydides 135
Thyard 63
Timon 136, 139
Tiraboschi 162, 163, 242
Trapezunt 43, 230
Traversari 26, 69
Trissino 36, 53, 242
Troeltsch 212
Turnebus 63

Ubaldo 15, 19
Valens 101
Valentinus 210
Valeriano 34, 242
Valla 25, 37, 41, 54, 55, 60, 67-69, 242
Varro 139
Vasari 15, 43, 174, 242
Vergil 43
Villadei 55
Villani 13, 27, 66
Virgil 47
Vitruvius 14, 51, 77, 91, 132
Voigt 25, 28, 41, 42, 48, 54, 56, 149, 242
Voltaire 119, 192, 242
Vossler 37, 242, 249
Vulcanus 50
Wallis 116, 117, 242
Weber xxxvi, 150, 212, 222, 228
White xxiv-xxvi, liii, 88, 185, 219
Whitehead xli
Wiedemann-Franz 204
William of Hessen 156
Wilson 64
Wimpeling 64
Windelband xxxvi, 206, 212, 222
Winternitz 10, 242
Wolf 64
Wren 116, 117, 242
Wright 15, 77, 82-85, 92, 242
Xenophanes 123, 139
Xilander 50, 64
Ympyn 14
Zannoni 163, 242
Zeno 100
Zeus 100
Zilsel 4, 5, 7-xiv, xvi-xxxi, xxxiii- lviii, lx, 10, 12, 15, 22, 34, 53, 65, 90, 114, 119, 128, 149, 154, 158, 159, 172, 173, 227, 233, 235, 243, 247, 249-252
Zinner 133, 242
Zoroaster 74, 78

INDEX OF TOPICS

Absolutism 121
academic xvi, xix, xxi, xxiii, xxv, xxxvi, xxxv,xxxviii, xliv, liii, lviii, lx, lxi,3, 5,11, 15, 17, 20, 24, 59, 62, 71, 84, 90, 92, 93,127,139, 154, 156, 158, 159, 174, 253
academy xxxvi, 20, 132, 133, 138, 139, 162-166, 168, 174, 242
  - Académie Française 17, 163, 231
  - Accademia del Cimento 163-166, 168, 241
  - Accademia del Disegno 5, 15, 174
  - Accademia del Lincei 163
  - Accademia della Crusca 163, 235, 242
  - Aldine Academy 162
  - Prussian Academy 166
administration xxvii, 6, 9, 22, 27, 28, 31
alchemy 13, 52, 73, 75, 77, 89, 90, 107, 123, 162, 173
anatomy 4, 14, 90, 91, 158, 173, 186
animism 98, 179, 180, 183, 227
Antiquity xi, xiii, xxiii, xxxi, xxxii, lii, 4, 6, 8, 12, 16-19, 23, 27, 29, 33, 34, 36-39, 42, 43, 46, 48, 49, 56, 67, 78, 80, 90, 96, 97, 99, 101-103, 123, 125, 126, 128-131, 133-135, 137-139, 142-146, 156-159, 162, 167, 168, 171, 173, 175, 177, 182, 185, 205, 206, 227, 228, 230-232
anthropology 34, 184, 185, 187
architect, architecture liv, lvii, 3, 9, 13, 35, 39, 43, 49, 77, 91, 110, 151, 152, 153, 155, 173, 174, 205, 219, 229, 230
art xxxi, xxxii, xxxvi, 4, 7, 12, 14, 15, 17, 18, 20, 24, 27, 32, 38, 42, 43, 46-49, 51-53, 55, 59, 61-63, 67, 75, 82, 83, 86, 90-93,104, 105, 120, 125-127, 134,141, 149, 151-158, 174, 175, 193, 203, 204, 220, 221, 228, 239, 251
  - liberal arts 12, 24, 90, 127, 141, 174, 175
  - mechanical arts xxxi, xxxvi, 4, 7, 12, 14, 15, 20, 48, 51-53, 62, 63, 90, 92, 93, 158, 174, 175
artisans ix, xxxi, xxxii, lii, lvii, lviii, 7, 9, 12-14, 17-19, 22, 23, 31, 35, 36, 38, 42, 43, 46, 48, 58, 67, 110, 114, 119, 129, 133, 141, 142, 144,146, 148-150, 153, 154, 156, 158, 160, 161, 165, 167, 168, 173, 230-232, 251
  - superior artisans xxxi, xxxii, lvii, lviii, 13, 14, 18, 110, 114, 129, 149, 150, 168, 230-232
artists xxxi, xxxvi, lii, 4, 5, 7, 12-15, 18-20, 28, 31, 36, 38-40, 42, 43, 67, 77, 90, 91, 94, 104, 109, 150, 152, 158, 173, 174, 203, 205, 230, 231
artist-engineers xxxi, xxxvi, 4, 5, 7, 13-15, 20, 77, 90, 91, 94, 173, 174, 231
astronomer, astronomy xiii, xxxii-xxxiv, xxxvi, lx, 3, 5, 7-9, 11, 13-15, 17, 18, 20, 21, 24, 49, 51, 52, 63, 64, 71, 72, 75, 76, 80, 82, 84-88, 90, 91, 94, 101, 104, 106, 108, 111, 123, 125, 126, 131, 133, 137-139, 142-148, 150, 154, 156-159, 163, 164, 168, 174, 176, 185,191, 195, 200-203, 208, 217, 228, 235, 239, 240
astrophysics 176, 197
authority xii, xxx, xliv, 6, 7, 15-17, 36-38, 57, 68, 73, 76, 77, 123, 128, 144, 155, 228-230
Bible 60, 69, 97, 98, 103-107, 109, 112, 114
biology xi, xvi, xxvii, xxxiv, xxxix, xl, xlvii, lxii, 8, 18, 29, 51, 52, 103, 106,133, 144, 145, 179, 186-189, 195, 196, 203, 211, 216-220, 228, 234, 244
bookkeeping xxxvi, 3, 9, 14, 158, 229
Capitalism ix, xii, xxiv, xxxi, xxxii, 3, 5-10, 17-19, 20, 23, 27, 28, 33, 38, 39, 56, 81, 89, 91, 110, 112, 119, 121, 145, 150, 160,162,168, 167, 172, 174, 192, 201, 229- 231
cartography 15, 52, 82, 88, 159-161
causality ix, xxx, xxxix, xl, xlii, lii- lviii, 4, 6-8, 11, 13, 17-19, 36, 97, 102, 104, 115, 127, 130, 176, 177, 183-188, 191, 192, 194-196, 205, 211-213, 215, 221, 222, 228, 234
  - causal research lv, 4, 11, 13, 130, 176, 186, 195, 212, 213
  - causal thinking 6-8, 127
chemist, chemistry xi, xxxiv, 3, 4, 8, 14, 173, 176, 178, 185, 188, 189, 197

259

church 22, 38, 44, 45, 54, 68, 103, 109, 139, 146, 151, 164, 251
city, citizen xii, xvii, xix, xxiii-xxv, 2, 11, 22-25, 27, 28, 33, 34, 39, 58, 61, 63, 65, 66, 97, 132, 143, 144, 152, 158, 162, 171, 177, 207, 228, 237, 238
civilization xii, xiii, 31, 39, 56
college xix, xxv-xxvii, lv, lvi, 2, 59, 61, 69, 158, 165, 166, 174, 239
compass 12, 14, 15, 20, 72, 82, 84- 86, 88, 89, 94, 95, 152, 172, 173, 175
competition xxxi, 3, 6, 8, 9, 13, 35, 44, 81, 150, 172, 198, 204, 228, 229, 232
control xxx, 7, 29, 42, 44, 96, 146, 179, 192, 221, 222, 229
- of nature 7, 42, 44, 221
cooperation xii, xiii, xiv, xxxii, xxxiv, xxxvi, xlvi, 7, 9, 17, 20, 44, 45, 94, 128, 129, 142, 161, 162, 165, 208, 229, 231
crafts xxxiv, 29, 114, 150, 151, 155, 157
- craftsmen xxxii, xxxvi, 3-8, 10, 12-16, 20, 36, 88, 90-92, 94, 109, 127, 141, 149-151, 153, 155, 158, 160, 172-175, 180, 205, 251
criticism xliv, xlv, xlvii, xlviii, 3, 9, 34, 54, 55, 57, 60, 67, 75, 80, 137, 144, 181, 183-185, 206, 214, 223, 229
- critical ix, xxi, xxii, xxiv, xli, xlii, xlvii, 9, 56, 73, 75, 76, 89, 139, 185, 242, 251, 253
- critique 223, 249, 251
cultural ix, xxx, xxxi, xxxix, xli, xlvi, xlviii, xlix, lii, liv, lviii, 11, 23, 29, 33, 42, 61, 129, 143, 144, 146, 161, 171, 193, 194, 207, 231, 244, 251
culture xi-xiv, xxx, xxxi, xliii, xlvi, l, lii, liv, lvii, 4, 7, 8, 12, 18, 19, 23, 32, 33, 39, 56, 96, 114, 128, 131, 134, 136, 144, 145, 148, 168, 187, 201, 202, 205-208, 222, 223, 227, 228, 231, 232
- ancient 18
- American 227
- Arabic 8, 148, 227
- Chinese xi, xii, xiv
- classical 18, 32, 136, 145, 201
- European 1
- humanistic 228
- mercantile xiii
- oriental 56, 128
- primitive 206, 227

- scientific xxx, 8
- urban xxxi, 134, 232
- western xxx, 114
deduction xxxviii, xlii, lvii, 16, 102, 103, 109, 110, 115, 171, 179, 183, 186, 189, 193, 196, 199-202, 233
determinism liii, liv, 101, 102, 116, 177, 178, 202
development xii-xiv, xxi-xxiii, xxx, xxxi, xxxiii, xxxiv, xxxix-xli, xliii, xlvi, l, li, liii, liv, lviii, 3, 9-12, 14, 16-18, 28, 32, 33, 38, 39, 40, 49, 55, 56, 58, 60-62, 77, 81, 83, 91, 96, 99, 102, 103, 119, 129, 131, 132, 139, 140, 142, 143, 146, 159, 166, 179, 181, 183-185, 187, 188, 192-194, 196, 202, 204, 206, 208, 216, 217, 227-232, 240, 251
dialectic 24, 49, 138, 174, 193
discoverer 4, 12, 49, 82, 90, 117, 154, 157, 172, 230
discovery xlvi, xlviii, lii, 4, 5, 14, 15, 16, 19, 46, 48, 82, 85-88, 91, 92, 130, 131, 141, 154, 172, 174, 175, 185, 196, 201, 228, 229, 231, 235, 238
dualism xli, 114, 181, 182
economic x, xiv, xliv, xlvi, l, liii, 3, 6, 8, 9, 13, 29, 39, 41, 44, 65, 81, 91, 95, 134, 136, 138, 150, 168, 172, 192-194, 196, 198, 204, 208, 228230
economy xxxi, 3, 6, 9, 10, 17-20, 22, 36, 78, 120, 121, 176, 192, 193, 196, 198, 199, 204, 228, 230-232
education ix, xxiii, xxxi, 1, 6, 12, 13, 15, 18, 23, 25, 30, 32, 35, 37, 40, 46, 49, 52, 60, 62, 65, 66, 88, 89, 91-93, 103, 105, 110, 128, 132, 133, 141, 145, 149, 154, 155, 161, 162, 173, 174, 204, 205, 220, 231, 232
- higher 2, 30, 46, 145
- humanistic 13, 32, 46, 52
electric 73, 94, 179, 196, 198, 203, 204, 207, 221, 222
electricity xxxiv, 71, 118, 120, 196, 198, 218
empire 12, 23, 88, 100, 145, 201, 205, 206, 228
empiricism 7, x, xxx, xlvii-l, 171, 173, 180-184, 199
energy xxvi, 108, 116, 195, 197
engineer xiii, xxxi, xxxvi, lviii, 3-5, 7, 9,

INDEX OF TOPICS

10, 13-15, 18, 20, 36, 49, 50, 77, 90, 91, 93, 94, 96, 109, 110, 132, 133, 134, 138, 141, 142, 155, 158, 160, 161, 168, 173, 174, 176, 179, 199, 229-231
engineering xxxii, 4, 5, 13, 15, 19, 77, 83, 89-93, 132, 138, 153-155, 173, 174, 199
Enlightenment 184, 187, 189, 192, 194
entelechy 105, 188
ethics 115, 131, 146, 177, 180, 187-189, 193, 209
evolution xxxiii, xxxix, 7, 8, 27, 53, 57, 126, 145, 173, 184, 186, 187, 194, 216-220
 - astrophysical 217
 - biological 8, 216, 217-219, 220
 - geological 217
 - historical xxxiii, 126, 217, 218
 - human 7
 - phylogenetic 187, 216, 218
 - scientific 57
experience ix, xxxiv, xlii, lvi, 3, 10, 17, 32, 56, 75, 78, 81, 83, 87, 88, 91, 93, 102, 128, 140, 141, 144, 145, 155, 157, 161, 171-173, 176- 178, 181-184, 190-193, 199, 210, 211, 220
experiment xii, xiii, xxxiii-xxxv, lii, lviii, lx, 4-6, 15, 16, 18, 19, 57, 71-73, 75-82, 85, 86-93, 94, 102, 109, 110, 117, 133, 146, 155, 162, 164, 168, 171, 173-176, 186-189, 191, 193-197, 205, 209, 210, 223, 229-232, 241, 246, 247
experimental xii, xxvii, xxxii-xxxiv, liii, liv, lviii, lxi, 5-7, 15, 18, 19, 72, 73, 78-80, 86, 88-93, 95, 108, 109, 119, 126, 127, 130, 142, 166, 171- 176, 180, 186, 188, 189, 230, 231
falsification li, 54
fame 4, 5, 11, 17, 27-29, 39-45, 48-50, 52, 61, 62, 129, 134-138, 146, 148, 153, 158, 159, 163, 166, 167, 231
feudal 3, 8, 9, 22, 33, 36, 67, 112, 120, 162, 177, 229, 230
Feudalism xii, 8, 9, 11, 22, 23, 44, 120, 121, 172, 228
geographer, geography xii, xxxvi, 12, 15, 34, 48, 51, 72, 82-84, 88, 91, 92, 131, 133, 139, 143, 159, 172, 242
geometry 4, 9, 12-14, 24, 49, 53, 77, 91, 115, 123, 126, 151, 152, 158, 173, 174, 183, 229, 237, 240

geophysics 74, 99, 202, 233
goldsmith 13, 152, 173
grammarians 12, 18, 46, 65, 75, 76, 140, 228, 230
guild 3, 8, 12, 13, 42, 43, 81, 91, 149, 150-154, 156, 157, 159, 162, 172, 173, 229
handwork 228-232
historiographer, historiography xxxiv, 18, 23, 26, 38, 46, 51, 63, 67, 134, 135, 145, 92, 204, 208, 222, 223
history ix-xii, xvi, xxii-xxviii, xxxi, xxxiv, xxxv, xxxvii, xxxviii, xl, xli, xliii-xlvii, xlix-li, liii-lv, lvii, lx, lxi, 16, 18, 20, 21, 23, 25, 32, 51, 52, 59, 62, 63, 65, 69, 72, 82, 92, 93, 96, 103, 112, 120, 121, 129, 131, 133, 136, 140, 142, 147-149, 154, 155, 158, 161, 162, 166-168, 171, 172, 174, 179, 183, 185, 189, 192-194, 196, 198-208, 212, 216-221, 223, 227, 228, 233, 234, 237-241, 244, 251, 253
 - of ideas xvi, xxvi, xxxv, 20, 96, 103, 112, 120, 158, 166, 212, 237-239, 244, 251
 - of mankind 16, 171, 174, 216, 217
 - of science x, xii, xxvi-xxviii, xxxvii, xli, lvii, lx, lxi, 20, 93, 140, 147, 189, 241, 251, 253
Humanism xxx-xxxiii, xl, lvi, 4, 5, 7, 10, 11, 16, 20, 22, 24, 27-29, 32, 33-38, 40-48, 50-63, 65-69, 76, 89, 91, 128, 146, 152, 161, 227-231
humanistic xvi, xxxiv, xxxviii-xl, lv-lvi i, lxii, 4-8, 12, 13, 15-17, 22- 25, 27, 28, 32-34, 36, 37, 40, 43-48, 50-53, 55, 58-62, 67, 68, 71, 78, 90, 91, 149, 158, 162, 163, 167, 173, 174, 221-223, 228, 230
humanists ix, lii, 4, 5, 7, 11, 12, 14, 16, 17, 19, 22-28, 30-69, 77, 90, 128, 129, 148-150, 152, 155, 158, 159, 162, 163, 167, 172, 175, 185, 207, 222, 230, 231
idealism xxxviii, xliii, lx, 66
ideals ix, xxii, xxxiii, xl, liii, liv, 5, 7, 17, 19, 23, 27-29, 31-33, 37-40, 42-45, 52, 56, 57, 60, 62, 66, 97, 115, 129-131, 135-138, 141, 146, 149-151, 157, 158, 160, 167, 168, 175, 188, 207, 212, 228, 230, 231

- craft 157
- cultural 207, 231
- ethical 115, 188
- humanistic 7, 28, 40
- guild 42, 43
- intellectual 44
- professional lii, 5, 17, 19, 27-29, 31, 33, 37, 45, 52, 136, 137, 149, 150
- religious 23, 44
- research liv
- scientific 44

ideology xlix, lv, 38, 41-43, 50, 60, 62, 121, 167, 227, 229-231

individualism 6, 8, 44, 143, 144, 162, 204, 229

individualistic 9, 38, 44, 60, 131, 143, 145, 146, 148, 149, 158, 204, 229

induction xiii, xxii, xlii, 16, 56, 175, 183, 184

industry xii, lii, liv, 94, 131, 166, 198

instruments xxxi, xxxii, xxxvi, 4, 7, 13-15, 17, 19-21, 72, 80, 82, 84-89, 91, 116, 125, 133, 147, 148, 150, 153, 154, 158-160, 173, 210, 230, 235, 236
- astronomical 13, 21, 133, 147, 148
- measuring 7, 17, 72, 84, 153, 158-160
- musical xxxi, 4, 7, 13, 91, 173, 230
- nautical 80, 82, 84, 154, 160
- physical 7, 15, 72, 116, 210

intellect xxi, 5, 14, 36, 50, 148, 174, 189

intellectual xi, xiv, xxi, xlvi, lii, lvi, lix, 3, 5, 7, 10-12, 17-19, 22, 23, 28, 32, 38, 44, 57, 60, 62, 65, 78, 80, 87, 90, 91, 107, 129, 134, 138, 147, 162, 167, 168, 171, 187, 188, 202- 207, 231

intelligence 7, 33, 35, 56, 84, 118, 161, 179

invention xiii, xxvi, 25, 31, 47, 51, 54, 72, 82, 150-152, 159, 164, 172

Jewish xxiv, 10, 98, 139, 147, 205, 227, 253

knowledge ix, xiii, xiv, xxx, xl-xlii, xliv, xlix, liii, lv, lviii, lx, 4, 11-14, 22, 27, 29-35, 44-47, 49, 50, 53, 57, 66, 67, 75-77, 79-81, 83, 84, 87, 88, 92, 96, 107, 118, 119, 130, 132, 133, 135, 137, 140-145, 150, 151, 156-158, 160, 161, 163, 165-168, 173, 175, 179-181, 183, 185, 189, 194, 196, 197, 199-201, 212, 221, 228-230, 233, 235

labor xxxi-xxxiii, xxxvi, 4-7, 10, 12, 13, 17-19, 45, 48, 90-93, 103, 109, 131, 138, 155, 157, 160, 167, 171, 173, 208, 231, 232
- division of 5, 13, 17, 45, 91, 138, 160, 208
- manual xxxi-xxxiii, xxxvi, 4, 6, 7, 12, 48, 90, 92, 93, 103, 109, 155, 171, 173, 232

law xxii, xxxii, xxxviii, xxxix, xli- xliii, li, liii-lviii, 1, 5, 7-11, 13-16, 18, 23, 26, 27, 29, 30, 35, 38, 44, 57-59, 63, 65, 87, 94-122, 126, 132, 144, 149, 154, 161, 169, 173, 175, 177, 178, 184-189, 191-208, 216- 223, 227-229, 231, 233, 234, 238, 239, 242, 251, 253
- historical xxix, xxxviii, xlvi, 200- 203, 205-208, 218, 233
- historico-sociological 207
- mechanical liii, 35, 113, 196-198, 201
- natural lvi, 10, 96, 97, 100, 105- 107, 112, 113, 116-118, 121, 177, 178, 185, 187, 188, 199, 216, 217, 221
- of nature xlii, 101, 106, 109, 112, 113, 115, 119, 120, 228, 229, 233, 251
- physical 10,11, 14, 15, 18, 29, 35, 96, 99, 102, 104, 108-110, 113, 116, 119, 120, 184, 204, 205, 234
- sociological xxxviii, liv, 8, 194, 195, 200, 207, 233, 234

logic ix, xiv, xliii, xliv, xlvi-xlviii, 7, 17, 29, 45, 55, 57, 84, 103, 104, 106, 133, 171, 178, 180, 181, 183, 189, 199, 201, 208, 209, 213-215, 221, 223, 228

magic xxx-xxxii, 6-8, 11, 30, 31, 75, 90, 101, 116, 118, 172, 175, 189, 193, 204, 210, 227, 228, 231, 232

magnetism, magnetic xiii, xxxiv, 71- 75, 79-83, 85-89, 93, 94, 108, 112, 154, 196, 207

mariner 4, 12, 15, 80, 83-89, 92, 154, 175, 238

mathematical xxi, xxxix, xli, xlii, lviii, lxi, 3, 9, 14-16, 35, 36, 49-51, 77, 82, 85, 92, 94-96, 102, 103, 108- 112, 114, 118, 123, 126, 132, 136, 143, 149, 153, 154, 158, 159, 168, 174, 175, 177, 197-199, 215, 229, 234, 237, 238

mathematician xxii, xxxvi, 3, 8, 19, 50-52, 75, 77, 82-84, 87, 91, 131, 133, 140, 148

mathematics ix, xii, xiii, xxii, xxvii, xxxii,

INDEX OF TOPICS

xxxiii, 5, 9, 15, 18, 20, 21, 35, 49-51, 57, 62, 88, 90, 92-94, 103, 110, 131, 133, 145, 146, 148, 154, 155, 158-160, 167, 174, 178- 180, 183, 186, 193, 196, 199, 229
mechanics xiii, xxxi, xxxiii, xxxvi, liii-lv, 3-5, 7-9, 12, 14, 15, 20, 35, 48, 49, 51-53, 62, 63, 73, 77, 90, 92-94, 108, 110, 113, 117, 119, 120, 123-127, 149, 154, 158, 159, 167, 175, 178-180, 182, 189, 195- 199, 201, 202, 206, 228
 - classical 178, 195, 198
 - quantum 198, 228
mechanist 76, 113-115, 126, 179, 180, 182, 198
medicine xii, xiv, 5, 15, 30-32, 65, 131, 136, 138, 140, 141, 145, 157, 186, 238
medicine man 30-32
Mercantilism xiii, 120, 166
metallurgy 4, 13-15, 78, 80-82, 89, 91, 92, 95, 173, 175
metaphysics xxvii, xxxiv, xliii, xlix, liii, liv, lx, lxi, 4, 9, 17, 72-75, 78, 79, 88, 89, 91, 99, 108, 114, 116-119, 125, 126, 128, 131-133, 140, 144,145, 171, 177, 180-183, 186, 188-190, 193, 194, 199, 218, 228
method ix, x, xxxii-xxxiv, xxxvi, xxxvii, xli, xlii, xliv-xlviii, lii-lv, lviii, lxii, 3-11, 14-17, 20, 22, 32, 34, 35, 37-39, 48, 49, 52-58, 60, 65-69, 71-73, 79, 80, 85, 89, 90, 92-95, 102, 103, 109, 115, 118, 125, 127-130, 141-148, 154, 157, 159, 161, 162, 167, 171-176, 180-186, 188-191, 193-196, 200, 203, 205, 207, 209-215, 217, 222, 223, 227-229
 - empirical xlvi, 176, 186, 195
 - experimental xxxii, xxxiv, liii, liv, 5-7, 73, 79, 89, 92, 93, 127, 171, 175, 180, 186, 188
 - philosophical xlii, 37, 214
 - quantitative xxxvi, lii, 7, 9, 57, 58, 128, 130, 186, 189, 194
 - scholastic xxxvi, 7, 10, 56, 66
 - scientific xxxvii, xlv, 3, 4, 7, 9, 14, 15, 20, 56, 69, 71, 72, 79, 80, 118, 159, 172, 174, 205, 212
 - statistical lvii, 176, 193, 194, 223
Middle Ages xxvi, xxxvi, liv, 3, 6, 8, 23, 27, 38, 44, 46, 53, 58, 60-62, 65, 67, 72, 93, 96, 103, 113, 120, 126, 128, 142, 146, 147,171-174, 177, 204, 227-230, 241
 - medieval ix, xiii, l, 3, 8-11, 19, 20, 22, 24, 27, 31, 34, 35, 38, 40, 44, 49, 53-55, 57, 58, 61, 66, 68, 73, 74, 76, 79-81, 93, 103, 107, 112, 120, 121, 123-125, 147-150, 154, 157, 162, 171, 172, 176, 177, 180, 182, 184, 205, 206, 212, 250
military xii, xxii, xxxi, xxxii, 3, 5, 6, 8, 9, 13, 15, 16, 19, 23, 50, 83, 89, 90, 92, 93, 109, 112, 132, 133, 137, 141, 153-156, 161, 162, 168, 173, 174, 229, 231
miner, mining 12, 13, 15, 80-82, 84, 89, 92, 93, 172, 175
modern ix, xi, xii, xiv, xxi-xxiii, xxv- xxvii, xxix-xxxi, xxxiii-xli, xliii, l-lv, lvii, 1, 3, 6-9, 11-22, 25, 27, 28, 30, 34, 35, 37-39, 42-47, 49, 51, 52, 54-57, 59, 61-63, 66-68, 71-82, 88-91, 96-111, 114-121, 123-133, 135-151, 154-156, 159-168, 171, 173-177, 180-188, 192, 194, 197- 199, 201-203, 206, 208, 212, 213, 227-229, 233, 235, 238, 251
modernization 1
monarchy 9, 100, 121
moral, morals x, 23, 28, 40, 84, 96- 100, 103, 105, 131, 136, 138, 146, 149, 164, 189, 193
myths, mythology 34, 102, 144, 227
nature ix, xi-xiii, xxvii, xxxi, xxxiii, xxxiv, xxxviii, xl-xlii, xlv, lii, liii, lv, 7, 8, 29, 36, 42, 44, 47, 56, 74, 75, 88, 91, 93, 96- 107, 109, 112-115, 117-121, 124, 125, 130, 138, 143, 147, 152, 159, 163, 164, 176-182, 193, 196, 197, 199, 201, 209, 221, 228, 229, 231, 233, 236, 240, 251
navigation, navigator xii, xxxi, xxxii, xxxvi, 5, 7, 14-17, 20, 21, 36, 38, 48, 52, 72, 80, 82-89, 91, 92, 95, 131, 148, 153, 154, 158, 160, 172, 242
nobleman 4, 11, 16, 22-25, 39, 62-64, 66, 79, 93, 120, 164, 172, 177, 200
objective, objectivity 17, 44, 89, 138, 146, 180, 182, 197, 204
objectivity 89
observation ix, xxxiii, xlvii, lii, lviii, lx, 5, 13-15, 71, 72, 74, 83, 85, 87, 91-93, 98,

99, 108, 112, 116, 126, 133, 142, 143, 156, 164, 173-176, 178, 180-182, 185-188, 191-195, 201, 203, 207, 216, 227, 229, 236

painter, painting xxxii, 4, 12, 13, 15, 18, 39, 42, 43, 52, 53, 91, 145, 150-152, 159, 173, 174, 205, 229

phenomenology xvi, xxxix, xl, 8, 209-215, 244, 251

philology 4, 11, 25, 32, 34, 38, 53-56, 60, 67, 68, 145, 221, 223

philosopher x, xxi, xxxviii, xliii-xlvi, lviii, lxi, 12, 18, 37, 47, 48, 50, 51, 54, 62, 63, 74-80, 88, 97, 106, 114, 123, 128-136, 138-141, 144-146, 155, 165, 171, 175, 178, 184, 188, 189, 191, 199, 202, 203, 211, 214, 217, 218, 237

philosophy 2, xi, xvi, xxii-xxv, xxx, xxxviii, xl, xlii-xlvii, xlix-li, lviii, lx, lxi, 10, 18, 20, 22, 23, 26, 37, 38, 46-51, 56, 63-67, 74, 75, 78, 88, 99-102, 112, 114, 116, 119, 123, 128, 130-136, 138-141, 144, 145, 151, 160-162, 165, 175, 177, 179-184, 187, 191-197, 201-204, 209, 212-215, 219, 242, 245-249, 251

physical 7, xvi, xxxii, xxxiv, xxxv, xxxviii, xlv, lxi, 7, 10, 11, 13-15, 18, 20, 29, 35, 72, 80, 91, 96-100, 102-106, 108-123, 125, 132, 149, 154, 160, 161, 165, 169, 176, 178-186, 189, 191, 195-197, 199-205, 210, 212, 214, 215, 219, 221, 233, 234, 244, 251

physics, physician x, xiii, xiv, xxii, xxiii, xxvi, xxvii, xxix, xxxix, xliii, xlv, liii, liv, lvi, lvii, 3, 8, 15, 20, 24-26, 49, 52, 57, 60, 61, 64-66, 72, 73, 75, 77, 80, 86, 92, 94, 95, 100, 103, 104, 108, 114, 115, 118-120, 124, 126, 127, 132, 133, 140, 141, 145, 147, 148, 161, 165, 173, 175, 176, 178, 179, 182, 184-189, 193, 195-204, 221, 228, 233, 234
- atomic 197, 189
- experimental 80, 95, 108, 175
- laboratory 104, 165, 176, 201, 202
- mechanistic liv, 94, 114, 179, 182, 198, 199

poet, poetry 16, 18, 23, 24, 41, 51, 52, 53, 61, 63-65, 110, 133, 134, 139, 162, 204

politics, politician ix, x, xx, xxi, xxiv, xli, xlv, li, liv, lv, lxi, 1, 7, 9-11, 23-25, 27, 28, 35, 39, 44-46, 54, 55, 58-64, 100, 106, 115, 132, 134, 135, 138, 145, 148, 166, 191-193, 196, 203, 205, 206, 216, 220, 229

positivism xi, xxxiii, xxxviii, xlii, li, lx, 194

priests xxxiii, 10, 17, 19, 25, 31, 32, 41, 61, 64, 120, 126, 132, 139, 142, 143, 145, 168, 184, 200, 205, 212, 228

printing xx, 25, 47, 48, 51, 54, 62, 64, 69, 128, 149, 151, 163, 172, 179, 240

production xxx, 8, 18, 19, 131, 183

profession 4, 10-12, 22, 23, 25, 28, 30, 31, 39, 40, 53, 58, 87, 132, 138, 148, 154, 155, 167, 173, 228

professor xxvii, lxi, 5, 7, 15, 16, 24-26, 28, 44, 45, 48, 51, 54, 55, 58-66, 69, 88, 108, 132, 138, 139, 158, 159, 165, 174, 206, 247

profit xxviii, 136, 150, 153, 154, 172

progress ix, xiii, xiv, xvi, xx, xxvii, xxxii, xxxiv, xxxv, l, liv, lxii, 5-7, 17, 32, 33, 38, 42, 44, 66, 128, 129, 131, 133, 138-144, 146, 148, 150-155, 157, 158, 160, 161, 163, 165-168, 175, 184, 192, 216, 231, 236, 251, 253
- scientific xvi, xx, xxvii, xxxv, lxii, 5, 7, 17, 128, 129, 133, 141, 146, 148, 154, 157, 158, 161, 163, 167, 168, 216, 251

Protestantism 60, 62, 69, 98

psychology, psychologist xxxiii, liii, liv, 29, 67, 115, 126, 136, 137, 176, 181-191, 195, 196, 202, 203, 207, 209, 213-215, 217, 228, 233, 234

rational, rationalism x, xxx, xli, xlii, xliv, liv, 3, 5-7, 9-11, 22, 40, 47, 52-54, 56, 57, 100, 120, 121, 124, 128, 162, 168, 174, 176, 179, 181, 184, 186, 191-193, 196, 198, 205, 227-229, 253

Reformation xii, 52

relativity 197, 198

religion, religious x, xiii, xiv, xxii, xliv, xlvi, 4, 11, 17, 23, 31, 32, 39, 44, 52, 60, 62, 74, 77, 89, 104, 111, 114, 119, 132, 135, 138, 143, 144, 164, 182, 184, 185, 193, 203, 204, 220, 221, 239

Renaissance ix, xxx, xl, lii, 4, 6, 12, 13, 15, 17-21, 24, 25, 27, 30, 33-36, 38-48, 51-54, 56-63, 66, 69, 75-78, 89, 90, 112, 128, 129, 148, 149, 152, 174, 185, 204, 205, 207, 228, 230, 231, 236, 237,

INDEX OF TOPICS

239-241, 251
research ix, xi, xvi, xix, xxi, xxii, xxiv, xxv, xxix, xxxvi-xxxviii, xl, xlii, xliii, xlv-xlix, li-lv, lvii, lviii, 4, 5, 11, 13, 16, 19, 24, 42-45, 57, 94, 96, 109, 126, 130, 131, 133, 134, 136, 138, 145-147, 149, 155, 160-164, 166, 171, 173-176, 178, 184-186, 188-190, 193-195, 201, 206, 208, 209, 212-214, 217, 221, 227, 228, 231, 233, 244, 251
- economic 193
- empirical xlii, xlvii, xlviii, li, 133, 171, 173, 175, 178, 184, 185, 188, 201
- experimental 109, 126
- historical xlvi-xlviii, lvii, lviii, 214, 228
- scientific 5, 16, 45, 138, 146, 155, 166, 174
- social xxi, xxiv, liv, lv, 194, 227, 233
- sociological lii, 19
revolution xi, xii, xiv, xxxiv, liii, liv, lx, lxi, 23, 32, 48, 81, 125, 161, 172-174, 196, 198, 199, 216, 230
- physical 196, 199
- scientific xi, xii, xiv, lxi
- technological 48, 174, 198
- theoretical 138, 221
rhetoric 16, 18, 24, 45, 46, 49, 55, 65, 99, 104, 128, 123, 132, 145, 171, 174, 228, 230
sailing, sailor 82-84, 85, 172
scepticism xxxvi, lviii, 139
scholar ix, xiii, xvii, xxi, xxiv-xxvi, xxviii, xxxi, xxxii, xxxv, xxxviii, lii, lx, 3-7, 10, 12, 14-19, 24, 25, 30-33, 35, 37-39, 43, 44, 50, 52-55, 57-59, 61-63, 65, 71, 75, 77-80, 82, 88-93, 108-110, 127, 129, 133, 134, 137-139, 142, 143, 147-149, 152-160, 162-165, 167, 168, 172-176, 208, 212, 231, 251
 - university-scholar ix, 3-7, 10, 12, 14, 71, 90, 93, 148, 173, 174, 231
Scholasticism, scholastic xxxi-xxxiii, xxxvi, xxxviii, 3-7, 10, 11, 14-19, 20, 22, 24, 25, 27, 30-33, 35, 37-39, 43, 44, 48, 50, 52-63, 65, 66-69, 71-73, 75, 77-80, 82, 88-93, 104, 108-110, 127, 129, 133, 134, 137-139, 142, 146, 160, 161, 171, 172, 177, 184, 187, 205, 206, 212, 218, 227
science xiv, xvi, xxi, xxii, xxv-xxxiv, xxxvii, xxxviii, xl-xlii, xliv, xlvi-lviii, lx, lxi, 1-9, 11, 14-22, 29-35, 38, 42, 44-49, 51-53, 56-58, 60, 63, 66-68, 71, 74, 75, 78, 79, 82, 88-93, 95-98, 100-102, 106-110, 114-120, 123, 128-131, 133-137, 139, 140, 142-145, 147, 149, 154-156, 158-168, 171-177, 180, 181, 184-211, 214-218, 221-223, 227-233, 236, 238, 239, 241, 244, 246, 251-253
- classical 139, 144, 145
- empirical xxxiii, xlii, xlix, lii, 171, 175, 176, 184-186, 191, 193-196, 200, 206, 207, 218
- experimental 78, 79, 88-93, 119, 142, 166, 171-174, 176
- human liv, lv
- modern ix, xi, xii, xiv, xxi, xxii, xxv, xxvii, xxx, xxxi, xxxiii, xxxiv, xxxviii, liii, liv, lvii, 1, 5-8, 14, 16, 17, 22, 34, 42, 44, 46, 63, 66, 74, 75, 89, 100, 102, 107, 120, 130, 136, 145, 154, 171, 175, 233
- natural ix, xii, xxxv, xxxviii, xliii, xlviii, xlix, lii, lv, lviii, lx, 5, 11, 16, 38, 47, 49, 51, 56, 57, 71, 82, 89, 90, 93, 95-98, 101, 102, 108, 109, 115, 116, 130, 133, 144, 145, 163, 165, 167, 175, 180, 185, 186, 189, 194, 200, 201, 206-211, 214-2, 217, 221-223, 232
- physical 18, 85, 101, 196
- social xxv, xlix, lv, lviii, lxi, 57, 96, 176, 191-193, 204, 207, 238
- special 145, 184, 195
- theoretical 143
scientific ix-xiv, xvi, xx, xxi, xxvi-xxx, xxxii-xxxviii, xl, xli, xliv-xlvi, xlviii-lii, lviii, lx-lxii, 3-11, 14-17, 19, 20, 22, 29, 31, 32, 34-36, 38, 39, 42, 44-51, 53, 56, 57, 67-69, 71-73, 75, 77, 79, 80, 83, 84, 89, 92-94, 98, 102, 106, 110, 112, 114-119, 127-133, 136-139, 141-149, 154-168, 172-175, 178-181, 185, 187-197, 199, 200, 204-206, 209, 211, 212, 216, 217, 222, 223, 229, 231, 236, 240, 244, 251
scientist xi, xii, li, lvi, 4-6, 8, 11, 14, 17, 29, 32, 35, 36, 42, 44, 45, 48-52, 57, 68, 89, 91, 97, 104, 118, 119, 128-130, 134-139, 145, 151, 160, 166, 167, 171,

174, 175, 177, 178, 182, 199, 202, 203, 208, 211, 216, 219, 221, 223, 229, 231
sculptor 13, 18, 39, 43, 91, 152, 173, 205
slave, slavery xxxi, 18, 19, 99, 103, 131, 132, 146, 167, 168, 173, 200, 232
social ix, x, xiv, xvi, xxi, xxiii-xxvii, xxix, xxx, xxxii-xxxiv, xxxvi-xli, xliv, xlviii-lv, lviii, lx, lxi, 1, 3-5, 7, 9-12, 15, 16, 18, 27, 29-33, 38, 53, 57, 62, 79, 82, 84, 90, 93, 95, 96, 99, 109, 112, 121, 129, 131, 134, 138, 146, 150, 156, 162, 172, 174, 176, 180, 186, 191-195, 204, 207, 216, 218, 227-229, 231, 233, 238, 252, 253
society ix, xiii, xiv, xvi, xvii, xix, xxiii, xxv-xxvii, xxxi, xxxii, xxxvii, xliii, xlix, l- lii, liv, lvi, lvii, lx, 1, 3, 5-10, 12, 17-19, 20, 22, 23, 30-33, 37, 44, 56, 57, 66, 68, 71, 78, 81, 91, 114-117, 120, 121, 129, 131, 132, 134, 136, 142, 143, 146, 154, 161-168, 172, 174, 187, 191, 193, 195, 213, 218, 219, 228-231, 239, 240, 242, 251, 253
- agrarian 168
- aristocratic 164
- capitalistic 3, 5, 7, 10, 18, 19, 23, 91, 167
- classical 131, 132, 136, 142, 146
- contemporary 17, 167
- early capitalistic 3, 5, 7, 10, 18, 19, 91, 167
- feudal 3, 8, 9, 230
- human 8, 9, 30, 218, 219, 228, 230, 231
- learned xxxii, 6, 17, 68, 161
- modern xxiii, lvii, 12, 78, 132
- scientific 20, 129, 162-166, 240
sociology, sociologist vii-xi, xiv, xvi, xxi-xxiii, xxix, xxx, xxxii-xxxv, xxxviii-xli, xlvi, xlvii, xlix, l, lii-lv, lvii, lviii, lxii, 7, 8, 10, 11, 19, 21- 23, 28, 29, 32, 33, 39-41, 52, 58, 61-63, 65, 69, 78, 90, 93, 103, 119, 121, 122, 126, 127, 129, 134, 137, 138, 141, 143, 144, 146, 148, 153, 158, 160, 162, 166-168, 176, 185, 187, 192-196, 200, 202, 203, 205, 207, 209, 213-215, 220, 222, 227, 228, 233, 234, 244
statistics xxxviii, xxxix, xl, xlii, xliii, xlvi, li-liii, lv, lvii, lviii, 12, 13, 70, 167, 176, 188, 193, 194, 198, 202, 203, 207, 223, 228, 233
stonedresser 4, 152, 173, 205
superstition lviii, lx, 13, 77, 101, 130, 173
technology xii, xiv, xxxi, xxxii, xxxiv, liii, liv, lviii, lx, 3, 5-9, 11, 12, 15, 17, 18-20, 29, 30, 31, 33, 34, 36, 49, 51, 56, 67, 78, 80, 81, 90-94, 96, 103, 109, 114, 119, 131, 132, 138, 142, 150, 153, 156, 158, 160, 161, 165-168, 172, 174-176, 179, 185, 196, 195, 198, 199, 228, 231, 232, 239, 251
teleology 104-106, 114, 117, 124-127, 130, 176, 177, 186, 187, 195, 198
theology xi, xxx, 3, 10, 48, 60, 64, 68, 118, 119, 128, 130, 145, 182, 227- 229, 251
theory xxiv, xl-xlii, xlviii, l-lxi, 9, 20, 29, 41, 73, 74, 77, 85, 89, 93, 94, 96, 121, 123, 125, 126, 129, 130, 131, 133, 135, 139, 141-143, 146, 156, 167, 172, 177-181, 184-188, 191-194, 196-199, 201-203, 207, 210, 218
thermodynamics 204
trade, tradesmen xxiii, 3, 8, 10, 33, 94, 185, 192
tradition xxx, 6, 8-10, 13, 22-24, 32, 44, 53, 57, 58, 65, 81, 119, 120, 140, 141, 143, 144, 147, 150, 151, 153, 160, 162, 163, 166-168, 172, 182, 184, 186, 187, 193, 199, 205, 206, 218-220, 222, 227-230
training xi, xxxvi, 5-7, 14, 17, 22, 24, 27, 28, 30, 32, 48, 62, 68, 93, 132, 147, 154, 157, 159, 174
- academic xxxvi, 62, 93, 154, 159
truth xli, 37, 47, 48, 56, 75, 130, 131, 137-139, 142, 144, 147, 219
unity of science xvi, xxv, xl, xli, l, liv, lvi, 3, 221, 244, 252
universe 74, 78, 94, 99-106, 111, 112, 121, 123-125, 135, 144
university ix, xii, xiii, xvi, xvii, xix, xxi-xxvii, xxix, xxxvii-xlviii, l, lii, lvi, lviii, lx, lxi, 3-7, 10-12, 14, 15, 24-26, 35, 42, 44, 45, 48, 52, 59-61, 63, 65, 69, 71, 84, 90, 92, 93, 110, 132, 144, 147, 148, 155, 158-160, 164-166, 171, 173, 174, 221, 231, 244, 251, 253
verification 102, 109, 206
vernacular 12-16, 20, 35, 37, 40, 43, 46, 49, 51, 52, 62, 68, 83, 91, 92, 109, 150, 154, 158, 164, 173

Vienna Circle xvi, xvii, xxi-xxv, xli, xlii, xliv, xlvii-li, lvi, lx, lxi, 3, 253
warfare 8, 18, 89, 131, 134, 143
worker ix, l, 5, 14, 17, 18, 44, 82, 89, 91 141, 145, 150, 155, 158, 167, 174, 231, 232

Boston Studies in the Philosophy of Science

37. H. von Helmholtz: *Epistemological Writings*. The Paul Hertz / Moritz Schlick Centenary Edition of 1921. Translated from German by M.F. Lowe. Edited with an Introduction and Bibliography by R.S. Cohen and Y. Elkana. [Synthese Library 79] 1977
ISBN 90-277-0290-X; Pb 90-277-0582-8
38. R.M. Martin: *Pragmatics, Truth and Language.* 1979
ISBN 90-277-0992-0; Pb 90-277-0993-9
39. R.S. Cohen, P.K. Feyerabend and M.W. Wartofsky (eds.): *Essays in Memory of Imre Lakatos.* [Synthese Library 99] 1976 ISBN 90-277-0654-9; Pb 90-277-0655-7
40. Not published.
41. Not published.
42. H.R. Maturana and F.J. Varela: *Autopoiesis and Cognition.* The Realization of the Living. With a Preface to "Autopoiesis' by S. Beer. 1980 ISBN 90-277-1015-5; Pb 90-277-1016-3
43. A. Kasher (ed.): *Language in Focus: Foundations, Methods and Systems.* Essays in Memory of Yehoshua Bar-Hillel. [Synthese Library 89] 1976
ISBN 90-277-0644-1; Pb 90-277-0645-X
44. T.D. Thao: *Investigations into the Origin of Language and Consciousness.* 1984
ISBN 90-277-0827-4
45. F.G.-I. Nagasaka (ed.): *Japanese Studies in the Philosophy of Science.* 1997
ISBN 0-7923-4781-1
46. P.L. Kapitza: *Experiment, Theory, Practice.* Articles and Addresses. Edited by R.S. Cohen. 1980 ISBN 90-277-1061-9; Pb 90-277-1062-7
47. M.L. Dalla Chiara (ed.): *Italian Studies in the Philosophy of Science.* 1981
ISBN 90-277-0735-9; Pb 90-277-1073-2
48. M.W. Wartofsky: *Models.* Representation and the Scientific Understanding. [Synthese Library 129] 1979 ISBN 90-277-0736-7; Pb 90-277-0947-5
49. T.D. Thao: *Phenomenology and Dialectical Materialism.* Edited by R.S. Cohen. 1986
ISBN 90-277-0737-5
50. Y. Fried and J. Agassi: *Paranoia.* A Study in Diagnosis. [Synthese Library 102] 1976
ISBN 90-277-0704-9; Pb 90-277-0705-7
51. K.H. Wolff: *Surrender and Cath.* Experience and Inquiry Today. [Synthese Library 105] 1976
ISBN 90-277-0758-8; Pb 90-277-0765-0
52. K. Kosík: *Dialectics of the Concrete.* A Study on Problems of Man and World. 1976
ISBN 90-277-0761-8; Pb 90-277-0764-2
53. N. Goodman: *The Structure of Appearance.* [Synthese Library 107] 1977
ISBN 90-277-0773-1; Pb 90-277-0774-X
54. H.A. Simon: *Models of Discovery* and Other Topics in the Methods of Science. [Synthese Library 114] 1977 ISBN 90-277-0812-6; Pb 90-277-0858-4
55. M. Lazerowitz: *The Language of Philosophy.* Freud and Wittgenstein. [Synthese Library 117] 1977 ISBN 90-277-0826-6; Pb 90-277-0862-2
56. T. Nickles (ed.): *Scientific Discovery, Logic, and Rationality.* 1980
ISBN 90-277-1069-4; Pb 90-277-1070-8
57. J. Margolis: *Persons and Mind.* The Prospects of Nonreductive Materialism. [Synthese Library 121] 1978 ISBN 90-277-0854-1; Pb 90-277-0863-0
58. G. Radnitzky and G. Andersson (eds.): *Progress and Rationality in Science.* [Synthese Library 125] 1978 ISBN 90-277-0921-1; Pb 90-277-0922-X
59. G. Radnitzky and G. Andersson (eds.): *The Structure and Development of Science.* [Synthese Library 136] 1979 ISBN 90-277-0994-7; Pb 90-277-0995-5

Boston Studies in the Philosophy of Science

60. T. Nickles (ed.): *Scientific Discovery.* Case Studies. 1980
    ISBN 90-277-1092-9; Pb 90-277-1093-7
61. M.A. Finocchiaro: *Galileo and the Art of Reasoning.* Rhetorical Foundation of Logic and Scientific Method. 1980 ISBN 90-277-1094-5; Pb 90-277-1095-3
62. W.A. Wallace: *Prelude to Galileo.* Essays on Medieval and 16th-Century Sources of Galileo's Thought. 1981 ISBN 90-277-1215-8; Pb 90-277-1216-6
63. F. Rapp: *Analytical Philosophy of Technology.* Translated from German. 1981
    ISBN 90-277-1221-2; Pb 90-277-1222-0
64. R.S. Cohen and M.W. Wartofsky (eds.): *Hegel and the Sciences.* 1984 ISBN 90-277-0726-X
65. J. Agassi: *Science and Society.* Studies in the Sociology of Science. 1981
    ISBN 90-277-1244-1; Pb 90-277-1245-X
66. L. Tondl: *Problems of Semantics.* A Contribution to the Analysis of the Language of Science. Translated from Czech. 1981 ISBN 90-277-0148-2; Pb 90-277-0316-7
67. J. Agassi and R.S. Cohen (eds.): *Scientific Philosophy Today.* Essays in Honor of Mario Bunge. 1982 ISBN 90-277-1262-X; Pb 90-277-1263-8
68. W. Krajewski (ed.): *Polish Essays in the Philosophy of the Natural Sciences.* Translated from Polish and edited by R.S. Cohen and C.R. Fawcett. 1982
    ISBN 90-277-1286-7; Pb 90-277-1287-5
69. J.H. Fetzer: *Scientific Knowledge.* Causation, Explanation and Corroboration. 1981
    ISBN 90-277-1335-9; Pb 90-277-1336-7
70. S. Grossberg: *Studies of Mind and Brain.* Neural Principles of Learning, Perception, Development, Cognition, and Motor Control. 1982 ISBN 90-277-1359-6; Pb 90-277-1360-X
71. R.S. Cohen and M.W. Wartofsky (eds.): *Epistemology, Methodology, and the Social Sciences.* 1983. ISBN 90-277-1454-1
72. K. Berka: *Measurement.* Its Concepts, Theories and Problems. Translated from Czech. 1983
    ISBN 90-277-1416-9
73. G.L. Pandit: *The Structure and Growth of Scientific Knowledge.* A Study in the Methodology of Epistemic Appraisal. 1983 ISBN 90-277-1434-7
74. A.A. Zinov'ev: *Logical Physics.* Translated from Russian. Edited by R.S. Cohen. 1983 [*see also* Volume 9] ISBN 90-277-0734-0
75. G-G. Granger: *Formal Thought and the Sciences of Man.* Translated from French. With and Introduction by A. Rosenberg. 1983 ISBN 90-277-1524-6
76. R.S. Cohen and L. Laudan (eds.): *Physics, Philosophy and Psychoanalysis.* Essays in Honor of Adolf Grünbaum. 1983 ISBN 90-277-1533-5
77. G. Böhme, W. van den Daele, R. Hohlfeld, W. Krohn and W. Schäfer: *Finalization in Science.* The Social Orientation of Scientific Progress. Translated from German. Edited by W. Schäfer. 1983 ISBN 90-277-1549-1
78. D. Shapere: *Reason and the Search for Knowledge.* Investigations in the Philosophy of Science. 1984 ISBN 90-277-1551-3; Pb 90-277-1641-2
79. G. Andersson (ed.): *Rationality in Science and Politics.* Translated from German. 1984
    ISBN 90-277-1575-0; Pb 90-277-1953-5
80. P.T. Durbin and F. Rapp (eds.): *Philosophy and Technology.* [*Also* Philosophy and Technology Series, Vol. 1] 1983 ISBN 90-277-1576-9
81. M. Marković: *Dialectical Theory of Meaning.* Translated from Serbo-Croat. 1984
    ISBN 90-277-1596-3
82. R.S. Cohen and M.W. Wartofsky (eds.): *Physical Sciences and History of Physics.* 1984.
    ISBN 90-277-1615-3

Boston Studies in the Philosophy of Science

83. É. Meyerson: *The Relativistic Deduction.* Epistemological Implications of the Theory of Relativity. Translated from French. With a Review by Albert Einstein and an Introduction by Milič Čapek. 1985 ISBN 90-277-1699-4
84. R.S. Cohen and M.W. Wartofsky (eds.): *Methodology, Metaphysics and the History of Science.* In Memory of Benjamin Nelson. 1984 ISBN 90-277-1711-7
85. G. Tamás: *The Logic of Categories.* Translated from Hungarian. Edited by R.S. Cohen. 1986 ISBN 90-277-1742-7
86. S.L. de C. Fernandes: *Foundations of Objective Knowledge.* The Relations of Popper's Theory of Knowledge to That of Kant. 1985 ISBN 90-277-1809-1
87. R.S. Cohen and T. Schnelle (eds.): *Cognition and Fact.* Materials on Ludwik Fleck. 1986 ISBN 90-277-1902-0
88. G. Freudenthal: *Atom and Individual in the Age of Newton.* On the Genesis of the Mechanistic World View. Translated from German. 1986 ISBN 90-277-1905-5
89. A. Donagan, A.N. Perovich Jr and M.V. Wedin (eds.): *Human Nature and Natural Knowledge.* Essays presented to Marjorie Grene on the Occasion of Her 75th Birthday. 1986 ISBN 90-277-1974-8
90. C. Mitcham and A. Hunning (eds.): *Philosophy and Technology II.* Information Technology and Computers in Theory and Practice. [*Also* Philosophy and Technology Series, Vol. 2] 1986 ISBN 90-277-1975-6
91. M. Grene and D. Nails (eds.): *Spinoza and the Sciences.* 1986 ISBN 90-277-1976-4
92. S.P. Turner: *The Search for a Methodology of Social Science.* Durkheim, Weber, and the 19th-Century Problem of Cause, Probability, and Action. 1986. ISBN 90-277-2067-3
93. I.C. Jarvie: *Thinking about Society.* Theory and Practice. 1986 ISBN 90-277-2068-1
94. E. Ullmann-Margalit (ed.): *The Kaleidoscope of Science.* The Israel Colloquium: Studies in History, Philosophy, and Sociology of Science, Vol. 1. 1986 ISBN 90-277-2158-0; Pb 90-277-2159-9
95. E. Ullmann-Margalit (ed.): *The Prism of Science.* The Israel Colloquium: Studies in History, Philosophy, and Sociology of Science, Vol. 2. 1986 ISBN 90-277-2160-2; Pb 90-277-2161-0
96. G. Márkus: *Language and Production.* A Critique of the Paradigms. Translated from French. 1986 ISBN 90-277-2169-6
97. F. Amrine, F.J. Zucker and H. Wheeler (eds.): *Goethe and the Sciences: A Reappraisal.* 1987 ISBN 90-277-2265-X; Pb 90-277-2400-8
98. J.C. Pitt and M. Pera (eds.): *Rational Changes in Science.* Essays on Scientific Reasoning. Translated from Italian. 1987 ISBN 90-277-2417-2
99. O. Costa de Beauregard: *Time, the Physical Magnitude.* 1987 ISBN 90-277-2444-X
100. A. Shimony and D. Nails (eds.): *Naturalistic Epistemology.* A Symposium of Two Decades. 1987 ISBN 90-277-2337-0
101. N. Rotenstreich: *Time and Meaning in History.* 1987 ISBN 90-277-2467-9
102. D.B. Zilberman: *The Birth of Meaning in Hindu Thought.* Edited by R.S. Cohen. 1988 ISBN 90-277-2497-0
103. T.F. Glick (ed.): *The Comparative Reception of Relativity.* 1987 ISBN 90-277-2498-9
104. Z. Harris, M. Gottfried, T. Ryckman, P. Mattick Jr, A. Daladier, T.N. Harris and S. Harris: *The Form of Information in Science.* Analysis of an Immunology Sublanguage. With a Preface by Hilary Putnam. 1989 ISBN 90-277-2516-0
105. F. Burwick (ed.): *Approaches to Organic Form.* Permutations in Science and Culture. 1987 ISBN 90-277-2541-1

Boston Studies in the Philosophy of Science

106. M. Almási: *The Philosophy of Appearances.* Translated from Hungarian. 1989
ISBN 90-277-2150-5
107. S. Hook, W.L. O'Neill and R. O'Toole (eds.): *Philosophy, History and Social Action.* Essays in Honor of Lewis Feuer. With an Autobiographical Essay by L. Feuer. 1988
ISBN 90-277-2644-2
108. I. Hronszky, M. Fehér and B. Dajka: *Scientific Knowledge Socialized.* Selected Proceedings of the 5th Joint International Conference on the History and Philosophy of Science organized by the IUHPS (Veszprém, Hungary, 1984). 1988
ISBN 90-277-2284-6
109. P. Tillers and E.D. Green (eds.): *Probability and Inference in the Law of Evidence.* The Uses and Limits of Bayesianism. 1988
ISBN 90-277-2689-2
110. E. Ullmann-Margalit (ed.): *Science in Reflection.* The Israel Colloquium: Studies in History, Philosophy, and Sociology of Science, Vol. 3. 1988
ISBN 90-277-2712-0; Pb 90-277-2713-9
111. K. Gavroglu, Y. Goudaroulis and P. Nicolacopoulos (eds.): *Imre Lakatos and Theories of Scientific Change.* 1989
ISBN 90-277-2766-X
112. B. Glassner and J.D. Moreno (eds.): *The Qualitative-Quantitative Distinction in the Social Sciences.* 1989
ISBN 90-277-2829-1
113. K. Arens: *Structures of Knowing.* Psychologies of the 19th Century. 1989
ISBN 0-7923-0009-2
114. A. Janik: *Style, Politics and the Future of Philosophy.* 1989
ISBN 0-7923-0056-4
115. F. Amrine (ed.): *Literature and Science as Modes of Expression.* With an Introduction by S. Weininger. 1989
ISBN 0-7923-0133-1
116. J.R. Brown and J. Mittelstrass (eds.): *An Intimate Relation.* Studies in the History and Philosophy of Science. Presented to Robert E. Butts on His 60th Birthday. 1989
ISBN 0-7923-0169-2
117. F. D'Agostino and I.C. Jarvie (eds.): *Freedom and Rationality.* Essays in Honor of John Watkins. 1989
ISBN 0-7923-0264-8
118. D. Zolo: *Reflexive Epistemology.* The Philosophical Legacy of Otto Neurath. 1989
ISBN 0-7923-0320-2
119. M. Kearn, B.S. Philips and R.S. Cohen (eds.): *Georg Simmel and Contemporary Sociology.* 1989
ISBN 0-7923-0407-1
120. T.H. Levere and W.R. Shea (eds.): *Nature, Experiment and the Science.* Essays on Galileo and the Nature of Science. In Honour of Stillman Drake. 1989
ISBN 0-7923-0420-9
121. P. Nicolacopoulos (ed.): *Greek Studies in the Philosophy and History of Science.* 1990
ISBN 0-7923-0717-8
122. R. Cooke and D. Costantini (eds.): *Statistics in Science.* The Foundations of Statistical Methods in Biology, Physics and Economics. 1990
ISBN 0-7923-0797-6
123. P. Duhem: *The Origins of Statics.* Translated from French by G.F. Leneaux, V.N. Vagliente and G.H. Wagner. With an Introduction by S.L. Jaki. 1991
ISBN 0-7923-0898-0
124. H. Kamerlingh Onnes: *Through Measurement to Knowledge.* The Selected Papers, 1853-1926. Edited and with an Introduction by K. Gavroglu and Y. Goudaroulis. 1991
ISBN 0-7923-0825-5
125. M. Čapek: *The New Aspects of Time: Its Continuity and Novelties.* Selected Papers in the Philosophy of Science. 1991
ISBN 0-7923-0911-1
126. S. Unguru (ed.): *Physics, Cosmology and Astronomy, 1300–1700.* Tension and Accommodation. 1991
ISBN 0-7923-1022-5

Boston Studies in the Philosophy of Science

127. Z. Bechler: *Newton's Physics on the Conceptual Structure of the Scientific Revolution.* 1991
    ISBN 0-7923-1054-3
128. É. Meyerson: *Explanation in the Sciences.* Translated from French by M-A. Siple and D.A. Siple. 1991
    ISBN 0-7923-1129-9
129. A.I. Tauber (ed.): *Organism and the Origins of Self.* 1991   ISBN 0-7923-1185-X
130. F.J. Varela and J-P. Dupuy (eds.): *Understanding Origins.* Contemporary Views on the Origin of Life, Mind and Society. 1992   ISBN 0-7923-1251-1
131. G.L. Pandit: *Methodological Variance.* Essays in Epistemological Ontology and the Methodology of Science. 1991   ISBN 0-7923-1263-5
132. G. Munévar (ed.): *Beyond Reason.* Essays on the Philosophy of Paul Feyerabend. 1991
    ISBN 0-7923-1272-4
133. T.E. Uebel (ed.): *Rediscovering the Forgotten Vienna Circle.* Austrian Studies on Otto Neurath and the Vienna Circle. Partly translated from German. 1991   ISBN 0-7923-1276-7
134. W.R. Woodward and R.S. Cohen (eds.): *World Views and Scientific Discipline Formation.* Science Studies in the [former] German Democratic Republic. Partly translated from German by W.R. Woodward. 1991   ISBN 0-7923-1286-4
135. P. Zambelli: *The Speculum Astronomiae and Its Enigma.* Astrology, Theology and Science in Albertus Magnus and His Contemporaries. 1992   ISBN 0-7923-1380-1
136. P. Petitjean, C. Jami and A.M. Moulin (eds.): *Science and Empires.* Historical Studies about Scientific Development and European Expansion.   ISBN 0-7923-1518-9
137. W.A. Wallace: *Galileo's Logic of Discovery and Proof.* The Background, Content, and Use of His Appropriated Treatises on Aristotle's *Posterior Analytics.* 1992   ISBN 0-7923-1577-4
138. W.A. Wallace: *Galileo's Logical Treatises.* A Translation, with Notes and Commentary, of His Appropriated Latin Questions on Aristotle's *Posterior Analytics.* 1992   ISBN 0-7923-1578-2
    Set (137 + 138) ISBN 0-7923-1579-0
139. M.J. Nye, J.L. Richards and R.H. Stuewer (eds.): *The Invention of Physical Science.* Intersections of Mathematics, Theology and Natural Philosophy since the Seventeenth Century. Essays in Honor of Erwin N. Hiebert. 1992   ISBN 0-7923-1753-X
140. G. Corsi, M.L. dalla Chiara and G.C. Ghirardi (eds.): *Bridging the Gap: Philosophy, Mathematics and Physics.* Lectures on the Foundations of Science. 1992   ISBN 0-7923-1761-0
141. C.-H. Lin and D. Fu (eds.): *Philosophy and Conceptual History of Science in Taiwan.* 1992
    ISBN 0-7923-1766-1
142. S. Sarkar (ed.): *The Founders of Evolutionary Genetics.* A Centenary Reappraisal. 1992
    ISBN 0-7923-1777-7
143. J. Blackmore (ed.): *Ernst Mach – A Deeper Look.* Documents and New Perspectives. 1992
    ISBN 0-7923-1853-6
144. P. Kroes and M. Bakker (eds.): *Technological Development and Science in the Industrial Age.* New Perspectives on the Science–Technology Relationship. 1992   ISBN 0-7923-1898-6
145. S. Amsterdamski: *Between History and Method.* Disputes about the Rationality of Science. 1992   ISBN 0-7923-1941-9
146. E. Ullmann-Margalit (ed.): *The Scientific Enterprise.* The Bar-Hillel Colloquium: Studies in History, Philosophy, and Sociology of Science, Volume 4. 1992   ISBN 0-7923-1992-3
147. L. Embree (ed.): *Metaarchaeology.* Reflections by Archaeologists and Philosophers. 1992
    ISBN 0-7923-2023-9
148. S. French and H. Kamminga (eds.): *Correspondence, Invariance and Heuristics.* Essays in Honour of Heinz Post. 1993   ISBN 0-7923-2085-9
149. M. Bunzl: *The Context of Explanation.* 1993   ISBN 0-7923-2153-7

Boston Studies in the Philosophy of Science

150. I.B. Cohen (ed.): *The Natural Sciences and the Social Sciences*. Some Critical and Historical Perspectives. 1994 ISBN 0-7923-2223-1
151. K. Gavroglu, Y. Christianidis and E. Nicolaidis (eds.): *Trends in the Historiography of Science*. 1994 ISBN 0-7923-2255-X
152. S. Poggi and M. Bossi (eds.): *Romanticism in Science*. Science in Europe, 1790–1840. 1994 ISBN 0-7923-2336-X
153. J. Faye and H.J. Folse (eds.): *Niels Bohr and Contemporary Philosophy*. 1994 ISBN 0-7923-2378-5
154. C.C. Gould and R.S. Cohen (eds.): *Artifacts, Representations, and Social Practice*. Essays for Marx W. Wartofsky. 1994 ISBN 0-7923-2481-1
155. R.E. Butts: *Historical Pragmatics*. Philosophical Essays. 1993 ISBN 0-7923-2498-6
156. R. Rashed: *The Development of Arabic Mathematics: Between Arithmetic and Algebra*. Translated from French by A.F.W. Armstrong. 1994 ISBN 0-7923-2565-6
157. I. Szumilewicz-Lachman (ed.): *Zygmunt Zawirski: His Life and Work*. With Selected Writings on Time, Logic and the Methodology of Science. Translations by Feliks Lachman. Ed. by R.S. Cohen, with the assistance of B. Bergo. 1994 ISBN 0-7923-2566-4
158. S.N. Haq: *Names, Natures and Things*. The Alchemist Jābir ibn Ḥayyān and His *Kitāb al-Aḥjār* (Book of Stones). 1994 ISBN 0-7923-2587-7
159. P. Plaass: *Kant's Theory of Natural Science*. Translation, Analytic Introduction and Commentary by Alfred E. and Maria G. Miller. 1994 ISBN 0-7923-2750-0
160. J. Misiek (ed.): *The Problem of Rationality in Science and its Philosophy*. On Popper vs. Polanyi. The Polish Conferences 1988–89. 1995 ISBN 0-7923-2925-2
161. I.C. Jarvie and N. Laor (eds.): *Critical Rationalism, Metaphysics and Science*. Essays for Joseph Agassi, Volume I. 1995 ISBN 0-7923-2960-0
162. I.C. Jarvie and N. Laor (eds.): *Critical Rationalism, the Social Sciences and the Humanities*. Essays for Joseph Agassi, Volume II. 1995 ISBN 0-7923-2961-9
    Set (161–162) ISBN 0-7923-2962-7
163. K. Gavroglu, J. Stachel and M.W. Wartofsky (eds.): *Physics, Philosophy, and the Scientific Community*. Essays in the Philosophy and History of the Natural Sciences and Mathematics. In Honor of Robert S. Cohen. 1995 ISBN 0-7923-2988-0
164. K. Gavroglu, J. Stachel and M.W. Wartofsky (eds.): *Science, Politics and Social Practice*. Essays on Marxism and Science, Philosophy of Culture and the Social Sciences. In Honor of Robert S. Cohen. 1995 ISBN 0-7923-2989-9
165. K. Gavroglu, J. Stachel and M.W. Wartofsky (eds.): *Science, Mind and Art*. Essays on Science and the Humanistic Understanding in Art, Epistemology, Religion and Ethics. Essays in Honor of Robert S. Cohen. 1995 ISBN 0-7923-2990-2
    Set (163–165) ISBN 0-7923-2991-0
166. K.H. Wolff: *Transformation in the Writing*. A Case of Surrender-and-Catch. 1995 ISBN 0-7923-3178-8
167. A.J. Kox and D.M. Siegel (eds.): *No Truth Except in the Details*. Essays in Honor of Martin J. Klein. 1995 ISBN 0-7923-3195-8
168. J. Blackmore: *Ludwig Boltzmann, His Later Life and Philosophy, 1900–1906*. Book One: A Documentary History. 1995 ISBN 0-7923-3231-8
169. R.S. Cohen, R. Hilpinen and R. Qiu (eds.): *Realism and Anti-Realism in the Philosophy of Science*. Beijing International Conference, 1992. 1996 ISBN 0-7923-3233-4
170. I. Kuçuradi and R.S. Cohen (eds.): *The Concept of Knowledge*. The Ankara Seminar. 1995 ISBN 0-7923-3241-5

Boston Studies in the Philosophy of Science

171. M.A. Grodin (ed.): *Meta Medical Ethics*: The Philosophical Foundations of Bioethics. 1995
    ISBN 0-7923-3344-6
172. S. Ramirez and R.S. Cohen (eds.): *Mexican Studies in the History and Philosophy of Science.* 1995 ISBN 0-7923-3462-0
173. C. Dilworth: *The Metaphysics of Science.* An Account of Modern Science in Terms of Principles, Laws and Theories. 1995 ISBN 0-7923-3693-3
174. J. Blackmore: *Ludwig Boltzmann, His Later Life and Philosophy, 1900–1906* Book Two: The Philosopher. 1995 ISBN 0-7923-3464-7
175. P. Damerow: *Abstraction and Representation.* Essays on the Cultural Evolution of Thinking. 1996 ISBN 0-7923-3816-2
176. M.S. Macrakis: *Scarcity's Ways: The Origins of Capital.* A Critical Essay on Thermodynamics, Statistical Mechanics and Economics. 1997 ISBN 0-7923-4760-9
177. M. Marion and R.S. Cohen (eds.): *Québec Studies in the Philosophy of Science.* Part I: Logic, Mathematics, Physics and History of Science. Essays in Honor of Hugues Leblanc. 1995
    ISBN 0-7923-3559-7
178. M. Marion and R.S. Cohen (eds.): *Québec Studies in the Philosophy of Science.* Part II: Biology, Psychology, Cognitive Science and Economics. Essays in Honor of Hugues Leblanc. 1996
    ISBN 0-7923-3560-0
    Set (177–178) ISBN 0-7923-3561-9
179. Fan Dainian and R.S. Cohen (eds.): *Chinese Studies in the History and Philosophy of Science and Technology.* 1996 ISBN 0-7923-3463-9
180. P. Forman and J.M. Sánchez-Ron (eds.): *National Military Establishments and the Advancement of Science and Technology.* Studies in 20th Century History. 1996
    ISBN 0-7923-3541-4
181. E.J. Post: *Quantum Reprogramming.* Ensembles and Single Systems: A Two-Tier Approach to Quantum Mechanics. 1995 ISBN 0-7923-3565-1
182. A.I. Tauber (ed.): *The Elusive Synthesis: Aesthetics and Science.* 1996 ISBN 0-7923-3904-5
183. S. Sarkar (ed.): *The Philosophy and History of Molecular Biology: New Perspectives.* 1996
    ISBN 0-7923-3947-9
184. J.T. Cushing, A. Fine and S. Goldstein (eds.): *Bohmian Mechanics and Quantum Theory: An Appraisal.* 1996 ISBN 0-7923-4028-0
185. K. Michalski: *Logic and Time.* An Essay on Husserl's Theory of Meaning. 1996
    ISBN 0-7923-4082-5
186. G. Munévar (ed.): *Spanish Studies in the Philosophy of Science.* 1996 ISBN 0-7923-4147-3
187. G. Schubring (ed.): *Hermann Günther Graßmann (1809–1877): Visionary Mathematician, Scientist and Neohumanist Scholar.* Papers from a Sesquicentennial Conference. 1996
    ISBN 0-7923-4261-5
188. M. Bitbol: *Schrödinger's Philosophy of Quantum Mechanics.* 1996 ISBN 0-7923-4266-6
189. J. Faye, U. Scheffler and M. Urchs (eds.): *Perspectives on Time.* 1997 ISBN 0-7923-4330-1
190. K. Lehrer and J.C. Marek (eds.): *Austrian Philosophy Past and Present.* Essays in Honor of Rudolf Haller. 1996 ISBN 0-7923-4347-6
191. J.L. Lagrange: *Analytical Mechanics.* Translated and edited by Auguste Boissonade and Victor N. Vagliente. Translated from the *Mécanique Analytique, novelle édition* of 1811. 1997
    ISBN 0-7923-4349-2
192. D. Ginev and R.S. Cohen (eds.): *Issues and Images in the Philosophy of Science.* Scientific and Philosophical Essays in Honour of Azarya Polikarov. 1997 ISBN 0-7923-4444-8

Boston Studies in the Philosophy of Science

193. R.S. Cohen, M. Horne and J. Stachel (eds.): *Experimental Metaphysics.* Quantum Mechanical Studies for Abner Shimony, Volume One. 1997 ISBN 0-7923-4452-9
194. R.S. Cohen, M. Horne and J. Stachel (eds.): *Potentiality, Entanglement and Passion-at-a-Distance.* Quantum Mechanical Studies for Abner Shimony, Volume Two. 1997 ISBN 0-7923-4453-7; Set 0-7923-4454-5
195. R.S. Cohen and A.I. Tauber (eds.): *Philosophies of Nature: The Human Dimension.* 1997 ISBN 0-7923-4579-7
196. M. Otte and M. Panza (eds.): *Analysis and Synthesis in Mathematics.* History and Philosophy. 1997 ISBN 0-7923-4570-3
197. A. Denkel: *The Natural Background of Meaning.* 1999 ISBN 0-7923-5331-5
198. D. Baird, R.I.G. Hughes and A. Nordmann (eds.): *Heinrich Hertz: Classical Physicist, Modern Philosopher.* 1999 ISBN 0-7923-4653-X
199. A. Franklin: *Can That be Right?* Essays on Experiment, Evidence, and Science. 1999 ISBN 0-7923-5464-8
200. D. Raven, W. Krohn and R.S. Cohen (eds.): *The Social Origins of Modern Science.* 2000 ISBN 0-7923-6457-0
201. Reserved
202. Reserved
203. B. Babich and R.S. Cohen (eds.): *Nietzsche, Theories of Knowledge, and Critical Theory.* Nietzsche and the Sciences I. 1999 ISBN 0-7923-5742-6
204. B. Babich and R.S. Cohen (eds.): *Nietzsche, Epistemology, and Philosophy of Science.* Nietzsche and the Science II. 1999 ISBN 0-7923-5743-4
205. R. Hooykaas: *Fact, Faith and Fiction in the Development of Science.* The Gifford Lectures given in the University of St Andrews 1976. 1999 ISBN 0-7923-5774-4
206. M. Fehér, O. Kiss and L. Ropolyi (eds.): *Hermeneutics and Science.* 1999 ISBN 0-7923-5798-1
207. R.M. MacLeod (ed.): *Science and the Pacific War.* Science and Survival in the Pacific, 1939-1945. 1999 ISBN 0-7923-5851-1
208. I. Hanzel: *The Concept of Scientific Law in the Philosophy of Science and Epistemology.* A Study of Theoretical Reason. 1999 ISBN 0-7923-5852-X
209. G. Helm; R.J. Deltete (ed./transl.): *The Historical Development of Energetics.* 1999 ISBN 0-7923-5874-0
210. A. Orenstein and P. Kotatko (eds.): *Knowledge, Language and Logic.* Questions for Quine. 1999 ISBN 0-7923-5986-0
211. R.S. Cohen and H. Levine (eds.): *Maimonides and the Sciences.* 2000 ISBN 0-7923-6053-2
212. H. Gourko, D.I. Williamson and A.I. Tauber (eds.): *The Evolutionary Biology Papers of Elie Metchnikoff.* 2000 ISBN 0-7923-6067-2
213. S. D'Agostino: *A History of the Ideas of Theoretical Physics.* Essays on the Nineteenth and Twentieth Century Physics. 2000 ISBN 0-7923-6094-X
214. S. Lelas: *Science and Modernity.* Toward An Integral Theory of Science. 2000 ISBN 0-7923-6303-5
215. E. Agazzi and M. Pauri (eds.): *The Reality of the Unobservable.* Observability, Unobservability and Their Impact on the Issue of Scientific Realism. 2000 ISBN 0-7923-6311-6

Also of interest:
R.S. Cohen and M.W. Wartofsky (eds.): *A Portrait of Twenty-Five Years Boston Colloquia for the Philosophy of Science, 1960-1985.* 1985 ISBN Pb 90-277-1971-3

OHIO UNIVERSITY LIBRARY
Please return t'   ok as soon as you have
'inished with   er to avoid a fine it must